Meiosis: Key Concepts

Meiosis: Key Concepts

Edited by **Morgan Key**

New York

Published by Callisto Reference,
106 Park Avenue, Suite 200,
New York, NY 10016, USA
www.callistoreference.com

Meiosis: Key Concepts
Edited by Morgan Key

International Standard Book Number: 978-1-63239-457-6 (Hardback)

Printed in the United States of America.

Contents

Preface | VII

Part 1   Molecular Biology of Mammalian Meiosis | 1

Chapter 1   **Energy Metabolism Regulating Mammalian Oocyte Maturation** | 3
N. Songsasen

Chapter 2   **The Control of Meiotic Arrest and Resumption in Mammalian Oocytes** | 17
Sylvie Bilodeau-Goeseels and Nora Magyara

Chapter 3   **PKC Regulation of Gametogenesis and Early Development** | 39
James J. Faust, Madhavi Kalive,
Anup Abraham and David G. Capco

Chapter 4   **Role of Bcl2l10 in Regulation of Meiotic Cell Cycle in the Mouse Oocyte** | 59
Kyung-Ah Lee, Se-Jin Yoon, Eun-Young Kim,
Jeehyeon Bae, Hyun-Seo Lee, Su-Yeon Lee and Eun-Ah Kim

Chapter 5   **Nuage Components and Their Contents in Mammalian Spermatogenic Cells, as Revealed by Immunoelectron Microscopy** | 69
Yuko Onohara and Sadaki Yokota

Chapter 6   **OMICS for the Identification of Biomarkers for Oocyte Competence, with Special Reference to the Mare as a Prospective Model for Human Reproductive Medicine** | 93
Maria Elena Dell'Aquila, Yoon Sung Cho, Nicola Antonio Martino,
Manuel Filioli Uranio, Lucia Rutigliano and Katrin Hinrichs

Chapter 7   Meiotic Behaviour of Chromosomes Involved in
             Structural Chromosomal Abnormalities Determined by
             Preimplantation Genetic Diagnosis   119
             L. Xanthopoulou and H. Ghevaria

Chapter 8   Insight Into the Molecular Program of Meiosis   135
             Hiba Waldman Ben-Asher and Jeremy Don

Part 2   Molecular and Cytogenetic Studies of Plant Meiosis   153

Chapter 9   Meiotic Behavior in Intra- and Interspecific Sexual and
             Somatic Polyploid Hybrids of Some Tropical Species   155
             Maria Suely Pagliarini, Maria Lúcia Carneiro Vieira
             and Cacilda Borges do Valle

Chapter 10   Quantifying Meiosis: Use of the Fractal Dimension,
             D_f, to Describe and Predict Prophase I
             Substages and Metaphase I   173
             Cynthia Ross Friedman and Hua-Feng Wang

Chapter 11   Haploid Independent Unreductional Meiosis in
             Hexaploid Wheat   191
             Filipe Ressurreição, Augusta Barão,
             Wanda Viegas and Margarida Delgado

Chapter 12   Embryology of Flowering Plants Applied to Cytogenetic
             Studies on Meiosis   201
             Jorge E. A. Mariath, André L. L. Vanzela, Eliane Kaltchuk-Santos,
             Karen L. G. De Toni, Célia G. T. J. Andrade, Adriano Silvério,
             Erica Duarte-Silva, Carlos R. M. da Silva, Juca A. B. San Martin,
             Fernanda Nogueira and Simone P. Mendes

Chapter 13   Investigating Host Induced Meiosis in
             a Fungal Plant Pathogen   223
             B. J. Saville, M. E. Donaldson and C. E. Doyle

             Permissions

             List of Contributors

Preface

This book provides an in-depth analysis of the recent advancements in meiosis. Meiosis, the procedure of producing gametes in preparation for sexual reproduction, has long been a focal point of concentrated research. It has been researched at the cytological, hereditary, molecular and cellular stages. Researches in model systems have exposed universal essential mechanisms while parallel studies in various organisms have led to the discovery of variations in meiotic methods. This book discusses topics related to the molecular biology of mammalian meiosis. It also includes molecular and cytogenetic studies of meiosis in plants. The book collects various strands of examination into this enthralling and demanding field of biology.

The researches compiled throughout the book are authentic and of high quality, combining several disciplines and from very diverse regions from around the world. Drawing on the contributions of many researchers from diverse countries, the book's objective is to provide the readers with the latest achievements in the area of research. This book will surely be a source of knowledge to all interested and researching the field.

In the end, I would like to express my deep sense of gratitude to all the authors for meeting the set deadlines in completing and submitting their research chapters. I would also like to thank the publisher for the support offered to us throughout the course of the book. Finally, I extend my sincere thanks to my family for being a constant source of inspiration and encouragement.

Editor

Part 1

Molecular Biology of Mammalian Meiosis

Energy Metabolism Regulating Mammalian Oocyte Maturation

N. Songsasen

Department of Reproductive Sciences, Center for Species Survival,
Smithsonian Conservation Biology Institute, Front Royal,
Virginia

1. Introduction

Oocyte maturation is a final step of gamete development that follows a prolonged period of cell growth within a growing follicle. Unlike oocyte growth that takes several weeks or months, maturation process is a short (hours or days), dynamic process. The period of oocyte maturation varies among species, ranging from 10-13 hours in the mouse (Edwards & Gates, 1959) to 16-24 hours in the cow (Dominko & First, 1997) and 48 -72 hours in the dog (Reynaud et al., 2005, Songsasen & Wildt, 2007). For an oocyte to fully capable of fertilizing and developing into an embryo, the gamete needs to undergo nuclear and cytoplasmic maturation. Nuclear maturation encompasses events associated with the separation of homologous chromosomes during meiosis I and the segregation of sister chromatids during meiosis 2 (Albertini & Limback, 2009). Events occur during this process include nuclear envelope breakdown, rearrangement of the cortical cytoskeleton and meiotic spindle assembly. Cytoplasmic maturation includes events of post-transcriptional and post-translational processes, including mRNA synthesis, rearrangement of cytoplasmic organelles and glutathione production that are essential for successful fertilization and subsequent embryonic development (Albertini & Limback, 2009, Watson, 2007). For an oocyte to appropriately progress to these dynamic process of nuclear and cytoplasmic maturation, it requires enormous energy from various substrates, including glucose, amino acids and lipids (Sutton et al., 2003). In addition, tight regulation of reactive oxygen species (ROS) and calcium homeostasis are important during this process. This chapter will review current knowledge on mammalian oocyte development, the roles of mitochondria on cell functions and energy metabolism and its impact on gamete maturation.

2. Mammalian oocyte development

The oocyte arises from the primordial germ cells (PGCs) developed during the embryogenesis (Edson et al., 2009). Once formed, the PGCs proliferate and migrate to the undifferentiated gonad that later becomes the ovary. Within the gonad, PGCs enter mitosis with incomplete cytokinesis to form clusters of germ cell nest consisting of oogonia connected to each other by intercellular bridges (Tingen et al., 2009), and the syncytia units are surrounded by pre-granulosa and stromal mesenchymal cells of the ovary. Oogenesis occurs *in utero* in rodents, ruminants and primates, whereas that of the cat, dog and ferret takes place after birth (Peter & McNatty, 1980). During this process, oogonia within germ cell nests transform into the oocytes by, albeit asynchronously, entering the first meiotic prophase and being arrested at the late diplotene stage. The interval between the appearance of the first and the last oocytes within the ovary varies greatly among species, ranging from 2 days in the rat to 5 months in the human (Peter & McNatty, 1980).

Formation of the primordial follicle begins during the fetal life in the human, monkey, horse, cow and pig, but not until during neonatal period in the mouse and rat or later in the second or third weeks after birth in cats, dogs, ferrets, rabbits and minks (Peter & McNatty, 1980). Primordial follicle formation begins with the breakdown of germ cell nest involving the degeneration of vast numbers of oocytes and the invasion of pre-granulosa cells into the germ cell syncytia (Tingen et al., 2009). The loss of oocytes during germ cell nest breakdown is substantial and is believed to be part of quality control processes to ensure that only healthy oocytes are enclosed inside primordial follicles (Tingen et al., 2009). It has been suggested that genetic defects or the failure of germ cells to produce mitochondria are responsible for selective loss of oocytes during the nest breakdown process (Tingen et al., 2009). The surviving oocytes are individually surrounded by squamous pre-granulosa cells and the entire unit is referred to as the primordial follicle (Edson et al., 2009). To date, it has been suggested that several factors produced by the oocytes and somatic cells, including synaptonemal complex protein, Foxl2, NOBOX, members of Notch signaling pathway and transforming growth factor family play roles in germ cell nest breakdown and formation of primordial follicles (Tingen et al., 2009).

Most primordial follicles (90%) leave the resting pool via apoptosis, while the remainders are activated by poorly understood mechanisms to enter the growing follicle pool and develop into the primary, secondary and antral stage (Gougeon, 2010, Picton, 2001). During the early stage of follicle development, the oocyte rapidly increases in size; however, as folliculogenesis proceeds, the growth rate significantly decline and the gamete reaches the maximum size shortly after antral formation (Griffin et al., 2006, Reynaud et al., 2009, Songsasen et al., 2009). During oocyte growth, several organelles including, mitochondria, endoplasmic reticulum and golgi complexes become more abundant (Peter & McNatty, 1980), indicating that energy production and protein synthesis are essential during this process. In most mammalian species, the oocyte resumes meiosis in the preovulatory follicle shortly before ovulation. However, meiotic resumption occurs after the dog oocyte is released from the follicle (Songsasen & Wildt, 2007). At the initiation of meiotic resumption, the nucleus of the oocyte migrates from the central position to the periphery of the cell, and the nuclear membrane and nucleoli disappear as the chromosomes become condensed. This stage is referred to as germinal vesicle breakdown (GVBD). The GVBD oocyte progresses to the metaphase I (MI) stage as the chromosomes are firmly attached to the meiotic spindle. This stage is followed by the separation of homologous chromosomes; one set remain in the secondary oocyte and another set move into a small portion of cytoplasm that is extruded as the second polar body. The chromosomes within the secondary oocyte resume metaphase configuration (i.e., metaphase II [MII] stage) which remains until fertilization. Completion of meiosis occurs after a spermatozoon penetrating the oocyte. During this process, the gamete extrudes the secondary polar body. The remaining chromosomes are enclosed in the nuclear membrane and forming female pronucleus that later fuses with the male counterpart to become the zygote. During its developmental process, the oocyte is coupled to the surrounding granulosa cells through trans-zonal processes (Eppig et al., 1996). This intimate physical connection is crucial for follicle and oocyte development as it facilitates bi-directional communication between the gamete and the somatic cells via gap junction and paracrine signaling (Eppig et al., 1996).

3. Mitochondria are the key organelles regulating cell functions

The mitochondria are characterized by having double membranes and their own DNA (mtDNA) inherited from the maternal origin (Van Blerkom, 2004). These organelles play fundamental roles in cell functions, including, providing energy, regulating apoptotic pathway and Ca^{2+} homeostasis (Van Blerkom, 2004, Ramalho-Santos et al., 2009). Dysfunction of mitochondria has been linked to several pathologies, such as heart diseases (Facecchia et al., 2011; Perrelli, et al. 2011), neurodegenerative (Facecchia et al., 2011) and infertility (Ramalho-Santos et al., 2009). The mitochondria provide ATP to cells via oxidative phosphorylation. During this process, high energy electrons derived from oxidation are carried by NADH + H^+ and $FADH_2$ to the inner mitochondrial membrane and transferred through cascades of electron transport chain that convert the electrons into ATP (Ramalho-Santos et al., 2009; Voet & Voet, 2004).

Mitochondria also participate in apoptotic pathways (Ramalho-Santos et al., 2009). Because mitochondria consume ~85% of cell's oxygen, these organelles are major producers of ROS (Ramalho-Santos et al., 2009). Excessive production of ROS can lead to DNA damage (especially mtDNA), oxidation of proteins and lipid, as well as release of cytochrome B into the extramitochondrial milieu (Ott et al., 2007). The release of cytochrome B into the cytosol triggers a series of events leading to proteolytic enzyme activation, including caspase 3, 6 and 7 that regulate cell death (Ott et al., 2007). Therefore, cells critically depend on the tight regulation of mitochondrial redox balance to maintain viability and proper functions. It has been shown that glutathione (GSH) and GSH-linked antioxidant enzymes, including Gpx1 and 4 play important roles in maintaining mitochondrial redox balance (Ott et al., 2007). Specifically, these enzymes catalyze the reduction of hydrogen peroxide and lipid peroxide (into H_2O) with GSH serves as the electron donor. Furthermore, mitochondrial thioredoxin has been shown to play roles in maintaining mitochondria protein in their reduce state, that in turn counter the reaction of ROS (Ott et al., 2007).

During folliculogenesis and oogenesis, mitochondria propagate simultaneously with the increase in cytoplasmic volume. Pre-migratory PGCs have < 10 mitochondria (Van Blerkom, 2004). However, the numbers of mitochondria within the germ cells increase 10-fold by the time PGCs reach the ovary and additional 2-fold after transformation into the oocyte. Primordial follicle oocytes contain 10,000 mitochondria which increase to 100,000 in mature gametes (Ramalho-Santos et al., 2009). The increase in number of mitochondria also coincides with changes in the distribution of the organelles (Rmalho-Santos et al., 2009). Specifically, mitochondria surround the nucleus in the primary follicle. As the follicle developing into secondary stage, mitochondria within the oocyte distribute throughout the cytoplasm (Peter & McNatty, 1980). In fully grown-germinal vesicle oocyte, the mitochondria homogenously distribute throughout the cell with some localization at the periphery (Sun et al., 2001). As the oocyte progresses through meiotic maturation, mitochondria relocate into the perinuclear region and aggregate into cluster (Yu et al., 2010; Van Blerkom et al., 2002; Sturmey et al., 2006); this event coincides with the rise in ATP levels (Yu et al., 2010). Therefore, the change in localization of mitochondria during oocyte growth is probably in response to energy demand during a given stage of development (Ramalho-Santos et al., 2009).

The oocyte contains two populations of mitochondria which differ in polarization (i.e., electrical and chemical gradient $[\Delta\Psi_M]$); one which is more abundant has low $\Delta\Psi_M$ and the smaller population is highly polarized (Ramalho-Santos et al., 2009; Acton et al., 2004, Van Blerkom et al., 2003). In mouse and human oocytes, highly polarized mitochondria are clustered in pericortical cytoplasm probably to (1) maintain sufficient ATP production in the sub-plasmalemma region in the preparation for fertilization and (2) play role in Ca^{2+} regulation during oocyte activation (Ramalho-Santos et al., 2009; Van Blerkom et al., 2003; Van Blerkom & Davis, 2007). It also has been shown that mitochondrial polarity is associated with developmental capacity of the oocyte and embryo (Van Blerkom, 2004; Acton et al., 2004; Van Blerkom & Davis, 2007; Fujii & Funahashi, 2009). Specifically, mitochondrial membrane potential (MMP) increases as oocytes progress through meiotic maturation (Fujii & Funahashi, 2009), and inhibition of MMP rise decreases the ability of the gamete to form pronucleus and impairs embryonic development (Fujii & Funahashi, 2009).

4. Glucose is a key substrate for providing energy during oocyte maturation

Glucose metabolism is crucial for oocyte maturation and development post-fertilization in many mammalian species (Sutton-McDowall et al., 2010; Krisher et al., 2007). *In vitro* culture of oocytes in sub-optimal concentrations of glucose results in delayed meiotic maturation, fertilization and embryonic development (Sutton-McDowall et al., 2010; Sato et al., 2007; Zheng et al., 2001). Delayed resumption of meiosis in prepubertal cattle oocytes is associated with retarded glucose metabolism (Steeves & Gardner, 1999). Furthermore, pharmacological stimulation of glucose metabolism enhances the developmental competence of cow (Krisher & Bavister, 1999) and pig (Herrick et al., 2006) oocytes *in vitro*. Finally, it has been shown that diabetic mice experience abnormal cellular metabolism, mitochondrial dysfunction and meiotic defect (Wang et al., 2010; Colton, et al., 2002).

Glucose uptake into the oocyte occurs via facilitative glucose transporters (GLUT) in the mouse (Purcell & Moley, 2009), cow (Augustin et al., 2001), sheep (Pisani et al., 2008), human (Dan Goor et al., 1997) and rhesus monkey (Zheng et al., 2007). But mammalian oocytes also have low capacity to utilize this substrate (Sutton-McDowall et al., 2010; Steeves & Gardner, 1999; Purcell & Moley, 2009, Brinster, 1971), possibly due to having limited amount of a glycolytic enzyme phosphofructokinase (Cetica et al., 2002). Thus, most mammalian species appear to rely on cumulus cells that contain an additional GLUT with high affinity to this substrate and high phosphofructokinase activity to convert glucose into readily utilized substrates (i.e., pyruvate, NADPH; (Sutton-McDowall et al., 2010; Biggers, et al., 1967). However, we have recently found that dog oocytes utilize glucose at a much higher rate than that of other species (Songsasen et al., 2012). This finding indicates that dog gamete may contain additional GLUT or high levels of glycolytic enzyme compared to those in other species.

The cumulus-oocyte complexes (COCs) have been found to metabolize glucose through four pathways, including glycolysis, the pentose-phosphate- (PPP), hexoxamine (HBP)- and polyol pathways (Sutton-McDowall et al., 2010), with the first two known to affect nuclear and cytoplasmic maturation of mouse (Downs, 1995), pig (Herrick et al., 2006; Krisher et al.,

2007), cow (Krisher et al., 2007; Steeves & Gardner, 1999, Rieger & Loskutoff, 1994) and cat (Spindler et al., 2000) oocytes.

COCs utilize substantial amount of glucose via glycolytic pathway produces energy (ATP), pyruvate and lactate. Pyruvate and lactate then enter the oocyte and are metabolized via the tricarboxylic pathway (TCA) followed by oxidative phosphorylation within the mitochondria that produce substantial amount of ATP essential for oocyte development. Glycolytic metabolism has been shown to play key roles in developmental competence of cow (Steeves & Gardner, 1999, Rieger & Loskutoff, 1994), cat (Spindler et al., 2000) and pig oocytes (Herrick et al., 2006). Specifically, glucose metabolism increases as immature cat oocytes progress through meiotic maturation from the GV to MII stage (Spindler et al., 2000). These same investigators also demonstrated that developmental competence post-fertilization, including the ability to advance to the blastocyst embryos, directly depended on glycolytic rate.

Although PPP does not generate ATP, this pathway generates NADPH that is essential for cytoplasmic integrity and GSH production (Sutton-McDowall et al., 2010; Garcia, et al., 2010). Furthermore, PPP produces ribose-5-phosphate that is critical for DNA and RNA syntheses (Sutton-McDowall et al., 2010; Krisher et al., 2007; Newsholme, et al., 2003). It has been shown the mouse oocyte preferentially metabolize glucose via this pathway. Stimulation of the PPP in mouse oocytes significantly increases GVBD due to conversion of ribose-5 phosphate to phosphoribosyl pyrophosphate that is involved in the production of purine nucleotides, important precursors of DNA and RNA synthesis (Downs et al., 1998). PPP also plays roles in completion of meiosis after fertilization via production of NADPH followed by generation of ROS essential for signaling pathways (Urner & Sakkas, 2005). In the pig, PPP plays critical role in resumption of meiosis and transition of GVBD to MII stage oocytes (Sato et al., 2007; Herrick et al., 2006; Funahashi et al., 2008). Inhibiting PPP activity decreases glycolysis and production of GSH that, in turn compromise oocyte developmental potential in this species (Herrick et al.,2006).

Follicle stimulating hormone (FSH) is the primary regulator of ovarian folliculogenesis (Gougeon, 2010). FSH has been shown to support meiotic resumption by promoting glucose metabolism in mouse oocyte. Specifically, FSH increases glucose uptake (Roberts et al., 2004), as well as promotes glycolysis and PPP in the mouse (Downs & Utecht, 1999), and augment TCA in the cow (Zuelke & Brackett, 1992). Luteinizing hormone surge (LH) plays significant roles in meiotic maturation of mammalian oocytes (Gougeon, 2010; Son et al., 2011, Hsieh et al., 2011). LH surge causes a significant decline in gap junctions leading to dissociation of granulosa cells from the gamete and expansion of the cumulus cells. LH also activates its G-protein-coupled receptor on theca and granulosa cells, which in turn leads to elevation of intracellular cAMP that subsequently triggers multiple downstream pathways regulating meiotic maturation and ovulation (Hsieh et al., 2011, Sun et al., 2009). Finally, LH promotes cow oocytes maturation by modifying gamete's nutritional microenvironment via increasing glucose utilization through glycolysis and TCA cycle (Zuelke & Brackett, 1992).

5. Amino acids influence cell functions

Amino acids play important roles in cellular functions, as they serve as substrates for protein synthesis, energy production, organic osmolytes and intracellular buffer (Sutton et al., 2003; Bae & Foote, 1974). Furthermore, cysteine, proline and glutamine are precursors of GSH (Sutton et al., 2003). It has been shown that there are differences in amino acid composition within follicular fluid among species. Hystinine, phynylalanine, asparagines and glutamate are present in high concentrations in human follicular fluid (Jozwik et al., 2006; Jimena, et al., 1993). However, glycine, glutamine and alanine are most abundant in the cow (Orsi et al., 2005). Finally, glycine, alanine and serine, and glutamine, glycine and aspartate are found in high concentration in horses (Engle et al., 1984) and mice (Harris et al., 2005), respectively. These apparent differences in follicular amino acids composition among species suggest that there are variations in amino acid requirement during oocyte maturation among various taxa.

Amino acid concentrations fluctuate with ovarian follicle development. Specifically, in the pig, most amino acids are present in lower concentration in large follicles than in small counterparts (Hong & Lee, 2007). This may be attributed to the decrease in the amount of specific amino acids following preferred consumption by follicular cells or oocytes during follicular development. Culturing pig oocytes in the presence of glutamine, aspargine and valine enhances cytoplasmic maturation based on the improvement of monospermic fertilization and embryonic development (Hong & Lee, 2007). Furthermore, addition of essential amino acids to chemically defined maturation media enhances oocyte maternal mRNA synthesis, embryo developmental rates and cell numbers in blastocyst embryos in the cow (Watson et al., 2000).

Mammalian oocytes utilize amino acids via the cumulus cells. Cumulus cells contain high concentration of two essential enzymes for amino acid metabolism, alanine aminotransferase [ALT] and aspartate aminotransferase [AST] (Cetica, et al., 2003), suggesting that the somatic cells supplies the oocyte with amino acid or oxidative substrate. Because oocytes also possess ALT and AST activities, albeit lower than cumulus cells, they continue to utilize amino acids after being dissociated from the somatic cells during maturation (Cetica et al., 2003).

Glutamine is the most widely studied amino acid among those found in follicular fluid. This amino acid is recognized as a key substrate for GSH synthesis, protein translation and gluconeogenesis (Bae & Foote 1974; Newsholme et al. 2003). In cattle, glutamine metabolism through the TCA cycle increases steadily during oocyte maturation and reaches maximum at 18 to 24 of *in vitro* culture (Steeves & Gardner, 1999; Rieger & Loskutoff, 1994). This finding suggests that this amino acid is critical for promoting final nuclear maturation in this species (Rieger & Loskutoff, 1994). Addition of glutamine to maturation medium increases the number of cow oocytes completing nuclear maturation (Bilodeau-Goeseels, 2006). However, supplement culture medium with glutamine alone does not improve developmental competence of rhesus monkey oocytes (Zheng et al., 2002). Nevertheless, combination of glutamine and 20 amino acids enhances nuclear maturation in this species (Zheng et al., 2002). It has been shown that glutamine is an effective energy substrate to support rabbit oocyte maturation (Bae & Foote, 1974). Our earlier study demonstrates that glutamine uptake peaks at about 12 h after the onset of culturing dog oocytes (Songsasen et al., 2007), suggesting its role in meiotic resumption in this species. However, addition of glutamine to a culture medium does not improve nuclear maturation rate, suggesting that this amino acid plays role in cytoplasmic maturation in the dog (Songsasen et al., 2007). In the mouse, the use of glutamine as a sole energy source (i.e., without carbohydrate supplementation) is sufficient to initiate meiotic resumption, but not enough to support the transition of GVBD to MII oocytes (Downs & Hudson, 2000). Gonadotropins have been shown to promote glutamine metabolism. Specifically, in the presence of FSH, glutamine promotes nuclear maturation of hamster oocytes (Kito & Bavister, 1997). LH stimulates cumulus cells to convert glutamine to α-ketoglutarate, which is then oxidized through TCA cycle to generate ATP (Zuelke & Brackett, 1993).

6. Lipids are important endogenous energy source during oocyte maturation

Intracellular lipids play essential roles in oocyte development (Sturmey et al., 2009, 2006; Sturmey & Leese, 2003; Ferguson & Leese, 2006). Lipids serve as reservoir of endogenous energy source, substrates for water production during blastocoel development and precursors of second messengers modulating cell functions and purine synthesis (McEvoy et al., 2000). Furthermore, membrane bound phospholipids regulate various cell functions associated with calcium flux (e.g., cortical granule exocytosis and fertilization (McEvoy et al., 2000).

Mammalian oocytes contain an endogenous lipid reserve that varies in quantity among species (ranging from 4 ng total lipid per oocyte (Loewenstein & Cohen, 1964) in the mouse to 63 ng (McEvoy et al., 2000) in the cow to 161 ng the pig (Sturmey & Leese, 2003; McEvoy et al., 2000). Triglyceride is the main lipid found in mammalian oocytes (range for 30% in sheep to 60% of total lipid in pig) (Sturmey & Leese, 2003; Sturmey et al., 2009; Ferguson & Leese, 1999). It has been shown that triglyceride stores decrease as cows (Leese & Ferguson, 1999) and pigs (Sturmey & Leese, 2003) oocytes progress through meiotic maturation, and this process is coincides with increased lipolysis (Cetica et al., 2002). These findings indicate that triglyceride provides a rich energy supply for oocyte maturation (Cetica et al., 2002) and sustaining pre-implantation embryo growth (Sturmey et al., 2006, Ferguson & Leese, 2006). Sturmey et al. (2006) utilized the fluorescence resonance energy transfer technology to examine the relationship between mitochondria and cytoplasmic lipid droplets during pig oocyte maturation and reported that the two organelles co-localize in a molecular level and form 'metabolic units' at the periphery of the gamete. There are evidence indicating that free fatty acids cleaved from triglyceride molecules stored in lipid droplets can be directly transported across the mitochondrial membrane and oxidized via β-oxidation that results in the production of acetyl CoA, a substrate for the TCA cycle (Sturmey et al., 2006; Dunning et al., 2010). Culturing pig and cow oocytes (Ferguson & Leese, 2006) in the presence of an inhibitor of carnitine palmitoyl transferase, the enzyme responsible for the transport of free fatty acids into the mitochondrial matrix compromises embryo development post-fertilization (Sturmey et al., 2006). Enhancing lipid metabolism and mitochondrial activity by addition of L-carnitine to culture medium decreases ROS levels, as well as improves MII and cleavage rates (Somfai et al., 2011), although the treatment does not impact blastocyst formation. In the mouse, β-oxidation is significantly up-regulated during both *in vivo* and *in vitro* oocyte maturation (Dunning et al., 2010). Inhibition of β-oxidation compromises meiotic resumption (Downs et al., 2009) and impairs embryonic development formation following fertilization (Dunning et al., 2010), suggesting that lipid metabolism plays essential roles in both nuclear and cytoplasmic maturation in the mouse. Furthermore, enhancing β-oxidation by supplement L-carnitine to the culture medium during in vitro follicle culture and *in vitro* maturation significantly improves mouse oocyte developmental competence (Dunning et al., 2011, 2010).

7. Species specificity in preferentially differences in energy substrate

To date, studies have revealed that there are species-specific energy requirement for oocyte maturation. Mouse oocytes require exogenous energy substrate (especially pyruvate) to complete nuclear maturation (Downs & Hudson, 2000) probably due to the limited amount of intracellular lipid (Loewenstein & Cohen, 1964). Mouse oocytes cultured in the absence of glucose and pyruvate fail to progress to the MII stage (Downs & Hudson, 2000). In contrast, cow oocytes which contain 20 fold of total lipid mass compared to the mouse are able to complete nuclear maturation in the absence of exogenous energy substrates (Sturmey et al., 2009). Furthermore, inhibition of triglyceride metabolism compromises nuclear maturation in this species (Ferguson & Leese, 2006).

Mouse (Sturmey et al., 2009; Downs et al., 2002; Downs & Mastropolo, 1994; Biggers et al., 1967), cow (Steeves & Gardner, 1999) and cat oocytes (Spindler et al., 2000) appear to prefer pyruvate to glucose as an energy source. Pyruvate supports maturation, maintains viability and promotes cleavage of the fertilized oocyte in the absence of cumulus cells (Downs et al., 2002; Downs & Mastropolo, 1994; Biggers et al., 1967). The metabolism of pyruvate is related to the stage of meiosis, with arrested oocytes (GV or MII) metabolizing less pyruvate than oocytes progressing through meiosis (Downs et al., 2002). Cow oocytes utilize pyruvate produced by the cumulus cell via glycolysis (Cetica et al., 2002). Furthermore, cow oocytes have more G6PDH than phosphofructokinase, suggesting that the gamete preferentially metabolize glucose via PPP pathway rather than glycolysis (Cetica et al., 2002).

Pig and dog oocytes utilize glucose as their primary energy substrate (Krisher et al., 2007; Songsasen & Wildt, 2007). Recent study has shown that lipid metabolism plays significant roles in oocyte maturation in the pig (Sturmey & Leese, 2003). This is not surprising since pig oocytes contain a substantial amount of triglyceride compared to other species (Sturmey & Leese, 2003). To date, there is no information on roles of intracellular lipid in dog oocyte development.

8. Summary and future perspectives

Energy metabolism is critical for oocyte maturation. Development of appropriate systems for *in vitro* oocyte maturation requires a great understanding of factors, including energy metabolism involved in the acquisition of oocyte developmental competence during folliculogenesis and during the maturation period. To date, studies have been focused on the impact of exogenous substrates, especially glucose and pyruvate on meiotic and cytoplasmic maturation. Very little attention has been centered on roles of intracellular lipid in oocyte development. Fatty acids are several-fold more energy rich than glucose. For example, 130 mols of ATP result from the oxidation of one mol of palmitic acid (C16:0), as compared to 38 mols of ATP from one mol of glucose. Thus, energy-dense lipids have potential ability to support and promote oocyte maturation and embryo development. Indeed, there has been increasing evidence that enhancing mitochondrial activity and β-oxidation lipid plays important role improve oocyte developmental competence in the mouse, a species that have small amount of intracellular lipids (Dunning et al., 2010). Furthermore, increase β-oxidation during *in vitro* folliculogenesis has been shown to promote mouse oocyte developmental competence (Dunning et al., 2011). Therefore, further research is needed to understand how the oocyte and early embryo utilize lipid substrates for energy production and whether this form of metabolism is developmentally beneficial.

Certainly, there are species-specificities in oocyte metabolism. An *in vitro* condition developed for oocytes of a certain species cannot be directly applied to another. Thus, one of future research priorities should be advance our understanding about oocyte metabolic requirement for understudied species, especially carnivores. At last, current and future development of *in vitro* culture conditions for mammalian oocytes will certainly benefit from comparative studies conducted in different animal species.

9. References

Acton, B. M., Jurisicova, A. & Casper, R. F. (2004) Alterations in mitochondrial membrane potential during preimplantation stages of mouse and human embryo development. *Molecular Human Reproduction*, 10, 23-32.

Albertini, D. F. & Limback, S. D. (2009) The natural life cycle of the mammalian oocyte. In Borini, A. and Coticchio, G. (eds), *Preservation of Human Oocytes*. Informa Healthcare, London. pp. 83-94

Augustin, R., Pocar, P., Navarrete-Santos, A., Wrenzycki, C., Gandolfi, F., Niemann, H. & Fischer, B. (2001) Glucose transporter expression is developmentally regulated in *in vitro* derived bovine preimplantation embryos. *Molecular Reproduction and Development*, 60, 370-376.

Bae, I.-H. & Foote, R. H. (1974) Utilization of glutamine for energy and protein synthesis by cultured rabbit follicular oocytes. *Experimental Cell Research*, 90, 432-436.

Biggers, J. D., Whittingham, D. G. & Donahue, R. P. (1967) The pattern of energy metabolism in the mouse oocyte and zygote. *Proceedings of National Academy of Sciences USA*, 58, 560-567.

Bilodeau-Goeseels, S. (2006) Effects of culture media and energy sources on the inhibition of nuclear maturation in bovine oocytes. *Theriogenology*, 66, 297-306.

Brinster, R. L. (1971) Oxidation of pyruvate and glucose by oocytes of the mouse and rhesus monkey. *Journal of Reproduction and Fertility*, 24, 187-191.

Cetica, P., Pintos, L., Dalvit, G. & Beconi, M. (2002) Activity of key enzymes involved in glucose and triglyceride catabolism during bovine oocyte maturation *in vitro*. *Reproduction*, 124, 675-681.

Cetica, P., Pintos, L., Dalvit, G. & Beconi, M. (2003) Involvement of enzymes of amino acid metabolism and tricarboxylic acid cycle in bovine oocyte maturation *in vitro*. *Reproduction*, 126, 753-763.

Colton, S. A., Pieper, G. M. & Downs, S. M. (2002) Altered meiotic regulation in oocytes from diabetic mice. *Biology of Reproduction*, 67, 220-231.

Dan Goor, M., Sasson, S., Davarashvili A. & Almagor, M. (1997) Expression of glucose transporter and glucose uptake in human oocytes and preimplantation embryos. *Human Reproduction*, 12, 2508-2510.

Dominko, T. & First, N. L. (1997) Timing in meiotic progression in bovine oocytes and its effect on early embryo development. *Molecular Reproduction and Development*, 47, 456-467.

Downs, S. M. & Hudson, E. D. (2000) Energy substrates and the completion of spontaneous meiotic maturation. *Zygote*, 8, 339-351.

Downs, S. M. & Mastropolo, A. M. (1994) The participation of energy substrates in the control of meiotic maturation in murine oocytes. *Developmental Biology*, 162, 154-168.

Downs, S. M. & Utecht, A. M. (1999) Metabolism of radiolabeled glucose by mouse oocytes and oocyte-cumulus cell complexes. *Biology of Reproduction*, 60, 1146-1452.

Downs, S. M. (1995) The influences of glucose, cumulus cells, and metabolic coupling on ATP levels and meiotic control in the isolated mouse oocytes. *Developmental Biology*, 167, 502-512.

Downs, S. M., Humpherson, P. G. & Leese, H. J. (1998) Meiotic induction in cumulus cell-enclosed mouse oocytes: involvement of the pentose phosphate pathway. *Biology of Reproduction*, 58, 1084-1094.

Downs, S. M., Humpherson, P. G. & Leese, H. J. (2002) Pyruvate utilization by mouse oocytes is influenced by meiotic status and the cumulus oophorus. *Molecular Reproduction and Development*, 62, 113-123.

Downs, S. M., Mosey, J. L. & Klinger, J. (2009) Fatty acid oxidation and meiotic resumption mouse oocytes. *Molecular Reproduction and Development*, 76, 844-853.

Dunning, K. R., Akison, L. K., Russell, D. L., Norman, R. J. & Robker, R. L. (2011) Increased beta-oxidation and improved oocyte developmental competence in response to L-carnitine during ovarian *in vitro* follicle development in mice. *Biology of Reproduction*, 85, 548-555.

Dunning, K. R., Cashman, K., Russell, D. L., Thompson, J. G., Norman, R. J. & Robker, R. L. (2010) Beta oxidation is essential for mouse oocyte developmental competence. *Biology of Reproduction*, 83, 909-918.

Edson, M. A., Nagaraja, A. K. & Matzuk, M. M. (2009) The mammalian ovary from genesis to revelation. *Endocrine Review*, 30, 624-712.

Edwards, R. G. & Gates, A. H. (1959) Timing of the stages of the maturation divisions, ovulation, fertilization and the first cleavage of egg of adult mice treated with gonadotrophins. *Journal of Endocrinology*, 18, 292-304.

Engle, C. C., Foley, C. W., Plotka, E. D. & Witherspoon, D. M. (1984) Free amino acids and protein concentrations in reproductive tract fluids of the mare. *Theriogenology*, 21, 919-930.

Eppig, J. J., O'Brien, M. O. & Wigglesworth, K. (1996) Mammalian oocyte growth and development *in vitro*. *Molecular Reproduction and Development*, 44, 260-273.

Facecchia, K., Fochesato, L. A., Stohs, S. J. & Pandey, S. (2011) Oxidative toxicity in neurodegenerative dieseases: role of mitochondrial dysfunction and therapeutic strategies. *Journal of Toxicology*, 683728, 1-12.

Ferguson, E. M. & Leese, H. J. (1999) Triglyceride content of bovine oocytes and early embryos. *Reproduction*, 116, 373-378.

Ferguson, E. M. & Leese, H. J. (2006) A potential role for tryglyceride as an energy sources during bovine oocyte maturation and early embryo development. *Molecular Reproduction and Development*, 73, 1195-1201.

Fujii, W. & Funahashi, H. (2009) Exogenous adenosine reduces the mitochondrial membrane potential of murine oocytes during the latter half of *in vitro* maturation and pronuclear formation following chemical activation. *Journal of Reproduction and Development*, 55, 187-193.

Funahashi, H., Koike, T. & Sakai, R. (2008) Effect of glucose and pyruvate on nuclear and cytoplasmic maturation of porcine oocytes in a chemically defined medium. *Theriogenology*, 70, 1041-1047.

Garcia, J., Han, D., Sancheti, H., Yap, L. P., Kaplowitz, N. & Cadenas, E. (2010) Regulation of mitochondrial glutathione redox status and protein glutathionylation by respiratory substrates. *Journal of Biological Chemistry*, 285, 39646-39654.

Gougeon, A. (2010) Human ovarian follicular development: From activation of resting follicles to preovulatory maturation. *Annls d' Endocrinologie*, 71, 132-143.

Griffin, J., Emery, B. R., Huang, I., Peterson, C. M. & Carrell, D. T. (2006) Comparative analysis of follicle morphology and oocyte diameter in four mammalian species (mouse, hamster, pig and human). *Journal of Experimental Clinical and Assisted Reproduction*, 3, 2.

Harris, S. E., Gopichandran, N., Picton, H. M., Leese, H. J. & Orsi, N. M. (2005) Nutrient concentrations in murine follicular fluid and the female reproductive tract. *Theriogenology*, 64, 992-1006.

Herrick, J. R., Brad, A. M. & Krisher, R. L. (2006) Chemical manipulation of glucose metabolism in porcine oocytes: effects on nuclear and cytoplasmic maturation *in vitro*. *Reproduction*, 131, 289-298.

Hong, J. & Lee, E. (2007) Intrafollicular amino acid concentration and the effect of amino acids in a defined maturation medium on porcine oocyte maturation, fertilization and preimplantation development. *Theriogenology*, 68, 728-735.

Hsieh, M., Thao, K. & Conti, M. (2011) Genetic dissection of epidermal growth factor receptor signaling during luteinizing hormone-induced oocyte maturation. *PLoS One*, 6, e21574.

Jimena, P., Castilla, J., A., Peran, F., Ramirez, J. P., Gil, T. & Mozas, J. (1993) Distribution of free amino acids in human preovulatory follicles. *Hormone and Metabolism Research*, 25, 228-230.

Jozwik, M., Jozwik, M., Teng, C. & Battaglia, F. C. (2006) Amino acid, ammonia and urea concentration in human pre-ovulatory ovarian follicular fluid. *Human Reproduction*, 21, 2776-2782.

Kito, S. & Bavister, B. D. (1997) Gonadotropins, serum, and amino acids alter nuclear maturation, cumulus expansion, and oocyte morphology in hamster cumulus-oocyte complexes *in vitro*. *Biology of Reproduction*, 56, 1281-1289.

Krisher, R. L. & Bavister, B. D. (1999) Enhanced glycolysis after maturation of bovine oocytes *in vitro* is associated with increased developmental competence. *Molecular Reproduction and Development*, 53, 19-26.

Krisher, R. L., Brad, A. M., Herrick, J. R., Sparman, M. L. & Swain, J. E. (2007) A comparative analysis of metabolism and viability in porcine oocytes during *in vitro* maturation. *Animal Reproduction Science*, 98, 72-96.

Leese, H. J. & Ferguson, E. M. (1999) Triglyceride content of bovine oocytes and early embryos. *Journal of Reproduction and Fertility*, 116, 373-378.

Loewenstein, J. E. & Cohen, A. (1964) Dry mass, lipid content and protein content of the intact and zona--free mouse ovum. *Journal Embryology Experiment and Morphology*, 12, 113-121.

McEvoy, T. G., Coull, G. D., Broadbent, P. J., Hutchinson, J. S. & Speake, B. K. (2000) Fatty acid composition of lipids in immature cattle, pig and sheep oocytes with intact zona pellucida. *Journal of Reproduction and Fertility*, 118, 163-170.

Newsholme, P., Lima, M. M. R., Procopio, J., Pithon-Curi, T. C., Doi, S. Q., Bazotte, R. B. & Curi, R. (2003) Glutamine and glutamate as vital metabolites. *Brazilian Journal of Medical and Biological Research*, 36, 153-163.

Orsi, N. M., Gopichandran, N., Leese, H. J., Picton, H. M. & Harris, S. E. (2005) Fluctuations in bovine ovarian follicular fluid throughout the oestrous cycle: a comparison with plasma and a TCM-199-based maturation medium. *Reproduction*, 129, 229-334.

Ott, M., Gogvadze, V., Orrenious, S. & Zhivotovsky, B. (2007) Mitochondria, oxidative stress and cell death. *Apoptosis*, 12, 913-922.

Perrelli, M.-G., Pagliaro, P. & Penna, C. (2011) Ischemia/reperfusion injury and cardioprotective mechanisms: role of mitochondria and reactive oxygen species. *World Journal of Cardiology*, 3, 186-200.

Peter, H. & McNatty, K. P. (1980) *The Ovary*, University of California Press, Berkley and Los Angeles. 175 pp.

Picton, H. M. (2001) Activation of follicle development: the primordial follicle. *Theriogenology*, 55, 1193-1210.

Pisani, L., F., Antonini, S., Pocar, P., Ferrari, S., Brevini, T. A., Rhind, S. M. & Gandolfi, F. (2008) Effects of pre-mating nutrition on mRNA levels of developmentally relevant genes in sheep oocytes and granulosa cells. *Reproduction*, 136, 303-312.

Purcell, S. H. & Moley, K. H. (2009) Glucose transporters in gametes and preimplantation embryos. *Trends in Endocrinology and Metabolism*, 20, 483-489.

Ramalho-Santos, J., Varum, S., Amaral, S., Mota, P. C., Sousa, A. P. & Amaral, A. (2009) Mitochondrial functionality in reproduction: from gonads and gametes to embryos and embryonic stem cells. *Human Reproduction Update*, 15, 553-572.

Reynaud, K., Fontbonne, A., Marseloo, N., Thoumire, S., Chebrout, M., de Lesegno, C. V. & Chastant-Maillard, S. (2005) *In vivo* meiotic resumption, fertilization and early embryonic development in the bitch. *Reproduction*, 130, 193-201.

Reynaud, K., Gicquel, C., Thoumire, S., Chebrout, M., Ficheux, C., Bestandji, M. & Chastant-Maillard, S. (2009) Folliculogenesis and morphometry of oocyte and follicle growth in the feline ovary. *Repro Domes Anim*, 44, 174-179.

Rieger, D. & Loskutoff, N. M. (1994) Changes in the metabolism of glucose, pyruvate, glutamine and glycine during maturation of cattle oocytes *in vitro*. *Journal of Reproduction and Fertility*, 100, 257-262.

Roberts, R., Stark, J., Latropoulou, A., Becker, D. L., Franks, S. & Hardy, K. (2004) Energy substrate metabolism of mouse cumulus-oocyte complexes: response to follicle-stimulating hormone is mediated by the phosphatidylinositol 3-kinase pathway and is associated with oocyte maturation. *Biology of Reproduction*, 71, 199-209.

Sato, H., Iwata, H., Hayashi, T., Kimura, K., Kuwayama, T. & Monji, Y. (2007) The effect of glucose on the progression of the nuclear maturation of pig oocytes. *Animal Reproduction Science*, 99, 299-305.

Somfai, T., Kaneda, M., Akagi, S., Watanabe, S., Haraguchi, S., Mizutani, E., Dang-Nguyen, T. Q., Geshi, M., Kikuchi, K. & Nagai, T. (2011) Enhancement of lipid metabolism with L-carnitine during *in vitro* maturation improves nuclear maturation and cleavage ability of follicular porcine oocytes. *Reproduction Fertility and Development*, 23, 912-920.

Son, W. Y., Das, M., Shalom-Paz, E. & Holzer, H. (2011) Mechanisms of follicle selection and development. *Minerva Ginecologica*, 63, 89-102.

Songsasen, N. & Wildt, D. E. (2007) Oocyte biology and challenges in developing *in vitro* maturation systems in the domestic dog. *Animal Reproduction Science*, 98, 2-22.

Songsasen, N., Fickes, A., Pukazhenthi, B. S. & Wildt, D. E. (2009) Follicular morphology, oocyte diameter and localization of fibroblast growth factors in the domestic dog ovary. *Reproduction in Domestic Animals*, 44 (Suppl 2), 65-70.

Songsasen, N., Spindler, R. E. & Wildt, D. E. (2007) Requirement for, and patterns of, pyruvate and glutamine metabolism in the domestic dog oocyte *in vitro*. *Molecular Reproduction and Development*, 74, 870-877.

Songsasen, N., Wesselowski, S., Carpenter, J. W., Wildt, D. E. (2012) The ability to achieve meiotic maturation in the dog oocyte is linked to glycolysis and glutamine oxidation. *Molecular Reproduction and Development*, in press.

Spindler, R. E., Pukazhenthi, B. S. & Wildt, D. E. (2000) Oocyte metabolism predicts the development of cat embryos to blastocyst *in vitro*. *Molecular Reproduction and Development*, 56, 163-171.

Steeves, T. E. & Gardner, D. K. (1999) Metabolism of glucose, pyruvate and glutamine during the maturation of oocytes derived from pre-pubertal and adult cows. *Molecular Reproduction and Development*, 54, 92-101.

Sturmey, R. G. & Leese, H. J. (2003) Energy metabolism in pig oocytes and early embryos. *Reproduction*, 126, 197-204.

Sturmey, R. G., O'Toole, P. J. & Leese, H. J. (2006) Fluorescence resonance energy transfer analysis of mitochondrial lipid association in the porcine oocyte. *Reproduction*, 132, 829-837.

Sturmey, R. G., Reis, A., Leese, H. J. & McEvoy, T. G. (2009) Role of fatty acids in enery provision during oocyte maturation and early embryo development. *Reproduction of Domestic Animals*, 44, 50-58.

Sun, Q. Y., Miao, Y. L. & Schatten, H. (2009) Towards a new understanding on the regulation of mammalian oocyte meiotic resumption. *Cell Cycle*, 8, 2741-2747.

Sun, Q. Y., Wu, G. M., Lai, L., Park, K. W., Cabot, R., Cheong, H. T., Day, B. N., Prather, R. S. & Schatten, H. (2001) Translocation of active mitochondria during pig oocyte maturation, fertilization and early embryo development *in vitro*. *Reproduction*, 122, 155-163.

Sutton, M. L., Gilchrist, R. B. & Thompson, J. G. (2003) Effects of *in-vivo* and *in-vitro* environments on the metabolism of the cumulus-oocyte complex and its influence on oocyte developmental capacity. *Human Reproduction Update*, 9, 35-48.

Sutton-McDowall, M. L., Gilchrist, R. B. & Thompson, J. G. (2010) The pivotal role of glucose metabolism in determining oocyte developmental competence. *Reproduction*, 139, 685-695.

Tingen, C., Kim, A. & Woodruff, T. K. (2009) The primordial pool of follicles and nest breakdown in mammalian ovaries. *Molecular Human Reproduction*, 15, 795-803.

Urner, F. & Sakkas, D. (2005) Involvement of the pentose phosphate pathway and redox regulation in fertilization in the mouse. *Molecular Reproduction and Development*, 70, 494-503.

Van Blerkom, J. & Davis, P. (2007) Mitochondrial signaling and fertilization. *Molecular Human Reproduction*, 13, 759-779.

Van Blerkom, J. (2004) Mitochondria in human oogenesis and preimplantation embryogenesis: engines of metabolism, ionic regulation and developmental competence. *Reproduction*, 128, 269-280.

Van Blerkom, J., Davis, P. & Alexander, S. (2003) Inner mitochondrial membrane potential (DYm), cytoplasmic ATP content and free Ca^{2+} levels in metaphase II mouse oocytes. *Human Reproduction*, 18, 2429-2440.

Van Blerkom, J., Davis, P., Mathwig, V. & Alexander, S. (2002) Domains of high-polarized and low-polarized mitochondria may occur in mouse and human oocytes and early embryos. *Human Reproduction*, 17, 393-406.

Voet, D. & Voet, J. G. (2004) *Biochemistry*. 3 edn, John, Wiley & Sons, Inc., Hoboken, NJ. 1591 pp.

Wang, Q., Frolova, A. I., Purcell, S., Adastra, K., Schoeller, E., Chi, M. M., Schedl, T. & Moley, K. H. (2010) Mitochondrial dysfunction and apoptosis in cumulus cells of type I diabetic mice. *PLoS One*, 5, e15901.

Watson, A. J. (2007) Oocyte cytoplasmic maturation: A key mediator of oocyte and embryo developmental competence. *Journal of Animal Sciences*, 85 (E. Suupl.), E1-E3.

Watson, A. J., de Sousa, P., Caveney, A., Barcroft, L., Natale, D., Urquhart, J. & Westhusin, M. E. (2000) Impact of bovine oocyte maturation media on oocyte transcript levels, blastocyst development, cell number, and apotosis. *Biology of Reproduction*, 62, 355-364.

Yu, Y., Dumollard, R., Rossbach, A., Lai, F. A. & Swann, K. (2010) Redistribution of mitochondria leads to bursts of ATP production during spontaneous mouse oocyte maturation. *Journal of Cellular Physiology*, 224, 672-680.

Zheng, P., Bavister, B. D. & Ji, W. (2001) Energy substrate requirement for *in vitro* maturation of oocytes from unstimulated adult rhesus monkeys. *Molecular Reproduction and Development*, 58, 348-355.

Zheng, P., Bavister, B. D. & Ji, W. Z. (2002) Amino acid requirements for maturation of rhesus monkey (*Macacca mulatta*) oocytes in culture. *Reproduction*, 124, 515-522.

Zheng, P., Vassena, R. & Latham, K. E. (2007) Effects of *in vitro* oocyte maturation and embryo culture on the expression of glucose transporters, glucose metabolism and insulin signaling genes in rhesus money oocytes and preimplantation embryos. *Molecular Human Reproduction*, 13, 361-371.

Zuelke, K. A. & Brackett, B. G. (1992) Effects of luteinizing hormone on glucose metabolism in cumulus-enclosed bovine oocytes matured *in vitro*. *Endocrinology*, 131, 2690-2696.

Zuelke, K. A. & Brackett, B. G. (1993) Increased glutamine metabolism in bovine cumulus cell-enclosed and denuded oocytes after *in vitro* maturation with lutenizing hormone. *Biology of Reproduction*, 48, 815-820.

2

The Control of Meiotic Arrest and Resumption in Mammalian Oocytes

Sylvie Bilodeau-Goeseels and Nora Magyara
Agriculture and Agri-Food Canada, Lethbridge Research Centre,
Canada

1. Introduction

In mammals, following implantation of the embryo, a small number of cells from the epiblast eventually become the precursors of primordial germ cells (reviewed by Edson et al., 2009). After mitotic proliferation of the primordial germ cells, migration to the genital ridges and rapid proliferation again, meiosis is initiated in the oogonia at about day 13 of development in mice, at day 82 in bovine and during the 3rd month in humans. A last round of DNA synthesis occurs after which the oocytes enter a long meiotic prophase which consists of different stages (leptotene, zygotene, pachytene). By the time of birth, oocytes of most mammals have reached the diplotene stage, meiosis is arrested and the nucleus is referred to as a germinal vesicle (GV). The timing of progression of oocytes through meiosis in relation to birth varies in different mammals and in a few species meiotic prophase is initiated only after birth (i.e., rabbits and cats). At about the time of birth, oocytes become surrounded by follicle cells and these primordial follicles and oocytes represent a large stockpile from which, at any given time, a few are selected to grow and develop. The gradual depletion of primordial follicles through either growth or degeneration (atresia) continues until menopause.

As oocytes grow, they also acquire the competence to re-initiate meiosis. In mammals, re-initiation of meiosis in the pre-ovulatory follicle is induced by the luteinizing hormone (LH) surge. The germinal vesicle membrane breaks down [this stage is called germinal vesicle breakdown (GVBD)] and chromosomes separate from one another: one set is extruded into the first polar body, whereas the other set aligns at the second metaphase (MII) plate. In most mammals, meiosis is arrested again at this stage until activation by a spermatozoon. When mammalian oocytes that are competent to re-initiate meiosis are removed from their follicles and cultured, they undergo spontaneous resumption of meiosis with progression to MII in the absence of gonadotropins, demonstrating that a signal(s) from the follicle holds oocytes in prophase arrest. Spontaneous resumption of meiosis was first observed by Pincus & Enzmann (1935) in rabbit oocytes and was subsequently observed in oocytes from other mammalian species (Edwards, 1965). Spontaneous oocyte maturation allowed the development of *in vitro* maturation (IVM), a reproductive technology which involves artificial removal of cumulus-oocyte complexes (COC) from antral follicles and culturing them in standard cell culture conditions for 24-48 h until they reach metaphase II. A proportion of these oocytes are then competent to develop following *in vitro* fertilization (IVF).

The cellular and molecular mechanisms involved in maintaining oocyte meiotic arrest for prolonged intervals in the ovary are not fully understood. Similarly, the chain of molecular and cellular events that leads to meiotic resumption, either *in vivo* in response to the gonadotropin surge, or spontaneously *in vitro*, has not been completely elucidated even though dramatic advances have been made in recent years in understanding the control of meiosis in rodent oocytes. Research aimed at understanding the mechanisms that control meiosis has relevance for practical applications as the oocyte, in addition to its important role in determining fertility, is a major player in reproductive biotechnologies such as IVM, IVF, cloning and transgenesis. For example, in livestock species, the efficiency of *in vitro* embryo production remains low as only approximately 40% of the fertilized oocytes develop to the blastocyst stage. One reason for this inefficiency is that although the oocytes are at the correct nuclear maturation stage (MII) after IVM, several of them did not have sufficient time to complete cytoplasmic maturation; i.e., they did not have time to accumulate all the mRNAs and proteins required for early development as RNA synthesis ceased when meiosis resumed. Physiological inhibition of spontaneous nuclear maturation *in vitro* for a period of time sufficient to complete cytoplasmic maturation, will be required to improve developmental competence. Similarly, the cornerstone of any wild animal fertility preservation approach is the capacity for successful *in vitro* culture of gametes and embryos (Comizzoli et al., 2010). Moreover, IVM has the potential to exploit the large supply of oocytes available within ovaries in the case of ovariectomy or death of a wild female donor. In 2009, the International Committee for Monitoring Assisted Reproductive Technologies reported that over 200,000 babies are born annually from assisted reproductive technologies (de Mouzon et al., 2009). Currently, the use of IVM for the treatment of human infertility is not widespread due to its low efficiency (Suikkari, 2008); however, IVM represents an attractive alternative as it would reduce the use of gonadotropins for ovarian stimulation, thereby reducing side effects and costs for patients. This chapter reviews research aimed at understanding the signalling pathways involved in the control of meiotic cell cycle arrest at the diplotene (GV) stage as well as in the control of meiotic resumption (GVBD) in oocytes from selected mammalian species.

2. The control of meiosis in rodent oocytes

2.1 Early data on the role of cAMP

Cyclic adenosine monophosphate (cAMP) is a second messenger which is responsible for the transduction of hormonal signals in a wide range of organisms. Cyclic AMP is synthesized from ATP by the enzyme adenylate cyclase (AC) which is located on the inner side of the plasma membrane. Adenylate cyclase is activated by a range of signalling molecules through the activation of adenylate cyclase stimulatory G (Gs)-protein-coupled receptors and inhibited by agonists of adenylate cyclase inhibitory G (Gi)-protein-coupled receptors. Cyclic AMP is degraded to 5'-AMP by phosphodiesterase (PDE) enzymes.

Cyclic AMP was thought early on to play a critical role in the control of meiotic maturation as LH, human chorionic gonadotropin (hCG), follicle-stimulating hormone (FSH) and prostaglandin E_2 (PGE$_2$) were effective in inducing meiotic resumption in rat oocytes cultured in intact follicles (Tsafriri et al., 1972). Additionally, injection of dibutyryl cyclic AMP (dbcAMP) into the follicular antrum also induced GVBD while LH and PGE$_2$ increased cAMP levels over 20-folds in follicles suggesting that the stimulatory effect of the hormones on maturation was mediated by cAMP (Tsafriri et al., 1972). Subsequently, several other studies, performed with cumulus-oocyte complexes cultured without follicles, provided evidence that cAMP levels within the oocyte determine meiotic status such that high levels result in meiotic arrest while low levels lead to re-initiation of meiotic maturation (Aberdam et al., 1987; Dekel et al., 1984; Schultz et al., 1983a). Therefore, cAMP appeared to have a dual role in the control of meiosis: mediation of the stimulatory action of gonadotropins on the follicle cells (oocytes do not have gonadotropin receptors) and an inhibitory effect on the oocyte itself. Challenging questions being addressed were: whether or not the oocyte itself could synthesize cAMP and whether the amount of cAMP synthesized by the oocyte was sufficient for meiotic arrest. Several studies with rodent oocytes suggested that cAMP derived from follicle cells was responsible for maintaining meiotic arrest as cAMP-elevating agents such as AC stimulators, PDE inhibitors and membrane-permeable cAMP derivatives did not inhibit meiosis in oocytes denuded of their cumulus cells [DO, (Dekel & Beers, 1978, 1980; Dekel et al., 1984; Racowsky, 1984)] and cAMP transfer from the cumulus cells to the oocyte was demonstrated (Bornslaeger & Schultz, 1985). However, other studies showed that the oocyte could synthesize cAMP and that cAMP-elevating agents could prevent GVBD in DO (Cho et al., 1974; Olsiewski & Beers, 1983; Urner et al., 1983).

2.2 Events downstream of cAMP and the control of maturation promoting factor

Maturation promoting factor (MPF) is a key regulator of the mitotic and meiotic cell cycle and integrates the signals from several pathways to control mitosis and meiosis. It is a serine-threonine kinase protein heterodimer composed of a catalytic subunit, cyclin-dependent kinase 1 (CDK1, also known as p34^{cdc2}), and a regulatory subunit, cyclin B. The activated form of MPF involves dephosphorylation at Thr14 and Tyr15 of CDK1 and its association with cyclin B (Clarke & Karsenti, 1991). For meiotic arrest to be maintained at prophase I, MPF must be kept inactive. How do high levels of cAMP in the oocyte maintain MPF inactive? In general, cAMP acts via protein kinase A (PKA). Recent studies [reviewed by Han & Conti (2006)] in mouse and Xenopus oocytes indicate that PKA directly regulates the activities of a kinase (Wee1B) and a phosphatase (Cdc25) for CDK1. High cAMP levels in oocytes result in active PKA, active Wee1B, and the phosphorylation and inactivation of CDK1 while a decrease in oocyte cAMP levels leads to PKA inactivation, Cdc25 activation, dephosphorylation of CDK1 and MPF activation. Regulation of the level of cyclin B by synthesis and degradation is also involved in the control of MPF activity (Ledan et al., 2001).

2.3 New findings on the control of meiosis in rodent oocytes

It is no longer believed that cAMP from follicle cells contributes significantly to mouse oocyte meiotic arrest. Horner et al. (2003) showed that mRNA and protein for the adenylate cyclase 3 (AC3) isoform were detected in rat and mouse oocytes and that the regulation of cAMP levels in oocytes indicated that the main AC from rat oocytes possessed properties similar to AC3 of somatic cells. Moreover, more than 50% of early antral follicles from AC3-deficient mice contained oocytes that had resumed meiosis (10-15% in wild-type oocytes), providing evidence that rodent oocytes contain a functional AC involved in meiotic arrest (Horner et al., 2003).

The development of a method to inject mouse oocytes within antral follicles allowed experiments to be performed that would shed more light on the mechanisms involved in the control of meiosis. Microinjection of a Gs antibody caused resumption of meiosis in follicle-enclosed oocytes indicating that inhibition of Gs in the oocyte leads to meiotic resumption and conversely, Gs activity in the oocyte is required to maintain meiotic arrest in the follicle (Mehlmann et al., 2002). Similarly, microinjection of a dominant negative form of Gs into mouse oocytes (also in Xenopus oocytes) caused resumption of meiosis (Kalinowski et al., 2004). Since Gs by itself has no constitutive activity, it was suspected that a receptor is required to activate Gs. A potential receptor, the orphan receptor GPR3, was identified through searching an expressed sequence tag database derived from a cDNA library obtained from mouse oocytes (Mehlmann et al., 2004). This receptor was of interest because it elevated cAMP when expressed in various cell lines. *In situ* hybridization showed that GPR3 RNA was localized in oocytes and 82% of the oocytes in antral follicles of ovaries from prepubertal GPR3 knockout mice had resumed meiosis (0% in wild-type, Mehlmann et al., 2004). Injection of GPR3 RNA into GPR3-/- oocytes reversed the knockout phenotype indicating that the GPR3 of the oocyte itself maintained meiotic arrest rather than the GPR3 of the follicle cells (Mehlmann et al., 2004). From these results, it appeared that the meiosis inhibitory signal was a GPR3 agonist which activated Gs and AC3; however, removal of oocytes from their follicles did not decrease Gs activation by GPR3 (Freudzon et al., 2005). Therefore, although GPR3 and Gs are required to maintain meiotic arrest, the signal from follicle cells acts by a mechanism other than providing a GPR3-activating ligand to maintain meiotic arrest.

If the signal from follicle cells to inhibit meiosis is not a GPR3-activating ligand then what are the other potential signals? The signal may act via gap junctions as initially believed, since gap junction inhibitors caused a decrease in oocyte cAMP and meiotic resumption (Norris et al., 2008; Sela-Abramovich et al., 2006). This brings us back to the original hypothesis that cAMP from somatic cells is transferred to the oocyte and inhibits meiosis. However, because the major oocyte phosphodiesterase is PDE3A (Masciarelli et al., 2004) which is inhibited by cyclic guanosine monophosphate (cGMP), and various studies showed that cGMP could be involved in meiotic arrest (Sela-Abramovich et al., 2008; Törnell et al., 1990), it was hypothesized that cGMP from the somatic cells could reach the oocyte through gap junctions, inhibit PDE3A and maintain meiotic arrest. Using Förster resonance energy transfer-based cyclic nucleotide sensors in follicle-enclosed oocytes, Norris et al. (2009) showed that cGMP does pass through gap junctions into the oocyte to contribute to the maintenance of high cAMP levels by inhibiting PDE3A.

How does LH stimulation of follicles lead to oocyte meiotic resumption? In the study from Norris et al. (2009), LH stimulation lowered cGMP levels in follicle cells and closed gap junctions. As a result, cGMP levels in the oocyte also decreased, PDE3A activity increased approximately 5-folds, oocyte cAMP decreased also approximately 5-folds and meiosis resumed (Norris et al., 2009). It is possible that LH could regulate oocyte cAMP via other mechanisms; for example, activation of Gi family G protein can inhibit AC or stimulate PDE thereby lowering cAMP. However, injection of pertussis toxin (a Gi inhibitor) into follicle-enclosed mouse oocytes did not prevent LH-induced meiotic resumption (Mehlmann et al., 2006). Similarly, the LH response was not prevented by inhibition of a Ca^{2+} elevation by injection of EGTA into follicle-enclosed mouse oocytes (Mehlmann et al., 2006) indicating that LH does not regulate cAMP in the oocytes via a Gi family G protein or calcium. Further studies also showed that LH does not act by terminating receptor-G(s) signalling (Norris et al., 2007).

There are still a number of questions remaining to answer to complete this exciting story such as: whether there is a direct role for somatic cell cAMP in meiotic inhibition. What maintains high cGMP levels? How does LH decrease cGMP in somatic cells: through inhibition of a guanylate cyclase or stimulation of a cGMP-specific phosphodiesterase? Further studies showed that the epidermal growth factor (EGF) pathway is required for LH-induced gap junction closure and a portion or all of the cGMP decrease (Norris et al., 2010; Vaccari et al., 2009). Another important question is how do other signalling pathways that can affect meiosis fit into this model? For example, several studies demonstrated that the energy sensor adenosine monophosphate-activated protein kinase (AMPK) plays a role in controlling the resumption of nuclear maturation in mouse oocytes. Pharmacological activation of AMPK induced nuclear maturation in cumulus-enclosed and denuded oocytes that had been arrested with dibutyryl cAMP or hypoxanthine (Downs et al., 2002). Moreover, oocyte AMPK was also activated by hormones and stress; this activation preceded GVBD, and inhibiting AMPK activity blocked the effects of these stimuli (Chen et al., 2006; Chen & Downs, 2008; LaRosa & Downs, 2006, 2007). A recent study provided evidence that AMPK is also involved after GVBD to promote the completion of meiosis and to prevent premature activation (Downs et al., 2010). In conclusion, these studies show that the control of oocyte meiosis probably involves a complex network of cross-talk between several signalling pathways in the oocyte and also in cumulus cells.

3. The control of meiosis in bovine oocytes

Research on the control of meiosis in livestock species has to accommodate a number of variables not encountered in laboratory animals. For example, the majority of studies were performed with oocytes recovered from ovaries collected from slaughtered animals. Although most researchers collect oocytes from a narrow range of follicle sizes and further select the oocytes based on morphological criteria, it remains that the selected oocytes represent a mixed population originating from follicles at varying stages of development, dominance and/or atresia. Moreover, contrary to laboratory animals for which all animals in an experiment are synchronized and treated the same way, the livestock females from which the oocytes are recovered are at different stages of the estrus cycle and are exposed to different environments such as nutrition, temperature or stress level, all of which can potentially affect reproductive cells. In addition, the generation of knockout animals to study molecules of interest is currently not possible in livestock species; therefore, definitive conclusions on the role of a specific enzyme isoform in a biological process can sometimes not be reached. Our understanding of the control of oocyte meiosis in livestock species has relied on the study of the presence of mRNA and proteins, the study of protein phosphorylation and, to a large extent, on pharmacological studies.

3.1 Role of cAMP in the control of bovine oocyte meiosis

As in rodent oocytes, several early studies showed that cAMP-elevating agents could decrease meiotic maturation of bovine cumulus-enclosed oocytes (Jagiello et al., 1975, 1981). However, higher concentrations of the cAMP-elevating agents were necessary to transiently increase the percentage of bovine oocytes remaining at the GV stage after *in vitro* culture

(Jagiello et al., 1981, Sirard & First, 1988) compared to concentrations that were effective in rodent oocytes. Since AC was detected on their plasma membrane (Kuyt et al., 1988), it was concluded that bovine oocytes could possibly generate sufficient amount of cAMP for meiotic arrest. In support of this hypothesis, the AC stimulator forskolin (FSK) transiently inhibited GVBD in oocytes denuded of cumulus cells (Homa, 1988), or denuded of both cumulus cells and zona pellucida thereby eliminating the possibility that the AC in the cumulus cell projections embedded in the zona pellucida was contributing to the FSK effect (Bilodeau et al., 1993). However, transfer of cAMP from cumulus cells to the oocyte was indirectly demonstrated in a few studies. For example, the culture of COCs with FSK led to ~20 to 90-fold increases in intra-oocyte cAMP compared to oocytes from complexes cultured in control conditions, whereas the culture of denuded oocytes with FSK resulted in only 2 to 6-fold increases in cAMP levels (Bilodeau et al., 1993; Thomas et al., 2004).

The enzyme AC possesses two metal ion-binding sites in its active site and Mn^{2+} is an AC activator (Tesmer et al., 1999). Moreover, a Mn^{2+}-dependent soluble AC is present in sperm cells (Braun & Dods, 1975). Culture of bovine oocytes in the presence of $MnCl_2$ resulted in an increase in the percentage of cumulus-enclosed oocytes (CEO) remaining at the GV stage after 7 h of culture, a decrease in the percentage of oocytes reaching the MII stage after 22 h of culture, and a six fold cAMP increase in complexes (Bilodeau-Goeseels, 2001, 2003a). An increase in cAMP in oocytes from complexes in response to $MnCl_2$ was detected in the presence of FSK only. In contrast to CEO, Mn^{2+} increased the percentage of denuded oocytes resuming meiosis, but also increased cAMP in the presence of FSK only (Bilodeau-Goeseels, 2003a). These results could be explained by the fact that Mn^{2+} is an activator of AC; however, Mn^{2+} is also a cofactor for many enzymes and the addition of $MnCl_2$ to the culture medium could possibly alter the activity of other enzymes involved in meiotic progression. The fact that inhibition of protein kinase A (PKA) activity abrogated the inhibitory effect of $MnCl_2$ in bovine CEO (Bilodeau-Goeseels, 2003a) supports the former possibility. Moreover, stimulation of protein kinase C (PKC) also abrogated the inhibitory effect of Mn^{2+} on meiosis (Bilodeau-Goeseels, 2003a) suggesting that the inhibitory effect of Mn^{2+} can be due to activation of AC4, 6, or 9 as these isoforms are inhibited by PKC (Sadana & Dessauer, 2009). Taken together, these results indicate that bovine oocytes and cumulus cells may contain AC isoforms with different sensitivities to Mn^{2+}. Lastro et al., (2006) determined that bovine cumulus cells obtained from follicles 2-8 mm contained mRNA for AC isoforms 1, 3, 4, 6, and 9 and were enriched in PKC-inhibited isoforms 4 and 6 and the Ca^{2+}-stimulated isoform 1. However, the AC isoforms of bovine oocytes have not yet been determined.

3.2 Role of phosphodiesterases in the regulation of bovine oocyte meiosis

As was the case with dbcAMP, the non-specific PDE inhibitor 3-isobutyl-1-methylxanthine (IBMX) inhibited bovine oocyte meiotic resumption partially and transiently (Sirard, 1990; Sirard & First, 1988) when used at much higher concentrations than concentrations that were inhibitory in rodent oocytes (Dekel et al., 1988; Schultz et al., 1983b; Vivarelli et al., 1983). In bovine oocytes, specific inhibition of PDE3 in vitro delayed spontaneous meiotic maturation and increased cAMP levels in cumulus cells and in denuded oocytes (Bilodeau-Goeseels, 2003b; Mayes & Sirard, 2002; Thomas et al., 2002). A study of PDE isoforms present in bovine follicle components showed that PDE3 accounted for 80% of the PDE activity in the oocyte, while PDE8 activity accounted for the remaining 20 and 60% of PDE activity in oocytes and cumulus cells, respectively. Inhibition of PDE8 in bovine COCs increased cAMP levels in oocytes and delayed meiosis (Sasseville et al., 2009). Collectively, these results suggest that cAMP and PDE enzymes are involved in the control of meiosis in bovine oocytes. However, even when cAMP levels were increased substantially and/or PDE inhibitors specific for the isoforms present in oocytes were used; meiosis was generally delayed and not totally inhibited as in rodent oocytes. It is suspected that other pathways acting in synergy with cAMP are probably important for the control of meiotic arrest in bovine oocytes.

3.3 Role of AMPK in the control of bovine oocyte meiosis

Contrary to results obtained with mouse oocytes where AMPK activation led to meiotic resumption (see section 2.3), meiosis was inhibited when bovine oocytes were cultured with the AMPK activator 5'-aminoimidazole-4-carboxamide 1-β-D-ribofuranoside (AICAR, Bilodeau-Goeseels et al., 2007). The inhibitory effect of AICAR was observed in cumulus-enclosed and denuded oocytes, was reversible, increased the inhibitory effect of the AC activator FSK, was dependent on its phosphorylation by adenosine kinase and was not due to increased cAMP levels or to increased purine nucleotide synthesis (Bilodeau-Goeseels et al., 2007). Metformin [MET, one of the most widely used drugs for the treatment of type 2 diabetes and an AMPK activator (Zhou et al., 2001)], was also reported to inhibit GVBD in bovine CEO and DO by Bilodeau-Goeseels et al. (2007). In a subsequent study, however, the AMPK inhibitor Compound C (CC) did not reverse the effect of AICAR and MET and even had a significant inhibitory effect itself on bovine oocyte meiosis (Bilodeau-Goeseels et al., 2011). In mouse oocytes, CC prevented AICAR-induced maturation and on its own had a slight inhibitory effect on GVBD (Chen et al., 2006).

No changes in the ratio of Thr172 phosphorylation (phosphorylation of Thr172 of the α subunit of AMPK is required for activation) to total AMPK were detected in extracts of cumulus cells, CEO or DO that had been treated with AICAR or MET at concentrations that inhibited meiosis (Bilodeau-Goeseels et al., 2011) suggesting that the inhibitory effect of AICAR and MET on bovine oocyte meiosis was not due to AMPK activation in cumulus cells or the oocyte. Different results were obtained in another laboratory where MET inhibited meiosis in CEO but not in DO, CC accelerated GVBD and culture with 5 or 10 mM MET [2 mM was used by Bilodeau-Goeseels et al. (2007)] resulted in increased Thr172 phosphorylation in cumulus cells and oocytes from complexes (Tosca et al., 2007). The different results were likely due to the different MET concentrations used and/or the different culture conditions. In both studies, but especially in the Bilodeau-Goeseels et al. 2011 study, there was already a certain level of Thr172 phosphorylation at the onset of culture (t = 0) and this could have made small changes in Thr172 phosphorylation in response to MET (and also AICAR) difficult to detect. In conclusion, more studies are needed to determine the extent of AMPK involvement in the control of meiosis in bovine oocytes. More specifically, the determination of the effects of culture conditions on basal AMPK activation and on the effects of activators and inhibitors, as well as the identification of the upstream kinase(s) that phosphorylates Thr172 will be the next steps.

3.4 Other signalling pathways involved in the control of bovine oocyte meiosis

As mentioned above, AICAR and MET may not inhibit bovine oocyte meiosis through AMPK activation and therefore, may act through other inhibitory signalling molecules. Additionally, since the inhibitory effect of cAMP is transient and high concentrations are required, it is possible that cAMP acts in synergy with another signalling pathway(s) to inhibit meiosis in bovine oocytes. Is the cGMP pathway as involved in the control of bovine oocyte meiosis as it is in rodent oocytes? The cGMP analog 8-bromo-cGMP as well as the guanylate cyclase stimulators atrial natriuretic peptide and protoporphyrin 1X did not have any effect on bovine oocyte meiosis after 7h of culture (Bilodeau-Goeseels, 2007), while cGMP derivatives inhibited spontaneous nuclear maturation in rat denuded oocytes (Törnell et al., 1990). The cGMP pathway can also be activated by nitric oxide (NO), which is synthesized by NO synthase (NOS) and activates soluble cytoplasmic guanylate cyclase. Inducible NOS (iNOS) inhibition studies suggested that endogenous NO is necessary for spontaneous nuclear maturation (Bilodeau-Goeseels, 2007) and/or the MI to MII transition (Matta et al., 2009; Schwarz et al., 2010). Yet, NO donors also inhibited GVBD (Bilodeau-Goeseels, 2007; Schwarz et al., 2008; Viana et al., 2007) but the inhibitory effect of the NO donor sodium nitroprusside was not through the cGMP/protein kinase G (PKG) pathway (Bilodeau-Goeseels, 2007). In mouse oocytes, NO donors reversed the inhibitory effect of NOS inhibitor on meiosis (Bu et al., 2003), demonstrating yet again differences in the control of meiosis between bovine and rodent oocytes.

There are even fewer studies on the role of other signalling pathways in the control of bovine oocyte meiosis, and meiotic resumption has not been studied as extensively as meiotic arrest in oocytes from livestock species. The role of Ca^{2+} in bovine oocyte fertilization and activation of development has been studied extensively (Tosti et al., 2002) and it appears to also be necessary for GVBD and progression of meiosis (He et al., 1997; Homa, 1995). Similarly, few studies have examined the role of PKC in bovine oocytes. Protein kinase C activation accelerated GVBD (Mondadori et al., 2008; Rose-Hellekant & Bavister, 1996) while PKC inhibition prevented GVBD (Homa, 1991). Using *in vivo* and *in vitro* experiments, Barreta et al. (2008) provided evidence that angiotensin II mediates LH-induced meiotic resumption in bovine oocytes and this event is dependent on prostaglandin E_2 or $F_{2\alpha}$ from follicular cells.

4. The control of meiosis in porcine oocytes

4.1 Role of cAMP, phosphodiesterases and AMPK

As in rodent and bovine oocytes, cAMP-elevating agents such as dbcAMP, the AC activator forskolin and non-specific and specific PDE isoform inhibitors can prevent or delay GVBD in porcine oocytes (Fan et al., 2002; Kim et al., 2008; Kren et al., 2004; Laforest et al., 2005; Racowsky, 1985). Phosphodiesterase activity in DO was almost completely inhibited by cilostamide (a PDE3 inhibitor) and PDE3A mRNA was detected by RT-PCR, suggesting that PDE3 is the main PDE in porcine oocytes (Sasseville et al., 2006). Although PDE3 activity represented only 19% of the total PDE activity in porcine COC, it potentially has a functional role in meiotic resumption as its mRNA level and activity were upregulated in COC but not in oocytes (therefore, upregulation was in cumulus cells) after 4 h of IVM. The up-regulation was gonadotropin- and cAMP-dependent (Sasseville et al., 2007). Similarly, cGMP-specific PDE activity also increased in a gonadotropin-dependent manner in porcine cumulus cells after 24 and 48 h of IVM (Sasseville et al., 2008). This PDE activity increase could potentially be responsible for a cGMP decrease leading to meiotic resumption as described in mouse follicles (see section 2.3). Although the presence and roles of AC3 or other AC isoforms and the GPR3 receptor have not yet been determined in porcine oocytes, an inhibitory role of

Gsalpha for meiotic resumption has been demonstrated as injection of an anti-Gsalpha antibody into porcine oocytes maintained in meiotic arrest with IBMX promoted cyclin B synthesis, MPF activation and GVBD (Morikawa et al., 2007). Similarly, as in mouse and Xenopus oocytes, Wee1B, the kinase which catalyzes the inhibitory phosphorylation of CDK1, is involved in meiotic arrest of porcine oocytes (Shimaoka et al., 2009).

Porcine oocytes and cumulus cells contained transcripts for at least one isoform of each of the three AMPK subunits. Moreover, AMPK activators alone or in combination with PDE inhibitors maintained porcine cumulus-enclosed oocytes in meiotic arrest (Mayes et al., 2007). As in bovine oocytes, the AMPK inhibitor CC did not reverse the effect of AMPK activators on meiosis and was itself inhibitory to oocyte nuclear maturation (Bilodeau-Goeseels et al., unpublished). Therefore, the effects of AMPK modulators on porcine oocyte meiosis are more similar to their effects on bovine oocytes compared to mouse oocytes indicating that the level of involvement of AMPK (and/or other signalling pathways affected by the AMPK modulators) is probably similar in bovine and porcine oocytes and different from rodent oocytes.

4.2 The role of mitogen-activated protein kinase in the control of porcine oocyte meiosis

4.2.1 MAPK in oocytes

Mitogen-activated protein kinase (MAPK) is universally activated in oocytes during maturation in all vertebrates studied so far. However, several studies suggested that MAPK activation in porcine oocytes is not implicated in meiotic resumption as spontaneous meiotic resumption occurred normally in porcine DO cultured with the MAPK inhibitor U0126 (Fan et al., 2003) and MAPK activation in pig oocytes occurred after GVBD (Liang et al., 2005). Moreover, microinjection of porcine oocytes with c-mos antisense RNA (c-mos is an upstream activator of MAPK) completely inhibited phosphorylation and activation of MAPK but did not have an effect on spontaneous meiotic resumption (Ohashi et al., 2003). Studies of the timing of MAPK activation and studies using inhibitors suggested that MAPK is probably involved in the regulation of meiosis after GVBD or during the MI/MII transition (Inoue et al., 1995; Lee et al., 2000; Ye et al., 2003). Similar conclusions were obtained in rodent and bovine oocytes (reviewed by Liang et al., 2007).

4.2.2 MAPK in follicular cells

Contrary to the situation in oocytes where MAPK is not necessary for GVBD, MAPK in follicular somatic cells is required for gonadotropin-induced meiotic resumption. Gonadotropins induced early and rapid MAPK activation in cumulus cells, MAPK activation in oocytes then occurred later at around the time of GVBD (Ebeling et al., 2007). The selective MEK inhibitors PD98059 and U0126 blocked FSH-induced meiotic resumption in mouse and porcine CEO but not spontaneous meiotic resumption in DO (Liang et al., 2005; Meinecke & Krischek, 2003; Su et al., 2002).

The mediator(s) between LH stimulation and MAPK activation in follicular cells as well as the mechanisms for inducing oocyte meiotic resumption following MAPK activation in follicle cells are not completely elucidated. Activation of MAPK could potentially lead to the production of a putative meiosis-inducing factor (Downs et al., 1988). However, LH stimulation closed gap junctions (Norris et al., 2009) and inhibition of MAPK activation blocked oocyte GVBD (Liang et al., 2005) as well as LH-induced inhibition of gap junction protein Cx43 translation in rat follicle cells (Kalma et al., 2004), suggesting that a MAPK-dependent pathway mediates LH-induced breakdown of gap junction communication and thus leads to oocyte maturation. The mediator of LH-stimulated MAPK activation could potentially be PKA as increased cAMP levels in porcine cumulus cells (resulting from inhibition of cumulus cell-specific PDE) activated MAPK (Liang et al., 2005). Protein kinase C also stimulated MAPK and oocyte meiotic resumption. The epidermal growth factor (EGF) network has an important role to play in mediating LH function during oocyte meiotic resumption; therefore, EGF could potentially be a mediator of LH-stimulated MAPK activation. In porcine granulosa cells, MAPK can be activated after a transient treatment with EGF (Keel & Davis, 1999) and EGF can induce meiotic maturation in porcine oocytes (Ding & Foxcroft, 1994). Studies in other species also suggested that activation of EGF receptor triggers signalling via the MAPK pathway (reviewed by Liang et al., 2007).

In conclusion, several questions remain unanswered about the role of MAPK in oocyte meiosis. For example, if PKA and PKC activate MAPK, are there other proteins in the signalling cascade between PKA/PKC and MAPK? Similarly, what is the mediator(s) between EGF receptor and MAPK?

5. The control of meiosis in equine oocytes

An increase in follicle diameter has been positively linked with increased nuclear and cytoplasmic maturation rates in the horse; however, oocyte diameter is not directly linked to oocyte competence (Goudet et al., 1997). Similar to other domestic species such as bovine and porcine, equine oocytes used for research tend to be recovered from slaughtered animals; however, due to the large size of follicles, ultrasound guided follicular puncture is often performed on mares. Follicular punctures are not nearly as efficient as follicle scraping of slaughterhouse ovaries with only approximately 8-9 COC being recovered every 22 days (Goudet et al., 1997); however, it allows for repeated harvesting of oocytes from the same mare when its reproductive cycle is known, thereby increasing the repeatability of results.

Equine oocytes have the lowest in vitro nuclear maturation rates of all of the domestic species discussed in the present review (Del Campo et al., 1995); therefore, the majority of studies on equine oocyte meiosis have focused on how to improve meiosis in vitro rather than on how to inhibit it. The low maturation rates may be related to the fact that the LH surge that triggers in vivo maturation lasts 4-6 days versus a 6-8 hour period observed in ewes (Alexander & Irvine, 1987; Irvine & Alexander, 1994); therefore, equine oocytes probably have different requirements and culture conditions still need to be optimized.

5.1 The role of follicular cells in the control of equine oocyte meiosis

Follicular cells play a key role in the control of meiosis since removal of oocytes from their follicle environment leads to spontaneous meiotic resumption (Pincus & Enzmann, 1935). Equine ovaries are especially suitable for the study of the role of follicular cells due to the follicles' large size which allows the different cell types to be easily isolated. Equine oocytes cultured while still attached to the membrana granulosa resumed meiosis. However, in the presence of theca cells or in theca cell-conditioned medium, more oocytes (attached to membrana granulosa) remained at the GV stage indicating that theca cells secrete a meiosis-inhibiting factor (Tremolada et al., 2003). These results are similar to results obtained with bovine oocytes as theca cells but not granulosa cell monolayers maintained meiotic arrest *in vitro* (Richard & Sirard, 1996).

5.2 Role of MPF, MAPK and other signalling molecules in the control of equine oocyte meiosis

The two subunits of MPF, p34^{cdc2} and cyclin B protein, are present in both immature and mature oocytes before and after *in vitro* culture (Goudet et al., 1998). Therefore, equine oocytes have all the key components required for MPF formation, suggesting a lack of MPF regulators in incompetent oocytes. Mitogen-activated protein kinase (MAPK) is present at all nuclear stages after *in vitro* culture (Goudet et al., 1998). Of the two forms of MAPK, ERK1 and ERK2, only ERK2 is detected in equine oocytes during the GV stage with decreased electrophoretic mobility due to modification by phosphorylation in Metaphase I and Metaphase II. Thus MAPK remains non-phosphorylated (inactive) in incompetent and immature oocytes but becomes phosphorylated (activated) after GVBD in competent and preovulatory oocytes (Goudet et al., 1998). The reasons for the inability to phosphorylate MAPK in incompetent equine oocytes are unknown. An extracellular calcium sensing receptor agonist, NPS R-467, increased the activity of MAPK in equine cumulus cells and oocytes (De Santis et al., 2009).

Epidermal growth factor (EGF), insulin-like growth factor-I (IGF-I) and growth hormone had a positive effect on *in vitro* nuclear maturation of equine oocytes (Carneiro et al., 2001; Lorenzo et al., 2002; Pereira et al., 2011). Receptors for EGF have been localized in equine follicles particularly in cumulus cells and to some extent in mural granulosa cells. The addition of A-47, a specific tyrosine kinase inhibitor, inhibited maturation, suggesting that EGF has a physiological role in the regulation of equine oocyte maturation (Lorenzo et al., 2002). Contrary to other mammalian species discussed above, we are not aware of any studies on the role of cAMP in the control of equine oocyte meiosis, except for a preliminary study which concluded that the positive effect of growth hormone on equine oocyte meiosis was mediated by the PKA pathway (Lorenzo et al., 2005).

6. The control of meiosis in ovine oocytes

6.1 The role of cumulus cells in the control of ovine oocyte meiosis

The presence of cumulus cells during ovine oocyte culture may have more importance in the promotion of nuclear maturation compared to bovine oocytes: only 3.6% of ovine DO reached MII (81.3% for CEO) after 24 h of culture in a complex medium containing serum, pyruvate and hormones (Shi et al., 2009). In contrast, 40-80% of bovine DO reached the mature stage after culture without hormones (Bilodeau-Goeseels, 2001, 2003b).

A study of paracrine factors released by gonadotropin-stimulated ovine COC also highlighted some differences in the control of meiosis between rodent and ovine (and potentially other species) oocytes. Meiosis-inducing signals from gonadotropin-stimulated cumulus cells from competent oocytes acted on cumulus cells from incompetent oocytes to induce meiosis but they did not have any effect on incompetent DO (Cecconi et al., 2008). In contrast, meiosis-stimulating factors from cumulus cells can act directly on mouse denuded oocytes (Downs, 2001). Additionally, cumulus expansion appears to be regulated by the cumulus cells themselves rather than by the oocyte as in mice (Cecconi et al., 2008; Su et al., 2003).

6.2 Role of cAMP, MAPK in the control of meiosis in ovine oocytes

Similar to bovine oocytes, meiotic resumption was only partially inhibited by cAMP-elevating agents in ovine oocytes (Jagiello et al., 1981). Ovine oocytes possess an AC enzyme as cholera toxin, a Gs activator, stimulated cAMP synthesis in isolated sheep oocytes (Crosby et al., 1985). As in other mammals, MAPK is activated at the time of GVBD in ovine cumulus cells from competent oocytes in response to gonadotropins, then later in the oocyte (Cecconi et al., 2008). Cumulus cells from incompetent oocytes (from small follicles) did not show MAPK activation even after exposure to gonadotropins. However, when co-cultured with competent complexes (and gonadotropins), more oocytes from small follicles could resume meiosis and MAPK was activated in these oocytes but not in their cumulus showing that meiotic arrest in oocytes from small follicles could be due to the inability of their surrounding cumulus cells to respond to gonadotropins (Cecconi et al., 2008).

7. Conclusion

The results obtained in rodent models suggest that, similarly to several other biological processes, oocyte nuclear maturation involves a complex network of cross-talk between several signalling pathways in oocytes and follicular cells. This review highlighted some differences between rodent and livestock species and it is anticipated that more differences will be discovered in the mechanisms controlling meiotic arrest and meiotic resumption as research progresses in non-rodent species.

8. References

Aberdam, E., Hanski, E. & Dekel, N. (1987). Maintenance of meiotic arrest in isolated rat oocytes by the invasive adenylate cyclase of *Bordetella Pertussis*. *Biology of Reproduction*, Vol.36, No.3, (April 1987), pp. 530-535, ISSN 0006-3363

Alexander, S.L. & Irvine, C.H.G. (1987). Secretion rates and short-term patterns of gonadotrophin-releasing hormone, FSH and LH throughout the periovulatory period in the mare. *Journal of Endocrinology*, Vol.114, No.3, (September 1987), pp. 351-362, ISSN 0022-0795

Barreta, M.H., Oliveira, J.F.C., Ferreira, R., Antoniazzi, A.Q., Gasperin, B.G., Sandri, L. & Gonçalves, B.D. (2008). Evidence that the effect of angiotensin II on bovine oocyte nuclear maturation is mediated by prostaglandins E_2 and $F_{2\alpha}$. *Reproduction*, Vol.136, No.6, (December 2008), pp. 733-740, ISSN 1470-1626

Bilodeau, S., Fortier, M.A. & Sirard, M.A. (1993). Effect of adenylate cyclase stimulation on meiotic resumption and cyclic AMP content of zona-free and cumulus-enclosed bovine oocytes in vitro. *Journal of Reproduction and Fertility*, Vol.97, No.1, (January 1993), pp. 5-11, ISSN 0022-4251

Bilodeau-Goeseels, S. (2001). Manganese inhibits spontaneous nuclear maturation in bovine cumulus-enclosed oocytes. *Canadian Journal of Animal Science*, Vol.81, No.2, (June 2001), pp.223-228, ISSN 0008-3984

Bilodeau-Goeseels, S. (2003a). Manganese inhibits spontaneous nuclear maturation via the cAMP/protein kinase A pathway in bovine cumulus-enclosed oocytes. *Journal of Animal and Veterinary Advance*, Vol.2, No.1, (January 2003), pp. 12-21, ISSN 1680-5593

Bilodeau-Goeseels, S. (2003b). Effects of phosphodiesterase inhibitors on spontaneous nuclear maturation and cAMP concentrations in bovine oocytes. *Theriogenology*, Vol.60, No.9, (December 2003), pp. 1679-1690, ISSN 0093-691X

Bilodeau-Goeseels, S. (2007). Effects of manipulating the nitric oxide/cyclic GMP pathway on bovine oocyte meiotic resumption *in vitro*. *Theriogenology*, Vol.68, No.5, (September 2007), pp. 693-701, ISSN 0093-691X

Bilodeau-Goeseels, S., Panich, P.L. & Kastelic, J.P. (2011). Activation of AMP-activated protein kinase may not be involved in AICAR- and metformin-mediated meiotic arrest in bovine denuded and cumulus-enclosed oocytes *in vitro*. *Zygote*, Vol.19, No.2, (May 2011), pp. 97-106, ISSN 0967-1994

Bilodeau-Goeseels, S., Sasseville, M., Guillemette, C. & Richard, F.J. (2007). Effects of adenosine monophosphate-activated kinase activators on bovine oocyte nuclear maturation *in vitro*. *Molecular Reproduction and Development*, Vol.74, No.8, (August 2007), pp. 1021-1034, ISSN 1040-452X

Bornslaeger, E.A. & Schultz, R.M. (1985). Regulation of oocyte maturation: effect of elevating cumulus cell cAMP on oocyte cAMP levels. *Biology of Reproduction*, Vol.33, No.3, (October 1985), pp. 698-704, ISSN 0006-3363

Braun, T. & Dods, R.S. (1975). Development of a Mn-2+-sensitive, "soluble" adenylate cyclase in rat testis. *Proceedings of the National Academy of Science of the USA*, Vol.72, No.3, (March 1975), pp. 1097-1101, ISSN 0027-8424

Bu, S., Xia, G., Tao, Y., Lei, L. & Zhou, B. (2003). Dual effects of nitric oxide on meiotic maturation of mouse cumulus cell-enclosed oocytes in vitro. *Molecular and Cellular Endocrinology*, Vol.207, No.1-2, (September 2003), pp. 21-30, ISSN 0303-7207

Carneiro, G., Lorenzo, P., Pimentel, C., Pegoraro, M., Ball, B., Anderson, G. & Liu, I. (2001). Influence of insulin-like growth factor-I and its interaction with gonadotropins, estradiol, and fetal calf serum on *in vitro* maturation and parthenogenic development in equine oocytes. *Biology of Reproduction*, Vol.65, No.3, (September 2001), pp. 899-905, ISSN 0003-3363.

Cecconi, S., Mauro, A., Capacchietti, G., Berardinelli, P., Bernabò, N., Di Vincenzo, A.R., Mattiolo, M. & Barboni, B. (2008). Meiotic maturation of incompetent prepubertal oocytes is induced by paracrine factor(s) released by gonadotropin-stimulated oocyte-cumulus cell complexes and involves mitogen-activated protein kinase activation. *Endocrinology*, Vol.149, No.1, (January 2008), pp. 100-107, ISSN 0013-7227

Chen, J. & Downs, S.M. (2008). AMP-activated protein kinase is involved in hormone-induced mouse oocyte meiotic maturation in vitro. *Developmental Biology*, Vol.313, No.1, (January 2008), pp. 47-57, ISSN 0012-1606

Chen, J., Hudson, E., Chi, M.M., Chang, A.S., Moley, K.H., Hardie, D.G. & Downs, S.M. (2006). AMPK regulation of mouse oocyte meiotic resumption *in vitro*. *Developmental Biology*, Vol.291, No.2, (March 2006), pp. 227-238, ISSN 0012-1606

Cho, W.K., Stern, S. & Biggers, J.D. (1974). Inhibitory effect of dibutyryl cAMP on mouse oocyte maturation in vitro. *Journal of Experimental Zoology*, Vol.187, No.3, (March 1974), pp. 383-386, ISSN 0022-104X

Clarke, P.R. & Karsenti, E. (1991). Regulation of p34cdc2 protein kinase: new insights into protein phosphorylation and the cell cycle. *Journal of Cell Science*, Vol.100, No.Pt3, (November 1991), pp. 403-414, ISSN 0021-9533

Comizzoli, P., Songsasen, N. & Wildt, D.E. (2010). Protecting and extending fertility for females of wild and endangered mammals. *Cancer Treatment and Research*, Vol.156, pp. 87-100.

Crosby, I.M., Moor, R.M., Heslop, J.P. & Osborn, J.C. (1985). cAMP in ovine oocytes: localization of synthesis and its action on protein synthesis, phosphorylation, and meiosis. *Journal of Experimental Zoology*, Vol.234, No.2, (May 1985), pp. 307-318, ISSN 0022-104X

Dekel, N., Aberdam, E. & Sherizly, I. (1984). Spontaneous maturation *in vitro* of cumulus-enclosed rat oocytes is inhibited by forskolin. *Biology of Reproduction*, Vol.31, No.2, (September 1984), pp. 244-250, ISSN 0006-3363

Dekel, N. & Beers, W.H. (1978). Rat oocyte maturation in vitro: relief of cyclic AMP inhibition by gonadotropins. *Proceedings of the National Academy of Science of the USA*, Vol.75, No.9, (September 1978), pp. 4369-4373, ISSN 0027-8424

Dekel, N. & Beers, W.H. (1980). Development of the rat oocyte in vitro: inhibition and induction of maturation in the presence or absence of the cumulus oophorus. *Developmental Biology*, Vol.75, No.2, (March 1980), pp. 247-254, ISSN 0012-1606

Dekel, N., Galiani, D. & Sherizly, I. (1988). Dissociation between the inhibitory and stimulatory action of cAMP on maturation of rat oocytes. *Molecular and Cellular Endocrinology*, Vol.56, No.1-2, (March 1988), pp. 115-121, ISSN 0303-7207

Del Campo, M.R., Donoso, X., Parrish, J.J. & Ginther, O.J. (1995). Selection of follicles, preculture oocyte evaluation, and duration of culture for *in vitro* maturation of equine oocytes. *Theriogenology*, Vol. 43, No. 7, (May 1995), pp. 1141-1153, ISSN 0093-691X

De Mouzon, J., Lancaster, P., Nygren, K.G., Sulliva, E., Zegers-Hochschild, F., Mansour, R., Ishihara, O. & Adamson, D. (2009). World collaborative report on assisted reproductive technology, 2002. *Human Reproduction*, Vol.24, No.9, (September 2009), pp. 2310-2320, ISSN 0268-1161

De Santis, T., Casavola, V., Reshkin, S.J., Guerra, L., Ambruosi, B., Fiandanese, N., Dalbies-Tran, R., Goudet, G. & Dell'Aquila, M.E. (2009). The extracellular calcium-sensing receptor is expressed in the cumulus-oocyte complex in mammals and modulates oocyte meiotic maturation. *Reproduction*, Vol.138, No.3, (September 2009), pp. 439-452, ISSN 1470-1626

Ding, J. & Foxcroft, G.R. (1994). Epidermal growth factor enhances oocyte maturation in pigs. *Molecular Reproduction and Development*, Vol.39, No.1, (September 1994), pp. 30-40, ISSN 1040-452X

Downs, S.M. (2001). A gap-junction-mediated signal, rather than an external paracrine factor, predominates during meiotic induction in isolated mouse oocytes. *Zygote*, Vol.9, No.1, (February 2001), pp. 71-82, ISSN 0967-1994

Downs, S.M., Daniel, S.A. & Eppig, J.J. (1988). Induction of maturation in cumulus cell-enclosed mouse oocytes by follicle-stimulating hormone and epidermal growth factor: evidence for a positive stimulus of somatic cell origin. *Journal of Experimental Zoology*, Vol. 245, No.1, (January 1988), pp. 86-96, ISSN 0022-104X

Downs, S.M., Hudson, E.R. & Hardie, D.G. (2002). A potential role for AMP-activated protein kinase in meiotic induction in mouse oocytes. *Developmental Biology*, Vol. 245, No.1, (May 2002), pp.200-212, ISSN 0012-1606

Downs, S.M., Ya, R. & Davis, C.C. (2010). Role of AMPK throughout meiotic maturation in the mouse oocyte: evidence for promotion of polar body formation and suppression of premature activation. *Molecular Reproduction and Development*, Vol.77, No.10, (October 2010), pp. 888-899, ISSN 1040-452X

Ebeling, S., Schuon, C. & Meinecke, B. (2007). Mitogen-activated protein kinase phosphorylation patterns in pig oocytes and cumulus cells during gonadotropin-induced resumption of meiosis in vitro. *Zygote*, Vol.15, No.3, (August 2007), pp. 139-147, ISSN 0967-1994

Edson, M.A., Nagaraja, A.K. & Matzuk, M.M. (2009). The mammalian ovary from genesis to revelation. *Endocrine Reviews*, Vol.30, No.6, (August 2009), pp. 624-712, ISSN 0163-769X

Edwards, R.G. (1965). Maturation in vitro of mouse, sheep, cow, pig, rhesus monkey and human ovarian oocytes. *Nature*, Vol.208, No.5008, (October 1965), pp. 349-351, ISSN 0028-0836

Fan, H.Y., Li, M.Y., Tong, C., Chen, D.Y., Xia, G.L., Song, X.F., Schatten, H. & Sun, Q.Y. (2002). Inhibitory effects of cAMP and protein kinase C on meiotic maturation and MAP kinase phosphorylation in porcine oocytes. *Molecular Reproduction and Development*, Vol.63, No.4 (December 2002), pp. 480-487, ISSN 1040-452X

Fan, H.Y., Tong, C., Lian, L., Li, S.W., Gao, W.X., Cheng, Y., Chen, D.Y., Schatten, H. & Sun, Q.Y. (2003). Characterization of ribosomal S6 protein kinase p90rsk during meiotic maturation and fertilization in pig oocytes: mitogen-activated protein kinase-associated activation and localization. *Biology of Reproduction*, Vol.68, No.3, (March 2003), pp. 968-977, ISSN 0006-3363

Freudzon, L., Norris, R.P., Hand, A.R., Tanaka, S., Saeki, Y., Jones, T.L., Rasenick, M.M., Berlot, C.H., Mehlmann, L.M. & Jaffe, L.A. (2005). Regulation of meiotic prophase arrest in mouse oocytes by GPR3, a constitutive activator of the Gs G protein. *Journal of Cell Biology*, Vol.24, No.2 (October 2005), pp. 255-265, ISSN 0895-2388

Goudet, G., Belin, F., Bézard, J. & Gérard, N. (1998). Maturation-promoting factor (MPF) and mitogen activated protein kinase (MAPK) expression in relation to oocyte competence for *in-vitro* maturation in the mare. *Molecular Human Reproduction*, Vol. 4, No.6, (June 1998), pp. 563-570, ISSN 1360-9947

Goudet, G., Bézard, J., Duchamp, G., Gérard, N. & Palmer, E. (1997). Equine oocyte competence for nuclear and cytoplasmic *in vitro* maturation: effect of follicle size and hormonal environment. *Biology of Reproduction*. Vol.57, No.2, (August 1997), pp. 232-245, ISSN 0006-3363

Han, S.J. & Conti, M. (2006). New pathways from PKA to the Cdc2/cyclin B complex in oocytes. *Cell Cycle*, Vol.5, No.3, (February 2006), pp. 227-232, ISSN 1538-4101

He, L.C., Damiani, P., Parys, J.B. & Fissore, R.A. (1997). Calcium, calcium release receptors, and meiotic resumption in bovine oocytes. *Biology of Reproduction*, Vol.57, No.5, (November 1997), pp. 1245-1255, ISSN 0006-3363

Homa, S.T. (1988). Effects of cyclic AMP on the spontaneous meiotic maturation of cumulus-free bovine oocytes cultured in chemically defined medium. *Journal of Experimental Zoology*, Vol.248, No.2, (November 1988), pp. 222-231 ISSN 0022-104X

Homa, S.T. (1991). Neomycin, an inhibitor of phosphoinositide hydrolysis, inhibits the resumption of bovine oocyte spontaneous meiotic maturation. *Journal of Experimental Zoology*, Vol.258, No.1, (April 1991), pp. 95-103, ISSN 0022-104X

Homa, S.T. (1995). Calcium and meiotic maturation of the mammalian oocyte. *Molecular Reproduction and Development*, Vol.40, No.1, (January 1995), pp. 122-134, ISSN 1040-452X

Horner, K., Livera, G., Hinckley, M., Trinh, K., Storm, D. & Conti, M. (2003). Rodent oocytes express an active adenylyl cyclase required for meiotic arrest. *Developmental Biology*, Vol.258, No.2 (June 2003), pp. 385-396, ISSN 0012-1606

Inoue, M., Naito, K., Aoki, F., Toyoda, Y. & Sato, E. (1995). Activation of mitogen-activated protein kinase during meiotic maturation in porcine oocytes. *Zygote*, Vol.3, No.3, (August 1995), pp. 265-271, ISSN 0967-1994

Irvine, C.H.G. & Alexander, S.L. (1994). The dynamics of gonadotrophin-releasing hormone, LH and FSH secretion during the spontaneous ovulatory surge of the mare as revealed by intensive sampling of pituitary venous blood. *Journal of Endocrinology*, Vol.140, No.2, (February 1994), pp. 283-295, ISSN 0022-0795

Jagiello, G., Ducayen, M.B. & Goonan, W.D. (1981). A note on the inhibition of *in vitro* meiotic maturation of mammalian oocytes by dibutyryl cyclic AMP. *Journal of Experimental Zoology*, Vol.218, No.2, (November 1981), pp. 309-311, ISSN 0022-104X

Jagiello, G., Ducayen, M., Miller, W., Graffeo, J. & Fang, J.S. (1975). Stimulation and inhibition with LH and other hormones of female mammalian meiosis *in vitro*. *Journal of Reproduction and Fertility*, Vol.43, No.1, (April 1975), pp. 9-22, ISSN 0022-4251

Kalinowski, R.R., Berlot, C.H., Jones T.L., Ross, L.F., Jaffe, L.A. & Mehlmann, L.M. (2004). Maintenance of meiotic prophase arrest in vertebrate oocytes by a Gs protein-mediated pathway. *Developmental Biology*, Vol.267, No.1, (March 2004), pp. 1-13, ISSN 0012-1606

Kalma, Y., Granot, I., Galiani, D., Barash, A. & Dekel, N. (2004). Luteinizing hormone-induced connexin 43 downregulation: inhibition of translation. *Endocrinology*, Vol. 145, No.4, (April 2004), pp. 1617-1624, ISSN 0013-7227

Keel, B.A. & Davis, J.S. (1999). Epidermal growth factor activates extracellular signal-regulated protein kinases (ERK) in freshly isolated porcine granulosa cells. *Steroids*, Vol.64, No.9, (September 1999), pp. 654-658, ISSN 0039-128X

Kim, J.S., Cho, Y.S., Song, B.S., Wee, G., Park, J.S., Choo, Y.K., Yu, K., Lee, K.K., Han, Y.M. & Koo, D.B. (2008). Exogenous dibutyryl cAMP affects meiotic maturation via protein kinase A activation; it stimulates further embryonic development including blastocyst quality in pigs. *Theriogenology*, Vol.69, No.3, (February 2008), pp. 290-301, ISSN 0093-691X

Kren, R., Ogushi, S. & Miyano, T. (2004). Effect of caffeine on meiotic maturation of porcine oocytes. *Zygote*, Vol.12, No.1, (February 2004), pp. 31-38, ISSN 0967-1994

Kuyt, J.R., Kruip, T.A. & de Jong-Brink, M. (1988). Cytochemical localization of adenylate cyclase in bovine cumulus-oocyte complexes. *Experimental Cell Research*, Vol.174, No.1, (January 1988), pp. 139-145, ISSN 0014-4827

Laforest, M.F., Pouliot, E., Guéguen, L. & Richard, F.J. (2005). Fundamental significance of specific phosphodiesterases in the control of spontaneous meiotic resumption in porcine oocytes. *Molecular Reproduction and Development*, Vol.70, No.3, (March 2005), pp. 361-372, ISSN 1040-452X

LaRosa, C. & Downs, S.M. (2006). Stress stimulates AMP-activated kinase and meiotic resumption in mouse oocytes. *Biology of Reproduction*, Vol.74, No.3, (March 2006), pp. 585-592, ISSN 0006-3363

LaRosa, C. & Downs, S.M. (2007). Meiotic induction by heat stress in mouse oocytes: involvement of AMP-activated protein kinase and MAPK family members. *Biology of Reproduction*, Vol.76, No.3, (March 2007), pp.476-486, ISSN 0006-3363

Lastro, M., Collins, S. & Currie, W.B. (2006). Adenylyl cyclases in oocyte maturation: a characterization of AC isoforms in bovine cumulus cells. *Molecular Reproduction and Development*, Vol.73, No.9, (September 2006), pp. 1202-1210, ISSN 1040-452X

Ledan E., Polanski, Z., Terret, M.E. & Maro, B. (2001). Meiotic maturation of the mouse oocyte requires equilibrium between cyclin B synthesis and degradation. *Developmental Biology*, Vol.232, No.2, (April 2001), pp. 400-413, ISSN 0012-1606

Lee, J., Miyano, T. & Moor, R.M. (2000). Localisation of phosphorylated MAP kinase during the transition from meiosis I to meiosis II in pig oocytes. *Zygote*, Vol.8, No.2, (May 2000), pp. 119-125, ISSN 0967-1994

Liang, C.G., Huo, L.J., Zhong, Z.S., Chen, D.Y., Schatten, H. & Sun, Q.Y. (2005). Cyclic adenosine 3',5'-monophosphate-dependent activation of mitogen-activated protein kinase in cumulus cells is essential for germinal vesicle breakdown of porcine cumulus-enclosed oocytes. *Endocrinology*, Vol.146, No.10, (October 2005), pp. 4437-4444, ISSN 0013-7227

Liang, C.G., Su, Y.Q., Fan, H.Y., Schatten, H. & Sun, Q.Y. (2007). Mechanisms regulating oocyte meiotic resumption: roles of mitogen-activated protein kinase. *Molecular Endocrinology*, Vol.21, No.9, (September 2007), pp. 2037-2055, ISSN 0888-8809

Lorenzo, P.L., Liu, I.K., Carneiro, G.F., Conley, A.J. & Enders, A.C. (2002). Equine oocyte maturation with epidermal growth factor. *Equine Veterinary Journal*. Vol.34, No.4, (July 2002), pp. 378-382, ISSN 2042-3306

Lorenzo, P.L., Pereira, G.R., Bilodeau-Goeseels, S., Carneiro, G.F., Kastelic, J.P. & Liu, I.K.M. (2005). Effect of growth hormone and inhibitors on the *in vitro* maturation of equine oocytes. *Reproduction in Domestic Animal*, Vol.40, No.4, (August 2005), p. 378, ISSN 0936-6768

Masciarelli, S., Horner, K., Liu, C., Park, S.H., Hinckley, M., Hockman, S., Nedachi, T., Jin, C., Conti, M. & Manganiello, V. (2004). Cyclic nucleotide phosphodiesterase 3A-deficient mice as a model for female infertility. *Journal of Clinical Investigation*, Vol.114, No.2, (July 2004), pp. 196-205, ISSN 0021-9738

Matta, S.G., Caldas-Bussière, M.C., Viana, K.S., Faes, M.R., Paes de Carvalho, C.S., Dias, B.L. & Quirino, C.R. (2009). Effect of inhibition of synthesis of inducible nitric oxide synthase-derived nitric oxide by aminoguanidine on the *in vitro* maturation of oocyte-cumulus complexes of cattle. *Animal Reproduction Science*, Vol.111, No.2-4, (April 2009), pp. 189-201, ISSN 0378-4320

Mayes, M.A., Laforest, M.F., Guillemette, C., Gilchrist, R.B. & Richard, F.J. (2007). Adenosine 5'-monophosphate kinase-activated protein kinase (PRKA) activators delay meiotic resumption in porcine oocytes. *Biology of Reproduction*, Vol.76, No.4, (April 2007), pp. 589-597, ISSN 0006-3363

Mayes, M.A. & Sirard, M.A. (2002). Effect of type 3 and type 4 phosphodiesterase inhibitors on the maintenance of bovine oocytes in meiotic arrest. *Biology of Reproduction*, Vol.66, No.1, (January 2002), pp. 180-184, ISSN 0006-3363

Mehlmann, L.M., Jones, T.L.Z. & Jaffe, L.A. (2002). Meiotic arrest in the mouse follicle maintained by a Gs protein in the oocyte. *Science*, Vol.297, No.5585, (August 2002), pp. 1343-1345, ISSN 0036-8075

Mehlmann, L.M., Kalinowski, R.R., Ross, L.F., Parlow, A.F., Hewlett, E.L. & Jaffe, L.A. (2006). Meiotic resumption in response to luteinizing hormone is independent of a Gi family protein or calcium in the mouse oocyte. *Developmental Biology*, Vol.299, No.2, (November 2006), pp. 345-355, ISSN 0012-1606

Mehlmann, L.M., Saeki, Y., Tanaka, S., Brennan, T.J., Evsikov, A.V., Pendola, F.L., Knowles, B.B., Eppig, J.J. & Jaffe, L.A. (2004). The Gs-linked receptor GPR3 maintains meiotic arrest in mammalian oocytes. *Science*, Vol.306, No.5703, (December 2004), pp. 1947-1950, ISSN 0036-8075

Meinecke, B. & Krischek, C. (2003). MAPK/ERK kinase (MEK) signalling is required for resumption of meiosis in cultured cumulus-enclosed pig oocytes. *Zygote*, Vol.11, No.1, (February 2003), pp. 7-16, ISSN 0967-1994

Mondadori, R.G., Neves, J.P. & Gonçalves, P.B.D. (2008). Protein kinase C (PKC) role in bovine oocyte maturation and early embryo development. *Animal Reproduction Science*, Vol.107, No.1-2, (August 2008), pp. 20-29, ISSN 0378-4320

Morikawa, M., Seki, M., Kume, S., Endo, T., Nishimura, Y., Kano, K. & Naito, K. (2007). Meiotic resumption of porcine immature oocytes is prevented by ooplasmic Gsalpha functions. *Journal of Reproduction and Development*, Vol.53, No.6, (December 2007), pp. 1151-1157, ISSN 0916-8818

Norris, R.P., Freudzon, L., Freudzon, M., Hand, A.R., Mehlmann, L.M. & Jaffe, L.A. (2007). A G(s)-linked receptor maintains meiotic arrest in mouse oocytes, but luteinizing hormone does not cause meiotic resumption by terminating receptor-G(s) signaling. *Developmental Biology*, Vol.310, No.2, (October 2007), pp. 240-249, ISSN 0012-1606

Norris, R.P., Freudzon, M., Mehlmann, L.M., Cowan, A.E., Simon, A.M., Paul, D.L., Lampe, P.D. & Jaffe, L.A. (2008). Luteinizing hormone causes MAP kinase-dependent phosphorylation and closure of connexin 43 gap junctions in mouse ovarian follicles: one of two paths to meiotic resumption. *Development*, Vol.135, No.19, (October 2008), pp. 3229-3238, ISSN 0950-1991

Norris, R.P., Freudzon, M., Nikolaev, V.O. & Jaffe, L.A. (2010). Epidermal growth factor receptor kinase activity is required for gap junction closure and for part of the decrease in ovarian follicle cGMP in response to luteinizing hormone. *Reproduction*, Vol.140, No.5, (November 2010), pp. 655-662, ISSN 1470-1626

Norris, R.P., Ratzan, W.J., Freudzon, M., Mehlmann, L.M., Krall, J., Movsesian, M.A., Wang, H., Ke, H., Nikolaev, V.O. & Jaffe, L.A. (2009). Cyclic GMP from the surrounding somatic cells regulates cyclic AMP and meiosis in the mouse oocyte. *Development*, Vol.136, No.11, (June 2009), pp. 1869-1878, ISSN 0950-1991

Ohashi, S., Naito, K., Sugiura, K., Iwamori, N., Goto, S., Naruoka, H. & Tojo, H. (2003). Analysis of mitogen-activated protein kinase function in the maturation of porcine oocytes. *Biology of Reproduction*, Vol.68, No.2, (February 2003), pp. 604-609, ISSN 0006-3363

Olsiewski, P.J. & Beers, W.H. (1983). cAMP synthesis in the rat oocyte. *Developmental Biology*, Vol.100, No.2, (December 1983), pp. 287-293, ISSN 0012-1606

Pereira, G.R., Lorenzo, P.L., Carneiro, G.F., Ball, B.A., Gonçalves, P.B.D., Cantarelli Pegoraro, L.M., Bilodeau-Goeseels, S., Kastelic, J.P., Casey, P.J. & Liu, I.K.M. (2011).The effect of growth hormone (GH) and insulin-like growth factor-I (IGF-I) on *in vitro* maturation of equine oocytes. *Zygote*, Available on CJO 2011doi:10.1017/S0967199411000335

Pincus, G. & Enzmann, E.V. (1935). The comparative behaviour of mammalian eggs *in vivo* and *in vitro*. I. The activation of ovarian eggs. *Journal of Experimental Medicine*, Vol.62, No.5, (October 1935), pp. 665-677, ISSN 0022-1007

Rose-Hellekant, T.A. & Bavister, B.D. (1996). Roles of protein kinase A and C in spontaneous maturation and in forskolin or 3-isobutyl-1-methylxanthine maintained meiotic arrest of bovine oocytes. *Molecular Reproduction and Development*, Vol.44, No.2, (June 1996), pp. 241-249, ISSN 1040-452X

Racowsky, C. (1984). Effect of forskolin on the spontaneous maturation and cyclic AMP content of rat oocyte-cumulus complexes. *Journal of Reproduction and Fertility*, Vol.72, No.1, (September 1984), pp. 107-116, ISSN 0022-4251

Racowsky, C. (1985). Effect of forskolin on maintenance of meiotic arrest and stimulation of cumulus expansion, progesterone and cyclic AMP production by pig oocyte-cumulus complexes. *Journal of Reproduction and Fertility*, Vol.74, No.1, (May 1985), pp. 9-21, ISSN 0022-4251

Richard, F.J. & Sirard, M.A. (1996). Effects of follicular cells on oocyte maturation. II: Theca cell inhibition of bovine oocyte maturation *in vitro*. *Biology of Reproduction*, Vol.54, No.1, (January 1996), pp. 22-28, ISSN 0006-3363

Sadana, R. & Dessauer, C.W. (2009). Physiological roles for G protein-regulated adenylyl cyclase isoforms: insights from knockout and overexpression studies. *Neurosignals*, Vol.17, No.1, (October 2009), pp. 5-22, ISSN 1424-862X

Sasseville, M., Albuz, F.K., Côté, N., Guillemette, C., Gilchrist, R.B. & Richard, F.J. (2009). Characterization of novel phosphodiesterases in the bovine ovarian follicle. *Biology of Reproduction*, Vol.81, No.2, (August 2009), pp. 415-425, ISSN 0006-3363

Sasseville, M., Côté, N., Gagnon, M.C. & Richard, F.J. (2008). Up-regulation of 3'5'-cyclic guanosine monophosphate-specific phosphodiesterase in the porcine cumulus-oocyte complex affects steroidogenesis during *in vitro* maturation. *Endocrinology*, Vol.149, No.11, (November 2008), pp. 5568-5576, ISSN 0013-7227

Sasseville, M., Côté, N., Guillemette, C. & Richard, F.J. (2006). New insight into the role of phosphodiesterase 3A in porcine oocyte maturation. *BMC Developmental Biology*, Vol.6, No.6, (October 2006), p. 47, ISSN 1471-213X

Sasseville, M., Côté, N., Vigneault, C., Guillemette, C. & Richard, F.J. (2007). 3'5'-cyclic adenosine monophosphate-dependent up-regulation of phosphodiesterase type 3A in porcine cumulus cells. *Endocrinology*, Vol.148, No.4, (April 2007), pp. 1858-1867, ISSN 0013-7227

Schultz, R.M., Montgomery, R.R. & Belanoff, J.R. (1983a). Regulation of mouse oocyte meiotic maturation: implication of a decrease in oocyte cAMP and protein dephosphorylation in commitment to resume meiosis. *Developmental Biology*, Vol.97, No.2, (June 1983), pp. 264-273, ISSN 0012-1606

Schultz, R.M., Montgomery, R.R., Ward-Bailey, P.F. & Eppig, J.J. (1983b). Regulation of oocyte maturation in the mouse: possible roles of intercellular communication, cAMP, and testosterone. *Developmental Biology*, Vol.95, No.2, (February 1983), pp. 294-304, ISSN 0012-1606

Schwarz, K.R., Pires, P.R., Adona, P.R., Câmana de Bem, T.H. & Leal, C.L. (2008). Influence of nitric oxide during maturation on bovine oocyte meiosis and embryo development *in vitro*. *Reproduction Fertility and Development*, Vol.20, No.4, (April 2008), pp. 529-536, ISSN 1031-3613

Schwarz, K.R., Pires, P.R., de Bem, T.H., Adona, P.R. & Leal, C.L. (2010). Consequences of nitric oxide synthase inhibition during bovine oocyte maturation on meiosis and embryo development. *Reproduction in Domestic Animal*, Vol.45, No.1, (February 2010), pp. 75-80, ISSN 0936-6768

Sela-Abramovich, S., Edry, I., Galiani, D., Nevo, N. & Dekel, N. (2006). Disruption of gap junctional communication within the ovarian follicle induces oocyte maturation. *Endocrinology*, Vol. 147, No.5, (May 2006), pp. 2280-2286, ISSN 0013-7227

Sela-Abramovich, S., Galiani, D., Nevo, N. & Dekel, N. (2008). Inhibition of rat oocyte maturation and ovulation by nitric oxide: mechanism of action. *Biology of Reproduction*, Vol.78, No.6, (June 2008), pp. 1111-1118, ISSN 0006-3363

Shi, L., Yue, W., Zhang, J., Lv, L., Ren, Y. & Yan, P. (2009). Effect of ovarian cortex cells on nuclear maturation of sheep oocytes during in vitro maturation. *Animal Reproduction Science*, Vol.113, No.1-4, (July 2009), pp.299-304, ISSN 0378-4320

Shimaoka, T., Nishimura, T., Kano, K. & Naito, K. (2009). Critical effect of pig Wee1B on the regulation of meiotic resumption in porcine immature oocytes. *Cell Cycle*, Vol.8, No.15, (August 2009), pp. 2375-2384, ISSN 1538-4101

Sirard, M.A. (1990). Temporary inhibition of *in vitro* meiotic maturation by adenylate cyclase stimulation in immature bovine oocytes. *Theriogenology*, Vol.33, No.4, (April 1990), pp. 757-767, ISSN 0093-691X

Sirard, M.A. & First, N.L. (1988). *In vitro* inhibition of oocyte nuclear maturation in the bovine. *Biology of Reproduction*, Vol.39, No.2, (September 1988), pp. 229-234, ISSN 0006-3363

Su, Y.Q., Denegre, J.M., Wigglesworth, K.I., Pendola, F.L., O'Brien, M.J. & Eppig, J.J. (2003). Oocyte-dependent activation of mitogen-activated protein kinase (ERK1/2) in cumulus cells is required for the maturation of the mouse oocyte-cumulus cell complex. *Developmental Biology*, Vol.263, No.1, (November 2003), pp. 126-138, ISSN 0012 1606

Su, Y.Q., Wigglesworth, K., Pendola, F.L., O'Brien, M.J. & Eppig, J.J. (2002). Mitogen-activated protein kinase activity in cumulus cells is essential for gonadotropin-induced oocyte meiotic resumption and cumulus expansion in the mouse. *Endocrinology*, Vol.143, No.6, (June 2002), pp. 2221-2232, ISSN 0013-7227

Suikkari, A.M. (2008). *In vitro* maturation: its role in fertility treatment. *Current Opinion in Obstetrics & Gynecology*, Vol.20, No.3, (June 2008), pp. 242-248, ISSN 1040-872X

Tesmer, J.J.G., Sunahara, R.K., Johnson, R.A., Gosselin, G., Gilman, A.G. & Sprang, S.R. (1999). Two-metal-ion catalysis in adenylyl cyclase. *Science*, Vol.285, No.5428, (July 1999), pp. 756-760, ISSN 0036-8075

Thomas, R.E., Armstrong, D.T. & Gilchrist, R.B. (2002). Differential effects of specific phosphodiesterase isoenzyme inhibitors on bovine oocyte meiotic maturation. *Developmental Biology*, Vol.244, No.2, (April 2002), pp. 215-225, ISSN 0012-1606

Thomas, R.E., Armstrong, D.T. & Gilchrist, R.B. (2004). Bovine cumulus cell-oocyte gap junctional communication during *in vitro* maturation in response to manipulation of cell-specific cyclic adenosine 3',5'-monophosphate levels. *Biology of Reproduction*, Vol.70, No.3, (March 2004), pp. 548-556, ISSN 0006-3363

Törnell, J., Billig, H. & Hillensjö, T. (1990). Resumption of rat oocyte meiosis is paralleled by a decrease in guanosine 3', 5'-cyclic monophosphate (cGMP) and is inhibited by microinjection of cGMP. *Acta Physiologica Scandinavica*, Vol.139, No.3, (July 1990), pp.511-517, ISSN 0001-6772

Tosca, L., Uzbekova, S., Chabrolle, C. & Dupont, J. (2007). Possible role of 5'AMP-activated protein kinase in the metformin-mediated arrest of bovine oocytes at the germinal vesicle stage during *in vitro* maturation. *Biology of Reproduction*, Vol.77, No.3, (September 2007), pp. 452-465, ISSN 0006-3363

Tosti, E., Boni, R. & Cuomo, A. (2002). Fertilization and activation currents in bovine oocytes. *Reproduction*, Vol.124, No.6, (December 2002), pp. 835-846, ISSN 1470-1626

Tremoleda, J.L., Tharasanit, T., Van Tol, H.T.A., Stout, T.A.E., Colenbrander, B. & Bevers, M.M. (2003). Effects of follicular cells and FSH on the resumption of meiosis in equine oocytes matured *in vitro*. *Reproduction*, Vol.125, No.4, (April 2003), pp. 565-577, ISSN 1470-1626.

Tsafriri, A., Lindner, H.R., Zor, U. & Lamprecht, S.A. (1972). *In vitro* induction of meiotic division in follicle-enclosed rat oocytes by LH, cyclic AMP and prostaglandin E$_2$. *Journal of Reproduction and Fertility*, Vol.31, No.1, (October 1972), pp. 39-50, ISSN 0022-4251

Urner, F., Herrmann, W.L., Baulieu, E.E. & Schorderet-Slatkine, S. (1983). Inhibition of denuded mouse oocyte meiotic maturation by forskolin, an activator of adenylate cyclase. *Endocrinology*, Vol.113, No.3, (September 1983), pp. 1170-1172, ISSN 0013-7227

Vaccari, S., Weeks II, J.L., Hsieh, M., Menniti, F.S. & Conti, M. (2009). Cyclic GMP signalling is involved in the luteinizing hormone-dependent meiotic maturation of mouse oocytes. *Biology of Reproduction*, Vol.81, No.3, (September 2009), pp. 595-604, ISSN 0006-3363

Viana, K.S., Caldas-Bussière, M.C., Matta, S.G., Faes, M.R., de Carvalho, C.S. & Quirino, C.R. (2007). Effect of sodium nitroprusside, a nitric oxide donor, on the *in vitro* maturation of bovine oocytes. *Animal Reproduction Science*, Vol.102, No.3-4, (December 2007), pp. 217-227, ISSN 0378-4320

Vivarelli, E., Conti, M., De Felici, M. & Siracusa, G. (1983). Meiotic resumption and intracellular cAMP levels in mouse oocytes treated with compounds which act on cAMP metabolism. *Cell Differentiation*, Vol.12, No.5, (May 1983), pp. 271-276, ISSN 0045-6039

Ye, J., Flint, A.P., Luck, M.R. & Campbell, K.H. (2003). Independent activation of MAP kinase and MPF during the initiation of meiotic maturation in pig oocytes. *Reproduction*, Vol.125, No.5, (May 2003), pp. 645-656, ISSN 1470-1626

Zhou, G., Myers, R., Li, Y., Chen, Y., Shen, X., Fenyk-Melody, J., Wu, M., Ventre, J., Doebber, T., Furii, N., Musi, N., Hirshman, M.F., Goodyear, L.J. & Moller, D.E. (2001). Role of AMP-activated protein kinase in mechanism of metformin action. *Journal of Clinical Investigation*, Vol.8, No.8, (October 2001), pp. 1167-1174, ISSN 0021-9738

3

PKC Regulation of Gametogenesis and Early Development

James J. Faust, Madhavi Kalive, Anup Abraham and David G. Capco
Arizona State University,
USA

1. Introduction

The protein kinase C (PKC) family is comprised of 11 isotypes and many can exist in a single cell simultaneously. There are three general categories of PKC based on their cofactor requirements for activation: The conventional, novel, and atypical PKCs. The first category, the conventional PKC isotypes, are PKCα, -βI, -βII, and -γ and are activated by diacylglycerol (DAG), and negatively charged phospholipids in a calcium-dependent manner. The second category, the novel PKCs, are composed of PKCδ, -ε, -θ, and –η and require negatively changed phospholipids and DAG, but act in a calcium-independent manner. Finally, the atypical PKCs are human ι/mouse λ and PKCζ and only require negatively charged phospholipids for activation. Each PKC isotype phosphorylates serine/threonine residues on protein substrates. There are several known substrates for PKC and they include MARCKS proteins, RACK proteins, Dynamin, EGFR, MEK5, the two subunits of MPF, $p34^{cdc2}$ and cyclin B, among others (Yu et al., 2004; Steinberg 2008; Meier et al., 2009).

Many PKC isotypes have overlapping substrate specificity. Since many isotypes function in different cellular locations and have different cofactor requirements for activation they provide for differential function during various cellular events. Germ cells (i.e. eggs and sperm) are highly specialized cells that eventually give rise to a new organism through several developmental transitions, these include: a) The oocyte must become mature as a fertilization-competent egg arrested in meiotic metaphase II of the cell cycle; b) Post fertilization events follow a series of time-dependent changes both at the structural and biochemical level that convert the egg to the zygote, and c) The zygote to the embryo through pivotal developmental transitions. Disruption of these time-dependent changes, likely results in spontaneous abortion or abnormal development. Equally important is the maturation process the sperm undergoes. Sperm cells require first that capacitation and later the acrosome reaction occur before becoming fertilization-competent. If these gametes interact without first undergoing these initial changes, the orchestration of fertilization will not take place. Indeed the egg should be thought of as a cell programmed for death, but rescued by the key developmental transition of fertilization. This chapter examines the involvement of PKC isotypes in oocytes and eggs. The involvement of PKC in sperm development is next considered - followed by examining the involvement of PKC in fertilization and development of the early embryo. The concluding section addresses methods for the study of PKC in eggs/embryos.

2. Oocytes and eggs

With the exception of M-phase at which point the spindle apparatus reorganizes from the nucleus, the cell can be thought of as partitioned into nuclear and cytoplasmic compartments. As such, the nuclear envelope provides a physical barrier permitting access of different signaling molecules to these two compartments. At the onset of M-phase the nuclear envelope is disassembled and, accompanied by the disassembly of the interphase array of microtubules and replacement with the meiotic spindle apparatus. Despite the absence of the nuclear envelope in partitioning the cell, differential enzyme substrate specificity can continue to afford unique enzyme action at different cellular locations. Thus, in a sense, it can be thought that the nuclear-cytoplasmic partition remains.

In mammals, oocytes are arrested in the diplotene state of the meiotic cell cycle within the ovary awaiting maturity. At this point in the cell's development the oocyte nucleus, also known as the germinal vesicle, is still present. Rupture of the follicles trigger resumption of meiosis and there is a release of oocytes each surrounded by a few layers of cumulus cells. During this time the oocyte is released from arrest at the diplotene state and progresses to the next cell cycle arrest point, meiotic metaphase II. Briefly, in mammals cumulus cells are responsible for physical support of the developing gamete and maintenance of a so-called microenvironment as reviewed elsewhere (Huang & Wells, 2010).

The presence of cumulus cells surrounding the oocyte can cause problems during experimental analysis, most notably when agonists or antagonists of signal transduction pathways are applied because they are applied externally. When applied in the short term these pharmacological agents necessarily act first on the cumulus cells surrounding the oocyte and then later on the cumulus-enclosed oocyte. Under these conditions the exact concentration of the pharmacological agent within the oocyte is unlikely to be consistent since the membrane-permeate, pharmacological agent must first interact with the "sponge-like" cumulus cells that surround the oocyte before diffusing into the oocyte. In addition, many pharmacological agents are foreign to cellular metabolism and thus any paracrine communication between not only cumulus cells, but also the cumulus cell and the oocyte may be altered. Thus, for investigations employing agents applied to oocytes, it is best to first remove the cumulus cells prior to application of a PKC agonist or antagonist. However, when the natural, unaltered distribution of PKC or some other endogenous component is to be studied, the experiment is benefited by the retention of the cumulus cells, since that is the natural state (Downs et al., 2006).

One such study (Avazeri et al., 2004) shows that all of the conventional PKC isotypes, except for PKCγ, undergo a change from a cytoplasmic to nuclear localization just prior to germinal vesicle breakdown. The investigators also injected antibodies directed against each of the previously mentioned PKC isotypes to block their biological activity. When injected into the nucleus, blocking PKCα and βII had the greatest effect on inhibition of germinal vesicle breakdown, followed by a lesser effect of antibodies to PKCβI and finally antibodies to PKCγ had no effect. Other studies examined a single PKC isotype and confirm the presence of PKCα (Quan et al., 2003), and PKCβI (Denys et al., 2007) in the nucleus prior to germinal vesicle breakdown. One study (Luria et al, 2000) examined PKCα, βI, βII and found all three isotypes present in the cytoplasm of the germinal vesicle stage oocyte which is in agreement with the above studies at the early germinal vesicle stage (Avazeri et al., 2004). The focus of the study by Luria and colleagues (2000) was to apply PMA as an agonist to activate PKCs. This treatment caused resumption of the meiotic cell cycle and consequent disruption of the germinal vesicle. Through biochemical techniques it was shown that PKCδ is localized in the germinal vesicle as well as the cytoplasm when PKC activity is elevated. Later, during meiosis I, PKCδ was shown to be associated with the spindle apparatus (Viveiros et al., 2001, 2003), and that its active form, p-PKCδ, was enriched at the spindle poles and colocalized with both pericentrin and γ-tubulin (Ma et al., 2008).

The fertilization-competent egg arrested at meiotic metaphase II features notable differences in PKC isotypes and their individual localization. Total antibody, that is, antibody that detects both the phosphorylated and unphosphorylated form of the kinase, directed against PKCα, -γ, -δ and –ζ were found not only in the cytoplasm, but also enriched on the spindle apparatus of the fertilization-competent egg (Baluch et al., 2004). The tightness of this PKC binding was assessed by detergent extraction, a process which removes the soluble components of the cell and retains the egg's cytoskeleton and associated proteins. Total PKCζ remained behind in the cytoskeleton after the detergent extraction process; the closeness of this binding was confirmed by FRET analysis between α-tubulin of the meiotic spindle and the PKC isotypes. FRET analysis demonstrated a close molecular proximity not only between α-tubulin and PKCζ, but also between α-tubulin and PKCδ (Baluch et al., 2004, Baluch & Capco, 2008). Other investigations have shown that PKCβ exists in the cytoplasm of the fertilization-competent egg (Raz et al., 1998). In mouse eggs, p-PKCζ is enriched at the ends of the acentrosomal spindle, whereas total PKCζ decorates the length of the spindle apparatus.

Upon fertilization PKCα, -β, and –γ localize to the plasma membrane (Raz et al., 1998; Baluch et al., 2004, Halet et al., 2004). Here these conventional PKCs are poised to become active, as they are in a prime location to interact with their co-factors (i.e. DAG and phospholipids) for activation. Intriguingly, total PKCζ and the phosphorylated form of PKCζ (p- PKCζ) remain associated with the spindle apparatus, from the arrest at meiotic metaphase II through the anaphase II transition. This result suggested that p-PKCζ may play a role in its association with the spindle. In fact, inhibition of p-PKCζ with a specific, membrane-permeate, peptide inhibitor caused inactivation of the isotype and appeared to result in a disruption of the spindle apparatus. Inhibitors that would block most other active isotypes of PKC, but not p-PKCζ, did not result in spindle disassembly.

Evidence exists that also suggests a role for p-PKCδ in spindle stability (Ma et al., 2008) as others have noted p-PKCδ is enriched in the same region of the spindle pole as p-PKCζ in the metaphase egg. A targeted knockdown of PKCδ expression with siRNA disrupts the spindle apparatus and also decreases the expression of pericentrin. However, these data leave open the possibility that the knockdown may have several effects on the egg's spindle apparatus (Ma et al., 2008), as the reduction of pericentrin would provide no sites for p-PKCζ to bind at the poles. This alternative interpretation could explain the disruption of the spindle since p-PKCζ activity is required for spindle stability as previously described (Baluch et al., 2004, 2008). Moreover, both immunopurification and FRET analysis at the spindle poles demonstrates that total and p-PKCζ are in close molecular proximity with γ-tubulin, Par6, and ser9-GSK3β (Baluch & Capco, 2008). The Par6/PKCζ/GSK3β pathway may have a role in microtubule stability (Baluch & Capco, 2008). When GSK3β is active, microtubule networks can be destabilized (Zhou & Snider, 2005). However, when GSK3β is rendered inactive by phosphorylation on serine 9, spindle microtubules are stabilized. Ser9 p-GSK3β and p-PKCζ are localized to the spindle poles and kinetochore regions of the mouse egg and FRET analysis demonstrates their close molecular proximity (Baluch & Capco, 2008). A knockdown of the PKCδ gene in mice using gene targeting strategy shows that mice are viable and fertile, and thus suggests no deleterious effects on the spindle apparatus (Leitges et al., 2001). This may suggest that cells have a "plan-B" to protect the spindle in the event of a complete gene knockout. However, the aim of that study was to investigate vein arteriosclerosis and therefore the spindle was not specifically examined, so it remains unclear if the PKCδ gene knockout triggered any other signaling events to occur regarding the spindle.

3. Sperm

Using a combination of biochemical and immunocytochemical methods, Rotem and coworkers (1990) demonstrated that cPKCs are present in sperm in the equatorial segment, the principal piece of the tail and occasionally visualized in the centriole region. Application of staurosporine, 2-amino-4-octadecene-1,3-diol (sphingosine), or 1-(5-isoquinoylinylsulfonyl)-2-methylpiperazine (H7) all older, concentration-dependent PKC inhibitors blocked sperm motility, while application of a phorbol ester, phorbol 12-myristate 13-acetate (PMA) which activates PKC, increased motility.

For sperm to become fertilization competent two separate events must first take place that is, capacitation and subsequently the acrosome reaction. Capacitation is a series of physiological changes in the sperm that typically occur as the sperm traverses the female reproductive tract to reach the egg. During capacitation, there is an efflux of cholesterol from the sperm plasma membrane followed by an increase in the level of intracellular-free calcium, and cAMP levels increase to activate PKA (Breitbart, 2003). Capacitation also induces a rise in free-calcium that itself could lower the requirement for PKC activation even in the absence of increased levels of DAG. However, the calcium-dependent isotype of PLC (PLCγ) is activated by calcium which could also generate DAG by hydrolysis of PIP_2, leading to the activation of many of the PKC isotypes (Breitbart, 2003) at the time of capacitation.

Capacitation is accompanied by assembly of an actin network between the outer acrosomal membrane and the overlying sperm plasma membrane. PKC itself or in concert with PLCγ and/or phospholipase D could mediate the formation of this network as both have been shown to be associated with the underlying actin network in this area (Breitbart 2002; Cohen et al., 2004). In order for the acrosome reaction to take place this cortical actin network must subsequently disassemble. Evidence that the disassembly of the actin network is necessary for completion of the acrosome comes from the application of an actin-stabilizing agent such as phalloidin which was shown to inhibit the acrosome reaction (Spungin et al., 1995). The opening of calcium channels is thought to be mediated by the binding of sperm to ligands in the zona pellucidaa and initiates the acrosome reaction (Breitbart 2002). This calcium influx induced by the zona pellucidaa could activate PLC which in turn can produce DAG, all of which are required for PKC activation. Studies on human sperm (Rotem et al., 1992) demonstrated that capacitated sperm could be induced to undergo the acrosome reaction in the absence of intracellular-free calcium via the non-specific PKC agonist, PMA. Both PKCα and –βII were found in a sharp band at the equatorial segment, whereas PKCβI and PKCε were found at the principal piece of the tail (Rotem et al., 1992).

The acrosome reaction is a secretory event in the sperm where the large secretory vesicle effectively a large lysosomal vesicle of the sperm head, the acrosome, vesiculates to release digestive enzymes. This permits the fertilization-competent sperm to burrow its way between cumulus cells to reach the egg zona pellucida. Here, through the use of digestive enzymes and other proteins released from the acrosome, the sperm is able to penetrate the zona pellucida to fuse with the egg plasma membrane. The zona pellucida is a glycoprotein-rich network exterior to the plasma membrane of the egg. The acrosome reaction is dependent on extracellular-free calcium (Yanagimachi & Usui, 1974; Florman et al., 1989); however, utilizing PMA to activate PKC induced the acrosome reaction independent of extracellular-free calcium (Rotem et al. 1992). PMA is not part of normal cellular metabolism and therefore the possibility exists that the calcium-independent acrosome reaction was an artifact. The notion, however, that PKC is involved in the acrosome reaction is supported by application of a diacylglycerol, a natural PKC agonist that also induced the acrosome reaction in the absence of extracellular-free calcium. Moreover, it was speculated that the increase in intracellular-free calcium required during the earlier capacitation event supplied sufficient calcium for the PKC-induced acrosome reaction while inhibitors to PKC blocked this response (O'Toole et al., 1996).

Although progesterone secreting cumulus cells surrounding the mammalian egg have been shown to trigger the acrosome reaction (Garcia & Meizel, 1999), PKC inhibitors block the acrosome in human sperm (O'Toole et al., 1996). In addition, some have shown that components of the egg zona pellucida are responsible for the acrosome reaction, but similarly the older drug inhibitor of PKC, Staurosporine, blocks the acrosome reaction (Liu and Baker, 1997).

It has also been reported that PKC is essential for the maintenance of sea urchin sperm motility, although other kinases such as PKA and tyrosine kinase are present, but not key for motility (White et al., 2007). Intriguingly, this article also reported the presence of PKM, the nonmembrane-bound catalytic subunit of PKC in the sea urchin sperm. PKC is cleaved to form PKM which has different substrate specificity than PKC. Since PKM is not membrane-bound it phosphorylates substrates in the sperm interior. PKM also has been reported in mammalian eggs and has been shown to phosphorylate cytoskeletal substrates in the egg interior as discussed in the fertilization section (Gallicano & Capco 1995). Although activation of PKC in sperm seems to result in similar developmental fates in a diversity of animals, it is not to be considered a universal activator since PKC activation in fowl sperm blocks the acrosome reaction (Ashizawa et al., 2006). Some level of PKC activity also has been implicated in retaining the proper volume in sperm as it traverses fluids of different osmolarity during its passage from the testis to the oviduct (Petrunkina et al., 2007).

4. Fertilization

The process of fertilization induces a number of structural and biochemical changes within the egg as the single sperm cell penetrates the egg to subsequently form the zygote. This pivotal developmental transition appears to be dependent on a series of signal transduction events initiated by sperm penetration (Figure 1). Increasing evidence suggest that phospholipase C (PLC) ζ, supplied by the sperm, initiates this developmental transition within the egg (Saunders et al., 2002; Knott et al., 2005; Lee et al., 2006; Swann et al., 2006; Yu et al., 2008). Once PLCζ is within the egg, it then hydrolyzes phosphatidylinositol 4,5-bisphosphate (PIP$_2$) to produce inositol triphosphate (IP$_3$) and DAG. IP$_3$ subsequently diffuses and binds to IP$_3$ receptors to initiate a rise in calcium from sequestered stores in the egg, which potentially initiates calcium oscillations through a calcium-induced activation of PLC or PKC.

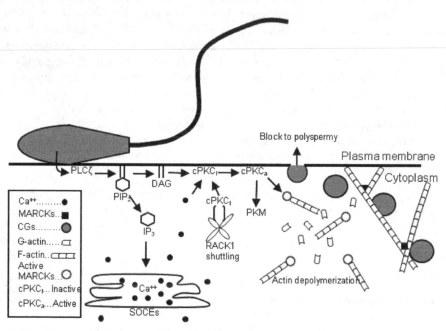

Fig. 1. Diagrammatic representation of PKC signaling events that may be involved in cortical granule exocytosis, the block to polyspermy, and subsequently fertilization.

Intracellular-free calcium along with DAG are essential cofactors for the activation of conventional PKCs. Translocation of the conventional PKC isotypes to the cell cortex indicates activation of the kinase and one study (Halet et al., 2004) has provided strong evidence that the cPKC, PKCα, is translocated to the egg plasma membrane after every calcium oscillation, and functions to induce calcium transients as part of a positive feedback loop that regulates store-operated calcium entry (SOCE). Others have shown the translocation of PKCs to the egg plasma membrane resulting from egg activation (Gallicano et al., 1995; Raz et al., 1998, Baluch et al., 2004; Carbone & Tatone, 2009) however, the investigation by Halet et al. (2004) clearly provided a link to calcium oscillations and shows that PKC has a role in the positive feedback loop that generates the calcium oscillations through SOCE.

After recruitment to the egg plasma membrane, perhaps in part by the proposed shuttling action of RACK1 (Haberman et al., 2010), activated PKCs are poised to phosphorylate MARCKS proteins. MARCKS proteins are substrates for PKC, and MARCKS have the ability to crosslink actin filaments into a network at the cell cortex (Eliyahu et al., 2005, 2006). However, once MARCKS are phosphorylated by PKC the actin network can then disassemble (Tsaadon et al., 2008). Disassembly of the actin network at the egg cortex could release cortical granules from the cortical actin network and allow them to reach the plasma membrane to exocytose (Tsaadon et al., 2008). Once cPKCs are activated, as part of their down regulation, the kinase is cleaved between its membrane-binding domain and catalytic subunit to form PKM (Jaken 1990; Hashimoto et al., 1991). PKM remains active in the cytosol and its altered folding provides for altered substrate specificities. Gallicano et al. (1995) demonstrated that PKM which forms as a consequence of fertilization acts to phosphorylate a specialized intermediate filament network known as cytoskeletal sheets (reviewed by Capco 2001). In addition, through a series of experiments it was also shown to remodel the cytoskeletal interior (Gallicano et al., 1995). The authors first determined that it was indeed PKC acting downstream of calcium responsible for the changes in the cytoskeletal interior by experimentally activating PKC while clamping calcium to low levels. Subsequently the distribution of PKC was mapped over time with the PKC reporter dye Rim-1 and PKC was found at the egg cortex indicating it was activated (Gallicano et al., 1995). Furthermore, it was noticed during a biochemical assay that utilized a monoclonal antibody to the catalytic subunit of PKCα that there was a change in the molecular weight of PKC from 80 to 50 kDa and that the kinase shifted compartments from the detergent-soluble fraction to the detergent-resistant, sheet-enriched fraction indicative of the formation of PKM. A further biochemical assay utilizing inhibitors to PKC, myosin light-chain kinase, and tyrosine kinase could not block the phosphorylation of exogenous substrate while an inhibitor to PKC/PKM blocked phosphorylation (Gallicano et al., 1995). Taken together these results suggest that the membrane-bound PKC downstream of calcium was cleaved forming PKM and that PKM bound to and phosphorylated the internal cytoskeletal elements. For these reasons the cPKCs were proposed to have a chronometer function that acts on the cell spatially and temporally. Through the action of PKC at the membrane the egg first blocks polyspermy by inducing cortical granule exocytosis and later, following each calcium oscillation, acts on substrates within the egg interior through the action of PKM as the egg remodels to form the cytoskeletal framework appropriate for the zygote.

There has been debate as to whether or not PKC acts alone to stimulate cortical granule exocytosis. Some reports suggest that another calcium-dependent kinase known as calcium/calmodulin-dependent protein kinase II (CaMKII) may be involved (Abbott & Ducibella, 2001; Abbott et al., 2001; Madgwick et al., 2005). In a series of experiments, Knott et al. (2006) injected constitutively active CaMKIIα into mouse eggs and it was reported that CaMKII was responsible for meiotic resumption and cortical granule exocytosis. Both PKC and CaMKII have several different isotypes, and the different isotypes have different substrate specifities. The δ subunit of CaMKII was shown to be effective in the resumption of mouse eggs (Johnson et al., 1998), while others have shown cell cycle resumption to be dependent on the γ subunit (Change et al., 2009; Backs et al., 2010). A plausible alternative explanation is upon injection the constitutively active CaMKIIα (Knott et al., 2006) could immediately phosphorylate substrates in the cytoplasm and alter the sequence of events leading to activation. As many others have noted, since the cortical granule block to polyspermy is key for continued development, redundant mechanisms regulating this important event are likely to be in place.

5. The early embryo

Various studies can demonstrate that the zygote and preimplantation embryo display a change in cellular localization of several PKC isotypes during postfertilization development. Both PKC ζ and -δ, which were enriched on the spindle apparatus prior to fertilization later appear enriched within the pronuclei at six hours postfertilization, while PKCα and -γ are absent from the pronuclei (Viveros et al. 2003; Baluch et al., 2004). By the two-cell stage PKCα, γ, δ, and ζ are enriched in the nuclei (Viveros et al. 2003; Baluch et al., 2004; Dehghani and Hahnel, 2005). Reports from Pauken and Capco (1999, 2000) have shown a marked reduction in the nuclear localization of these isotypes after the two-cell stage with the exception that PKCζ lines the nuclear periphery at the four-cell stage. However, these isotypes appear in unique locations poised to presumably interact during subsequent developmental transitions: At the time of compaction PKCα, γ, δ, and μ line the cell-cell boundaries to differing extents, and these isotypes are considerably absent from the nucleus (Pauken & Capco, 1999, 2000). During the late eight-cell stage just prior to compaction the ζ isotype is greatly enriched in the nuclei (Pauken & Capco, 2000). Moreover, when PKC was experimentally activated by the natural agonist (DiC8 a diacylglycerol and natural activator of PKC) it induces PKCα to localize at internal cell-cell boundaries. Then β-catenin becomes phosphorylated and accumulates at these internal cell-cell boundaries as the blastomeres begin to flatten out on each other during the process of embryonic compaction (Pauken & Capco, 1999). In addition, it was shown that immediately before compaction begins β-catenin becomes part of the detergent-resistant cytoskeleton at intercellular boundaries indicative of its association with the adherens junctions that are responsible for adhering and subsequently flattening of the blastomeres later during compaction (Pauken & Capco, 1999). Although PKCβI and -βII were not investigated in the aforementioned study, others (Dehghani and Hahnel, 2005) have shown that PKCβI accumulates in the nuclei of embryos at the four-cell stage and later during the postcompaction stage, while PKCβII appears to become uniformly distributed throughout the cytoplasm and nuclei from the four-cell stage until the blastocyst stage.

PKCδ and ε are reported to transiently enrich in nuclei of the four-cell embryo (Dehghani and Hahnel, 2005). In this study investigators blocked both PKCδ and ε with peptides that interfere with adapter sites to block movement of these isotypes, which subsequently altered transcription. This suggests that the location of the isotype may be involved in its activity and that both isotypes have a role in transcription at the four-cell stage.

6. Consideration of experimental procedures & methods

6.1 Pharmacological agents & PKC

To determine the role of a single type of kinase investigators often employ various pharmacological agents (Table 1).

Common name	Also known as	Target kinase(s)	IC$_{50}$
Inhibitors	N/A	PKA	7 nM
Staurosporine		CaMK	20 nM
		MLCK	1.3 nM
		PKC	700 pM
		PKG	8.5 nM
Chelerythrine Chloride	N/A	PKC	660 nM
Bisindolylmaleimide	BIM	PKC	10 nM
		GSK-3	360 nM
		GSK-3β	170 nM
		PKA	2 µM
Rottlerin	Mallotoxin	PKCδ/θ	3-6 µM
		PKCα/β/γ	30-40 µM
		PKCε/η/ζ	80-100 µM
		CaMKIII	5 µM
H-7	1-(5-Isoquinolinylsulfonyl)-2-methylpiperazine dihydrochloride, Isoquinoline-5-sulfonic 2-methyl-1-piperazide dihydrochloride	PKA PKC	Determine empirically
Myristoylated PKCζ Pseudosubstrate	N/A	PKCζ	Determine empirically
Myristoylated PKCθ Pseudosubstrate	N/A	PKCθ	Determine empirically
Myristoylated PKCη Pseudosubstrate	N/A	PKCη	Determine empirically
Activators			Determine
PMA	TPA, phorbol 12-myristate 13-acetate	PKC	empirically
OAG	1-(cis-9-Octadecenoyl)-2-acetyl-sn-glycerol, 2-Acetyl-1-oleoyl-sn-glycerol	PKC	Determine empirically
DiC8	1,2-Dioctanoyl-sn-glycerol	PKC PKC	Determine empirically

Table 1. A large variety of agonists/antagonists exist for the experimental study of PKC. Some PKC inhibitors have the ability to interact with multiple kinases since the concentration of the drug increases as it accumulates within the cell over time.

While their use is essential, it is important to understand their limitations and design experiments containing the proper controls when conclusions are drawn from their uses. Ideally, a log-fold dose response curve should be conducted prior to experimentation to determine the lowest working concentration to activate/inhibit a single type of kinase of interest. The lowest working concentration should be used in subsequent experiments as many agonists/antagonists have the ability to interact with multiple kinases as the cell continues to absorb the membrane-permeate pharmacological agent over time. Examples include the PKC inhibitors 3-[1-[3-(dimethylamino)propyl]-1H-indol-3-yl]-4-(1H-indol-3-yl)-1H-pyrrole-2,5-dione (BIM, also referred to as Gö 6850, or GF 10923X), Staurosporin, and Rottlerin, among others. An alternative to pharmacological inhibitors of PKC are membrane–permeant peptide inhibitors, which exist for some of the PKC isotypes and other kinases and have a greater degree of specificity since their peptide sequence is designed to block the active site of the specific kinase. In addition to membrane-permeant peptide inhibitors, it would be wise to employ two structurally and chemically distinct inhibitors to reduce the possibility of misinterpretation of experimental data (Stricker et al., 2010).

It also is wise to adopt natural agonists such as a diacylglycerol (DAG); for instance OAG or DiC8 can be routinely used in parallel when pharmacological agents are employed (Gallicano et al., 1995; Pauken & Capco, 1999; Eliyahu & Shalgi, 2002; Halet et al., 2004). In addition, it is useful to confirm the activity of the PKC isotype. A rapid method to determine whether a kinase is active is to employ an antibody specific to the phosphorylated/active form of the isotype during immunocytochemical or western blot analysis.

Some molecular strategies at the DNA or RNA level have been developed to remove or knockdown a specific PKC isotype. Knockout mice have been developed for PKCα, PKCβ, PKCγ, PKCδ, PKCε, PKCθ, and PKCζ and the effects of gene knockout have been studied (Abeliovich et al., 1993; Leitges et al., 1996, 2001a, 2001b, 2002; Sun et al., 2000; Meier et al., 2007). Care should be taken when interpreting experimental data from these models as PKC operates within many other signaling pathways.

6.2 Introduction to the handling and procurement of mouse eggs

There exist several technical challenges concerning the procurement and manipulation of mammalian eggs/embryos. For starters, the mouse egg is about 80 μm in diameter. This small size requires the use of a finely pulled, flame-polished Pasteur pipette attached to a pipetting apparatus; a 1 mL syringe attached to a 24" piece of flexible tubing through which suction can be created for aspiration or expulsion. Under optimal conditions, only 20-40 eggs can be obtained from a single female, and consequently any loss of sample during handling can significantly decrease the amount of material for study. In order to obtain 20-40 eggs/embryos female mice are superovulated by an injection regimen of gonadotropins. An outbred mouse strain CD-1 (Charles River Laboratories) is routinely used in this laboratory and maintained under a 14-hour light/ 10-hour dark schedule and given *ad libitum* access to food and water.

6.2.1 Procedure for the procurement of cumulus-free eggs

1.   Pregnant mare serum gonadotropin (PMSG) administration is accomplished by intraperitoneally (i.p.) injection of 5.0 IU PMSG in a 0.1 mL sterile volume of 100 mM Phosphate-buffered saline (PBS) at approximated 3:00 p.m.
2.   Approximately 48 hours later, inject sterile human chorionic gonadotropin (5 IU).
3.   Approximately 15 hours post-hCG injection, females are euthanized according to the established IUCUC protocol for the given institution.
4.   Oviducts are removed by gently cutting away the mesometrium while holding the uterine horn near the oviduct. Remove the fat that allows the ovary to adhere to the

abdominal viscera. Cut between the oviduct and ovary followed by a cut at the top of the uterine horn to remove the oviduct. Place the oviduct in a Petri dish with pre-equilibrated FHM-BSA. All subsequent steps are conducted under a dissection microscope equipped with a stage warmer set to 37°C and fiber-optic lighting. Locate the cumulus mass. Removal of the cumulus mass is accomplished by carefully immobilizing one side of the oviduct with Dumont no. 5 forceps while gently slicing the oviduct on the opposite side with a clean hypodermic needle. If the cumulus mass does not immediately "pop" out of the oviduct, gently tease it out with the blunt side of the needle. Alternatively, a 26 or 30G needle coupled to a syringe filled with medium can be inserted in the ostium and flushed with excess medium to release the cumulus mass from the opposite end of the oviduct.

5. Collect cumulus masses with aid from a fire polished unpulled, 5" glass Pasteur pipette and place in the pre-equilibrated medium FHM-BSA (Millipore, product number MR-024-D) containing 300 µg/mL hyaluronidase and observe to determine when cumulus cells are released.

6. Once denuded, immediately transfer eggs through several drops (approximately 1 mL) of FHM-BSA to wash away any remaining hyaluronidase. Proceed immediately to the next step in the experiment (e.g. treatment with peptide inhibitors or pharmacological agents, immobilization for cytological fixation, or production of a lysate for a biochemical kinase assay, etc.) as spontaneous activation will begin at approximately 9 A.M.

The above procedure details the procurement of cumulus-free mouse eggs for experimental purposes. However, this protocol can be modified to collect fertilized eggs and embryos. To produce fertilized eggs and embryos a male is placed with a female for copulation after hCG injection. The following morning the female is checked for a white copulation (vaginal) plug to indicate that copulation occurred. The day at which the plug is detected is 0.5 days post coitum (dpc), since fertilization is assumed to occur at midnight. Two-cell embryos can be collected at noon 1.5 dpc and four-cell embryos approximately 8-12 hours later. Eight-cell embryos are collected on the morning of 2.5 dpc, while morulas are collected in the evening.

It is most efficient to flush the oviducts with approximately 0.1 mL of FHM-BSA since, beginning at the two-cell stage and beyond, the cumulus cells sloughed off which makes them difficult to locate.

6.3 Detergent extraction

The cytoskeleton is the cellular framework composed of actin filaments, intermediate filaments, microtubules and associated protein that regularly reorganize in response to stimuli. Each filament has a subset of signaling proteins responsible for cytoskeletal reorganization. During meiosis, the egg undergoes a remodeling of the cytoskeleton in preparation of the developmental transition of fertilization. If one is interested in studying the cytoskeleton and its associated proteins at a morphological and biochemical level a technique known as detergent extraction can be employed.

Detergent extraction removes the soluble components of the cell through the use of a non-ionic detergent in an intracellular buffer (ICB) that mimics the intracellular milieu. As a result of this process approximately one-third of the egg's total protein is extracted into a cell fraction referred to as the soluble fraction. Two-thirds remains insoluble and constitutes the detergent-resistant, cytoskeletal fraction. This latter fraction (i.e. the cytoskeletal fraction) contains all three filament systems and those proteins that are tightly bound to each cytoskeletal framework. For eggs and embryos the detergent extraction medium is ICB (Aggeler et al., 1983; Webster and McGaughey, 1990) made 1% with Tween-20 and 200 µg/mL with the protease inhibitor 4-(2-Aminoethyl) benzenesulfonyl fluoride hydrochloride (AEBSF). ICB is composed of 100 mM KCl, 5 mM $MgCl_2$, 3 mM EGTA, 20 mM HEPES, (pH 6.8). Tween-20 is the non-ionic detergent used for eggs and embryos as Triton X-100 was shown to remove selective peptides from and destabilize the cytoskeletal network in mammalian eggs and embryos (Gallicano et al., 1991, 1994).

How long does it take to completely detergent extract cells? The answer to this question is dependent on the cell type the investigator employs and should be determined empirically. To determine the proper extraction time the investigator should detergent extract the cells and take aliquots of the medium at increasing time intervals. The protein containing medium can then be analyzed to determine protein concentration and the plateau at which the proteins are no longer rapidly released from cells. The time that corresponds to the establishment of a plateau should then be used in replicate experiments as the extraction end-point and clearly indicated in the methods section of the manuscript.

6.3.1 Detergent extraction procedure for eggs/embryos

1. Eggs/embryos are washed through 3 drops of PBS and placed in detergent extraction medium (ICB made 1% with Tween-20 and 200 µg/mL AEBSF).
2. Prepare the soluble-fraction and detergent-resistant cytoskeleton as in the detergent extraction section. For biochemical analysis (i.e. gel electrophoresis), eggs/embryos can be transferred into SDS-sample buffer executing caution not to transfer excess fluid with the cytoskeleton. The detergent-soluble fraction is isolated in a large volume and needs to be concentrated; this is done by precipitating the soluble-fraction in ice-cold 95% ethanol and subsequently incubating it at -20°C overnight. Centrifuge the precipitated soluble fraction into a pellet and decant the supernatant. The pellet containing the soluble-fraction can then be solubilized by the addition of SDS-sample buffer.
3. The fixative is largely dependent on the type of microscopic analysis being conducted. The detergent-resistant cytoskeleton of eggs/embryos is processed for observation by light or electron microscopy. For instance, if immunocytochemistry is the form of analysis, the detergent-extracted specimen is cytologically fixed by the addition of 0.05% glutaraldehyde 2% formaldehyde made in ICB for 5 minutes at room temperature and then transferred to 2% formaldehyde made in ICB for 25 minutes at room temperature. Any remaining free aldehyde groups are inactivated, and sticky sites are mitigated by transferring eggs/embryos through the blocking solution ICB-BSA (ICB made 1% with bovine serum albumin). This treatment allows the specimen to be transferred through antibody-containing solutions to view specific proteins.

6.4 Conventional and embedment-free electron microscopy

Conventional, resin-embedded transmission electron microscopy has been utilized by investigators as a powerful tool to interrogate the ultrastructural changes that accompany fertilization (Bement and Capco, 1989; 1991; Gallicano et al., 1991, 1994, 1995). However, mammalian eggs and embryos do not afford high contrast images in resin-embedded, ultrathin sections (silver to gold interference patterns corresponding to 60-70 nm). To mitigate the inherent low-contrast of eggs and embryos the authors routinely employ 0.1% tannic acid during glutaraldehyde fixation. Tannic acid acts as a mordant allowing OsO_4 to more effectively bind. Alternatively, *en block* staining in aqueous 2% uranyl acetate prior to dehydration can be used to further improve contrast of the biological material. The aforementioned steps produce marked electron density for resin-embedded ultrathin sections and provide detail unattainable at the level of the light microscope.

Further detail and three-dimensional information is afforded by removing the embedding medium entirely. This requires employing embedding medium that can be removed without damaging the specimen. By implementing the embedment medium, diethylene glycol disterate (DGD, also referred to as Pentament), Capco and coworkers (1984) refined a procedure for the complete removal of the DGD in HeLa and Madin-Darby Canine Kidney cultured cells for its use in transmission and scanning electron microscopy. Removing embedment not only provides high resolution images, but significantly increased electron density in biological material due to the absence of a resin, and also allowed for the use of thick sections (200 nm sections and greater corresponding to purple interference) at the level of the electron microscope. Gallicano and coworkers exploited the use of DGD in detergent extracted mammalian eggs and for the first time novel cytoskeletal elements known as "sheets" were characterized in mammalian eggs (Gallicano et al., 1991).

6.4.1 Procedure for embedment-free electron microscopy for use with eggs/embryos

1. Before obtaining eggs/embryos coverslips are prepared by cleaning no. 1 coverglass in a beaker containing acetone by sonication for 3x for 5 minutes each followed by sonication in distilled water for 3x for 5 minutes each. Allow coverslips to dry by placing them on Whatman no. 1 filter paper in a Petri dish protected from dust. Poly-L-lysine (10% w/v) is spread evenly over the surface of the coverslips and placed on a 37°C slide warmer until dry.
2. Obtain eggs/embryos (see Procurement section) and wash through 3 drops of PBS at 37°C.
3. In a glass Petri dish, eggs/embryos are added to prepared coverslips immersed in an excess volume of PBS. The specimens should adhere immediately. If specimens fail to adhere, they are reaspirated washed again and added to a different prepared coverslip. Care should be exercised during manipulation as poly-L-lysine will clog the tip of the micropipette in the event that the tip touches the poly-L-lysine and this is typically followed by a significant loss of sample. Using a disposable, fire polished, 5" glass Pasteur pipette, remove three-fourths total volume of PBS and slowly add copious amounts of extraction buffer. The density of the extraction buffer will displace the PBS. Repeat the controlled addition of extraction buffer 1x. Allow specimens to extract.
4. Remove three-fourths total volume of the extraction buffer and add fresh extraction buffer containing 2% (v/v) glutaraldehye and 1% formaldehyde for 30 minutes at room temperature.
5. Remove three-fourths total volume of the fixation medium for 3x 15 minutes with 100 mM sodium cacodylate.

6. Post fixation occurs by removal of three-fourths total volume of sodium cacodylate wash buffer and the addition of fresh 1% (v/v) OsO_4 in sodium cacodylate at 4°C in the dark for 1 hour. Upon completion of secondary fixation specimens are washed 3x for 15 minutes in sodium cacodylate containing no OsO_4.

7. Specimens are then dehydrated through a series of increasing ethanol concentrations starting with 10% ethanol. Remove three-fourths volume of sodium cacodylate wash buffer and add 10% ethanol. Repeat 1x and allow 10-15 minutes time for the specimen to equilibrate. This procedure should be repeated for 30, 50, 70, 90, and 100% ethanol. Exchange 100% ethanol 3x for 20 minutes each with 1:1 100% ethanol: 100% n-butyl alcohol. After the third change allow the specimens to equilibrate for 30 minutes. Remove three-fourths volume 1:1 100% ethanol: 100% n-butyl alcohol, and replace with 100% n-butyl alcohol 3x for 20 minutes each change. After the third change allow the specimens to equilibrate for 30 minutes. Place the Petri dish in a 70°C oven for 15 minutes. Pour off the n-butyl alcohol and immediately add prewarmed 1:1 100% n-butyl alcohol: 100% DGD. Repeat this 3x for 30 minutes each. Pour off 1:1 100% n-butyl alcohol: 100% DGD and add fresh DGD 3x for 30 minutes each.

8. Preheat the flat embedding mold (Ted Pella, Inc., Redding CA; product 10505) in an oven. Add molten DGD to each block of a Teflon-coated flat-embedding mold under a heat lamp. Carefully remove eggs/embryos from the poly-L-lysine with the blunt end of a prewarmed needle. Using a prewarmed, fire-polished pulled pipette, immediately transfer eggs/embryos to the embedding mold and allow the wax to solidify at room temperature. It is necessary to add additional DGD as the block begins to cool and contract. Remove blocks carefully once solidified, as they tend to stick if left for extended periods of time.

9. Trim the block face to produce a square. Blocks are sectioned on an ultramicrotome with a knife angle adjusted to 10° (Capco et al., 1993). 200 nm sections are collected on formvar-coated, carbon-stabilized grids precoated with poly-L-lysine and subsequently dried overnight in a vacuum desiccator.

10. Removal of the wax is accomplished by placing the formvar-coated, carbon-stabilized grids containing thick sections into excess 100% n-butyl alcohol for 1 hour at room temperature in a glass Petri dish. After 1 hour gently swirl the dish containing 100% n-butyl alcohol and let sit for an additional 15 minutes. Remove the grids and place into a fresh Petri dish containing the transition fluid 1:1 100% ethanol: 100% n-butyl alcohol for 15 minutes. Remove the grdis and place into 100% ethanol for 1 hour. After 1 hour, carefully swirl the 100% ethanol and let sit for an additional 15 minutes.

11. Dry the specimens through the CO_2 critical point and immediately view the sample with a conventional transmission electron microscope with an accelerating voltage of 60 kV.

7. Conclusion

The 11 isotype family of PKC has been identified within both the sperm and egg during key developmental transitions such as gametogenesis, fertilization, and early development, and this suggests that the PKC family may take on multiple, essential tasks. Differential regulation of individual isotypes can occur in a number of ways, most notably through different cofactor requirements for activation of individual isotypes, differential substrate specificity of individual isotypes, and localization or enrichment of individual iostypes.

These mechanisms likely serve to impart distinction among isotypes regarding the function of the kinase.

Elucidation of the possible involvement of PKC during each developmental transition will be expedited by the use of antibodies and inhibitors specific to individual isotypes. For instance, the use of a pan- (i.e. total) antibody directed against all of the PKC isotypes has been previously studied in great detail in this area, and is therefore counterproductive for use in a manuscript. In addition, care should be employed with the use of older, concentration-dependent PKC inhibitors as the interpretation of the results obtained by the use of these inhibitors can be difficult as these inhibitors can interact with multiple kinases. Furthermore, investigations focusing on localization or enrichment of specific PKC isotypes will likely reveal signaling-mediated changes during developmental transitions. These studies may expose the possible redundancy of specific isotypes, i.e. are there backup mechanisms in place in the event of a complete dysfunction of an individual isotype? Lastly, few have studied the potential role of PKM, the catalytic subunit of cPKCs that was shown to interact in the egg cytoskeletal interior. Studies directed towards PKM may reveal time-dependent changes in the egg interior that have long been ignored.

8. References

Abbott, A.L. & Ducibella, T. (2001). Calcium and the control of mammalian cortical granule exocytosis. *Froniers in Bioscencesi*, Vol. 6, No. D, July 2001, pp. 792-806.

Abbott, A.L., Fissore, R.A. & Ducibella, T. (2001). Identification of a translocation deficiency in cortical granule secretion in preovulatory mouse oocytes. *Biology of Reproduction*, Vol. 65, No. 6, December 2001, pp. 1640-1647.

Abeliovich, A., Paylor, R., Chen, C., Kim, J.J., Wehner, J.M. & Tonegawa, S. (1993). PKCγ mutant micr exhibit mild deficits in spatial and contextual learning. *Cell*, Vol. 75, No. 7, December 1993, pp. 1263-1271.

Aggeler, J., Takemura, R. & Werb, Z. (1983). High-resolution three-dimensional views of membrane-associated clatherin and cytoskeleton critical-point-dried macrophages. *Journal of Cell Biology*, Vol. 97, No. 5, November 1983, pp. 1452-1458.

Ashizawa, K., Wishart, G.J., Katayama, S., Takano, D., Ranasinghe, A.R., Narumi, K. & Tsuzuki, Y. (2006). Regulation of acrosome reaction of fowl spermatozoa: evidence for the involvement of protein kinase C and protein phosphatase-type 1 and/or -type 2A. *Reproduction*, Vol. 131, No. 6, June 2006, pp. 1017-1024.

Avazeri, N; Courtot, A.M. & Lefevre, B. (2004). Regulation of spontaneous meiosis resumption in mouse oocytes by various conventional PKC isozymes depends on cellular compartmentalization. *Journal of Cell Science*, Vol. 117, No. 21, October 2004, pp. 4969-4978.

Backs, J., Stein, P., Backs, T., Duncan, F.E., Grueter, C.E., McAnally, J., Qi, X., Schultz, R.M. & Olson, E.N. (2010). The γ isoform of CaM Kinase II controls mouse egg activation by regulating cell cycle resumption. *Proceedings from the National Academy of Science*, Vol. 107, No. 1, January 2010, pp. 81-86.

Baluch, D.P., Koeneman, B.A., Hatch, K.R., McGaughey, R.W. & Capco, D.G. (2004). PKC isotypes in post-activated and fertilized mouse eggs: association with the meiotic spindle. *Developmental Biology*, Vol. 274, No. 1, October 2004, pp. 45-55.

Baluch, D.P. & Capco, D.G. (2008). GSK3 beta mediates acentromeric spindle stabilization by activated PKC zeta. *Developmental Biology*, Vol. 317 No. 1, May 2008, pp. 46-58.

Bement, W.M. & Capco, D.G. (1989). Activators of protein kinase C triggers cortical granule exocytosis, cortical contraction, and cleavage furrow formation in Xenopus laevis oocytes and eggs. *Journal of Cell Biology*, Vol. 108, No. 3, March 1989, pp. 885-892.

Bement, W.M. & Capco, D.G. (1991). Parallel pathways of cell cycle control during Xenopus egg activation. *Proceedings from the National Academy of Science*, Vol. 88, No. 12, June 1991, pp. 5172-5176.

Breitbart, H. (2002). Intracellular calcium regulation in sperm capacitation and acrosomal reaction. *Mol Cell Endocrinol*, Vol. 187, No. 1-2, February 2002, pp. 139-144.

Breitbart, H. (2003). Signaling pathways in sperm capacitation and acrosome reaction. *Cell Mol Biol*, Vol. 49, No. 3, May 2003, pp. 321-327.

Capco, D.G. (2001). Molecular and biochemical regulation of early mammalian development. *International Review of Cytology*, Vol. 207, 2001, pp. 195-235.

Capco, D.G., Gallicano, G.I., McGaughey, R.W., Downing K.H. & Larabell, C.A. (1993). Cytoskeletal sheets of mammalian eggs and embryos: a lattice-like network of intermediate filaments. *Cell Motility and the Cytoskeleton*, Vol. 24, No. 2, 1993, pp. 85-99.

Capco, D.G., Krochmalnic, G. & Penman, S. (1984). A new method of preparing embedment-free sections for transmission electron microscopy: applications to the cytoskeletal framework and other three-dimensional networks. *Journal of Cell Biology*, Vol. 98, No. 5, May 1984, pp. 1978-1985.

Carbone, M.C. & Tatone, C. (2009). Alterations in protein kinase c signaling activated by a parthenogentic agent in oocytes from reproductively old mice. *Molecular Reproduction and Development*, Vol. 76, No. 2, February 2009, pp. 122-131.

Chang, H-Y., Minahan, K., Merriman, J.A. & Jones, K.T. (2009). Calmodulin-dependent protein kinase gamma 3 (CamKIIγ3) mediates the cell cycle resumption of metaphase II eggs in mouse. *Development*, Vol. 136, No. 24, September 2009, pp. 4077-4081.

Cohen, G., Rubinstein, S., Gur, Y. & Breitbart, H. (2004). Crosstalk between protein kinase A and C regulates phospholipase D and F-actin formation during sperm capacitation. *Developmental Biology*, Vol. 267, No. 1, pp. 230-241.

Dehghani, H. & Hahnel, A.C. (2005). Expression profile of protein kinase C isozymes in preimplantation mouse development. *Reproduction*, Vol. 130, No. 4, October 2005, pp. 441-451.

Dehghani, H., Reith, C. & Hahnel, A.C. (2005). Subcellular localization of protein kinase C delta and epsilon affects transcriptional and post-transcriptional processes in four-cell mouse embryos. *Reproduction*, 130, No. 4, October 2005, pp. 453-465.

Denys, A., Avazeri, N. & Lefevre, B. (2007). The PKC pathway and in particular its beta1 isoform is clearly involved in meiotic arrest maintenance but poorly in FSH-induced meiosis resumption of the mouse cumulus cell enclosed oocyte. *Molecular Reproduction and Development*, Vol. 74, No. 12, December 2007, pp. 1575-1580.

Downs, S.M., Gilles, R., Canderhoef, C., Humpherson, P.G. & Leese, H.J. (2006). Differential response of cumulus cell-enclosed and denuded mouse oocytes in a meiotic induction model system. *Molecular Reporduction and Development*, Vol. 73, No. 3, March 2006, pp. 379-389.

Eliyahu, E. & Shalgi, R. (2002). A role for protein kinase c during rat egg activation. *Biology of Reproduction*, Vol. 67, No. 1, July 2002, pp. 189-195.

Eliyahu, E., Shtraizent, N., Tsaadon, A. & Shalgi, R. (2006). Association between myristoylated alanin-rich C kinase substrate (MARCKS) translocation and cortical granule exocytosis in rat eggs. *Reproduction*, Vol. 131, No. 2, February 2006, pp. 221-231.

Eliyahu, E., Tsaadon, A., Shtraizent, N. & Shalgi, R. (2005). The involvement of protein kinase C and actin filaments in cortical granule exocytosis in the rat. *Reproduction*, Vol. 129, No. 2, February 2005, pp. 161-170.

Florman, H.M., Tombes, R.M., First, N.L. & Babcock, D.F. (1989). An adhesion-associated agonist from the zona pellucida activates G protein-promoted elevations of internal Ca2+ and pH that mediate mammalian sperm acrosomal exocytosis. *Developmental Biology*, Vol. 135, No. 1, September 1989, pp. 133-146.

Gallicano, G.I. & Capco, D.G. (1995). Remodeling of the specialized intermediate filament network in mammalian eggs and embryos during development: regulation by protein kinase C and protein kinase M. *Current Topics in Developmental Biology*, Vol. 31, 1995, pp. 277-320.

Gallicano, G.I., Larabell, C.A., McGaughey, R.W. & Capco, D.G. (1994). Novel cytoskeletal elements in mammalian eggs are composed of a unique arrangement of intermediate filaments. *Mechanisms of Development*. Vol. 45, No. 3, March 1994, pp. 211-226.

Gallicano, G.I., McGaughey, R.W. & Capco, D.G. (1995). Protein kinase M, the cytosolic counterpart of protein kinase C, remodels the internal cytoskeleton of the mammalian egg during activation. *Developmental Biology*, Vol. 167, No. 2, February 1995, pp. 482-501.

Gallicano, G.I., McGaughey, R.W. & Capco, D.G. (1991). The cytoskeleton of the mouse egg and embryo: Reorganization of planar elements. *Cell Motility and Cytoskeleton*, Vol. 18, No. 2, 1991, pp. 143-154.

Garcia, M.A. & Meizel, S. (1999). Progesterone-mediated calcium influx and acrosome reaction of human spermatozoa: pharmacological investigation of T-type calcium channels. *Biology of Reproduction*, Vol. 60, No. 1, January 1999, pp. 102-109.

Haberman, Y., Tsaadon, L., Eliyahu, E. & Shalgi, R., (2010). Receptor for activated c kinase (RACK) and protein kinase c (PKC) in egg activation. *Theriogenology*, Vol. 75, No. 1, January 2011, pp. 80-89.

Hashimoto, E., Takeuch,i F. & Yamamura, H. (1991). Studies on protein kinase C tightly-bound to rat liver plasma membrane and its protease-activated form. *The International Journal of Biochemistry*, Vol. 23, No. 4, 1991, pp. 395-403.

Halet, G., Tunwell, R., Parkinson, S.J. & Carroll, J. (2004). Conventional PKCs regulate the temporal pattern of Ca2+ oscillations at fertilization in mouse eggs. *The Journal of Cell Biology*, Vol. 164, No. 7, March 2004, pp. 1033-1044.

Huang, Z. & Wells, D. (2010). The human oocyte and cumuls cells relationship: new insights from the cumulus cell transcriptome. *Molecular human reproduction*, Vol. 16, No. 10, April 2010, pp. 715-725.

Jaken, S. (1990). Protein kinase C and tumor promoters. *Current opinion in Cell Biology*, Vol. 2, No. 2, April 1990, pp. 192-197.

Johnson, J., Bierle, B.M., Gallicano, G.I. & Capco, DG. (1998). Calcium/calmodulin-dependent protein kinase II and calmodulin: regulators of the meiotic spindle in mouse eggs. *Developmental Biology*, Vol. 204, No. 2, December 1998, pp. 464-477.

Knott, J.G., Gardner, A.J., Madgwick, S., Jones, K.T., Williams, C.J. & Schultz, R.M. (2006). Calmodulin-dependent protein kinase II triggers mouse egg activation and embryo development in the absence of Ca2+ oscillations. *Developmental Biology*, Vol. 296, No. 2, August 2006, pp. 388-395.

Knott, J.G., Kurokawa, M., Fissore, R.A., Schultz, R.M. & Williams, C.J. (2005). Transgenic RNA interference reveals role for mouse sperm phospholipase Czeta in triggering Ca2+ oscillations during fertilization. *Biology of Reproduction*, Vol. 72, No. 4, April 2005, pp. 992-996.

Lee, B., Yoon, S.Y. & Fissore, R.A. (2006). Regulation of fertilization-initiated [Ca2+]i oscillations in mammalian eggs: a multi-pronged approach. *Seminars in Cell & Developmental Biology*, Vol. 17, No. 2, April 2006, pp. 274-284.

Leitges, M., Mayr, M., Braun, U., Mayr, U., Li, C., Pfister, G., Ghaffari-Tabrizi, N., Baier, G., Hu, Y. & Xu, Q. (2001a). Exacerbated vein graft arteriosclerosis in protein kinase Cδ-null mice. *J Clin. Invest.*, Vol. 108, No. 10, November 2001, pp. 1505-1512.

Leitges, M., Plomann, M., Standaert, M.L., Bandyopadhyay, G., Sajan, M.P., Kanoh, Y. & Farese, R.V. (2002). Knockout of PKCα enhances insulin signaling through PI3K. *Molecular Endocrinology*. Vol. 16, No. 4, April 2002, pp. 847-858.

Leitges, M., Sanz, L., Martin, P., Duran, A., Braun, U., Garcia, J.F., Camacho, F., Diaz-Meco, M., Rennert, P.D. & Moscat, J. (2001b). Targeted disruption of the ζPKC gene results in the impairment of the NF-κB pathway. *Molecular Cell*, Vol. 8, No. 4, October 2001, pp. 771-780.

Leitges, M., Schmedt, C., Guinamard, R., Davoust, J., Schaal, S., Stabel, S. & Tarahovsky, A. Immunodeficiency in protein kinase cbeta-deficient mice. *Science*, Vol. 273, No. 5276, August 1996, pp. 788-791.

Liu, D.Y. & Baker, H.W. (1997). Protein kinase C plays an important role in the human zona pellucida-induced acrosome reaction. *Molecular Human Reproduction*, Vol. 3, No. 12, December 1997, pp. 1037-1043.

Luria, A., Tennenbaum, T., Sun, Q.Y., Rubinstein, S. & Breitbart, H. (2000). Differential localization of conventional protein kinase C isoforms during mouse oocyte development. *Biology of Reproduction*, Vol. 62, No. 6, June 2000, pp. 1564-1570.

Madgwick, S., Levasseur, M. & Jones, K.T. (2005). Calmodulin-dependent protein kinase II, and not protein kinase C, is sufficient for triggering cell-cycle resumption in mammalian eggs. *Journal of Cell Science*, Vol. 118, No. 17, August 2005, pp. 3849-3859.

Ma, W., Koch, J.A. & Viveiros, M.M. (2008). Protein kinase C delta (PKCdelta) interacts with microtubule organizing center (MTOC)-associated proteins and participates in meiotic spindle organization. *Developmental Biology*, Vol. 320, No. 2, August 2008, pp. 414-425.

Meier, M., Menne, J. & Haller, H. (2009). Targeting the protein kinase c family in diabetic kidney: lessons from analysis of mutant mice. *Diabetologia*, Vol. 52, No. 5, May 2009, pp. 765-775.

Meier, M., Menne, J., Holtz, M., Gueler, F., Kirsch, T., Schiffer, M., Mengel, M., Lindschau, C., Leitges, M. & Haller, H. (2007). Deletion of protein kinase c-epsilon signaling pathway induces glomerulosclerosis and tubelointerstitial fibrosis in vivo. *Journal of the American Society of Nephrology*, Vol. 18, No. 4, April 2007, pp. 1190-1198.

Odeon, M.M., Salatino, A.E., Rodriguez, C.B., Scolari, M.J. & Acosta, G.B. (2010). The response to postnatal stress: amino acids transporters and PKC activity. *Neurochem. Research*, Vol. 35, No. 7, July 2010, pp. 967-975.

O'Toole, C.M., Roldan, E.R. & Fraser, L.R. (1996). Protein kinase C activation during progesterone-stimulated acrosomal exocytosis in human spermatozoa. *Molecular Human Reproduction*, Vol. 2, No. 12, December 1996, pp. 921-927.

Pauken, C.M. & Capco, D.G. (1999). Regulation of cell adhesion during embryonic compaction of mammalian embryos: roles for PKC and beta-catenin. *Molecular Reproduction and Development*, Vol. 54, No. 2, October 1999, pp. 135-144.

Pauken, C.M. & Capco, D.G. (2000). The expression and stage-specific localization of protein kinase C isotypes during mouse preimplantation development. *Developmental Biology*, Vol. 223, No. 2, July 2000, pp. 411-421.

Petrunkina, A.M., Harrison, R.A., Tsolova, M., Jebe, E. & Topfer-Petersen, E. (2007). Signalling pathways involved in the control of sperm cell volume. *Reproduction*, Vol. 133, No. 1, January 2007, pp. 61-73.

Quan, H.M., Fan, H.Y., Meng, X.Q., Huo, L.J., Chen, D.Y., Schatten, H., Yang ,P.M. & Sun, Q.Y. (2003). Effects of PKC activation on the meiotic maturation, fertilization and early embryonic development of mouse oocytes. *Zygote*, Vol. 11, No. 4, November 2003, pp. 329-337.

Raz, T., Ben-Yose,f D. & Shalgi, R. (1998). Segregation of the pathways leading to cortical reaction and cell cycle activation in the rat egg. *Biology of Reproduction*, Vol. 58, No. 1, January 1998, pp. 94-102.

Rotem, R., Paz, G.F., Homonnai, Z.T., Kalina, M., Lax, J., Breitbart, H. & Naor, Z. (1992). Ca(2+)-independent induction of acrosome reaction by protein kinase C in human sperm. *Endocrinology*, Vol. 131, No. 5, November 1992, pp. 2235-2243.

Rotem, R., Paz, G.F., Homonnai, Z.T., Kalina, M. & Noar, Z. (1990). Protein kinase c is present in human sperm: possible role in flagellar motility. *Proceedings from the National Academy of Science*, Vol. 87, No. 18, September 1990, pp. 7305-7308.

Saunders, C.M., Larman, M.G., Parrington, J., Cox, L.J., Royse, J., Blayney, L.M., Swann, K. & Lai, F.A. (2002). PLC zeta: a sperm-specific trigger of Ca(2+) oscillations in eggs and embryo development. *Development*, Vol. 129, No. 15, August 2002, pp. 3533-3544.

Sun, Z., A, C.W., Ellmeier, W., Schaeffer, E.M., Sunshine, M.J., Gandhi, L., Annes, J., Petrzilka, D., Kupfer, A., Schwartzberg, P.L. & Littman, D.R. (2000). PKC-θ is required for TCR-induced NF-κB activation in mature but not immature T lymphocytes. *Nature*, Vol. 404, No. 6776, March 2000, pp. 402-407.

Spungin, B., Margalit, I. & Breitbart, H. (1995). Sperm exocytosis reconstructed in a cell-free system: evidence for the involvement of phospholipase C and actin filaments in membrane fusion. *Journal of Cell Science*, Vol. 108, No. 6, pp. 2525-2535.

Stricker, S., Escalona, J., Abernathy, S. & Marquardt, A. (2010) Pharmacological analysis of protein kinases regulating egg maturation in marine nemertean worms: A review and comparison with mammalian eggs. *Marine Drugs*, Vol. 2010, No. 8, August 2010, pp. 2417-2434, 1660-3397

Steinberg, S.F. (2008). Structural basis of protein kinase c isoform function. *Physiological Reviews*, Vol. 88, No. 4, October 2008, pp. 1341-1378.

Swann, K., Saunders, C.M., Rogers, N.T. & Lai, F.A. (2006). PLCzeta(zeta): a sperm protein that triggers Ca2+ oscillations and egg activation in mammals. *Seminars in Cell & Developmental Biology*, Vol. 17, No. 2, April 2006, pp. 264-273.

Tsaadon, L., Kaplan-Kraicer, R. & Shalgi, R. (2008). Myristoylated alanine-rich C kinase substrate, but not Ca2+/calmodulin-dependent protein kinase II, is the mediator in cortical granules exocytosis. *Reproduction*, Vol. 135, No. 5, May 2008, pp. 613-624.

Viveiros, M.M., Hirao, Y. & Eppig, J.J. (2001). Evidence that protein kinase C (PKC) participates in the meiosis I to meiosis II transition in mouse oocytes. *Developmental Biology*, Vol. 235, No. 2, July 2001, pp. 330-342.

Viveiros, M.M., O'Brien, M., Wigglesworth, K. & Eppig, J.J. (2003). Characterization of protein kinase C-delta in mouse oocytes throughout meiotic maturation and following egg activation. *Biology of Reproduction*, Vol. 69, No. 5, November 2003, pp. 1494-1499.

Webster, S.D. & McGaughey, R.W. (1990). The cortical cytoskeleton and its role in sperm penetration of the mammalian egg. *Developmental Biology*, Vol. 142, No. 1, November 1990, pp. 61-67.

White D., de Lamirande, E. & Gagnon, C. (2007). Protein kinase C is an important signaling mediator associated with motility of intact sea urchin spermatozoa. *J Exp Biol*, Vol. 210, No. 22, November 2007, pp. 4053-4064.

Xiao, J., Lui, C., Hou, J., Cui, C., Wu, C., Fan, H., Sun, X., Meng, J., Yang, F., Wang, E. & Yu, B. (2011). Ser[149] is another potential PKA phosphorylation target of Cdc25B in G_2/M transition of fertilized mouse eggs. *The Journal of Biological Chemistry*, Vol. 286, No. 12, January 2011, pp. 10356-10366.

Yanagimachi, R. & Usui, N. (1974). Calcium dependence of the acrosome reaction and activation of guinea pig spermatozoa. *Experimental Cell Research*, Vol. 89, No. 1, November 1974, pp. 161-174.

Yu, Y., Halet, G, Lai, F.A. & Swann, K. (2008). Regulation of diacylglycerol production and protein kinase C stimulation during sperm- and PLCzeta-mediated mouse egg activation. *Biol Cell*, Vol. 100, No. 11, November 2008, pp. 633-643.

Yu, B.Z., Zheng, J., Yu, A.M., Shi, W.Y., Liu, Y., Wu, D.D., Fu, W. & Yang, J. (2004). Effects of protin kinase c on M-phase promoting factor in early development of fertilized mouse eggs. *Cell Biochem. Funct*, Vol. 22, No. 5, September 2004, pp. 291-298.

Zhou, F.Q. & Snider, W.D. (2005). GSK-3beta and microtubule assembly in axons. *Science*, Vol. 308, No. 5719, April 2005, pp. 211-214.

4

Role of Bcl2l10 in Regulation of Meiotic Cell Cycle in the Mouse Oocyte

Kyung-Ah Lee[1], Se-Jin Yoon[2], Eun-Young Kim[1], Jeehyeon Bae[3],
Hyun-Seo Lee[1], Su-Yeon Lee[1] and Eun-Ah Kim[1]
[1]Department of Biomedical Science, College of Life Science, CHA University,
[2]Department of Genetics, Stanford University,
[3]Department of Pharmacy, College of Pharmacy, CHA University,
[1,3]Korea
[2]USA

1. Introduction

Bcl2l10, also known as Diva or Boo, is a member of the Bcl2 family. This factor is known to have opposing apoptotic functions in various tissues. That is, it can act as either a pro-apoptotic or an anti-apoptotic factor depending on the cellular milieu (Ke et al., 2001; Lee et al., 2001).

The pro-apoptotic factor Bad is expressed in various tissues including rat ovaries and testes (Kaipia et al., 1997). Bok, anti-apoptotic factor, has been detected in granulosa cells as well as in several reproductive tissues such as the ovaries, testes, and uterus (Hsu et al., 1997). Several anti-apoptotic Bcl2 homologs are expressed in the ovary. Mcl1, Bcl2, Bcl2l1 (Bcl-x), and other Bcl2 family members have also been detected in ovarian tissues (Hsu et al., 1997; Hsu and Hsueh, 2000; Kim and Tilly, 2004; Tilly et al., 1995). However, the expression patterns of these Bcl2 family members are quite different from those of Bcl2l10, which exhibits ovary- and oocyte-specific expression. Murine Bcl2l10 was first identified in expressed sequence tag clones from unfertilized, fertilized, and two-cell-stage mouse eggs (Inohara et al., 1998). Expression of this factor is restricted to the ovary and testis in adult mice (Inohara et al., 1998).

2. Expression of Bcl2l10 in the ovary

In a previous study, we used the annealing control primer-PCR method to investigate differentially expressed genes in the mouse oocytes. Using this approach, we found that Bcl2l10 was highly expressed in oocytes (Yoon et al., 2005). We confirmed that Bcl2l10 is constantly expressed in oocytes during oocyte maturation and found that its expression disappeared after the four-cell stage (Figure 1).

Since Bcl2l10 is a member of Bcl2 family, we first evaluated the relationship between Bcl2l10 expression and apoptosis in the ovary. Surprisingly, Bcl2l10 expression did not appear to be related to granulosa cell apoptosis in ovarian follicles (Figure 2). Analysis of serial ovarian sections using immunohistochemical labeling for Bcl2l10 and TUNEL assay revealed that BCL2L10 expression is oocyte-specific but that this expression is mismatched with the apoptotic death of follicular granulosa cells. Therefore, Bcl2l10 may have a function that is not related to the regulation of apoptosis in oocytes.

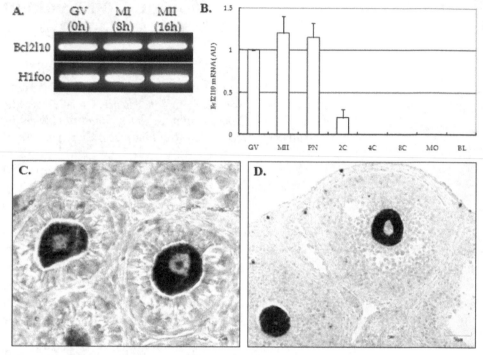

Fig. 1. Expression of Bcl2l10 mRNA in oocytes and preimplantational embryos. A) Semiquantitative RT-PCR analysis of Bcl2l10 expression during spontaneous oocyte maturation *in vitro*. GV, germinal vesicle-stage oocytes; MI, meiosis I oocytes collected after 8 h of culture; MII, meiosis II-stage oocytes collected after 16 h of in culture. B) Quantitative real-time RT-PCR of Bcl2l10 mRNA in oocytes and embryos. Messenger RNA isolated from oocytes and at various embryonic stages was reverse transcribed. For PCR, the cDNA from a single oocyte or an embryo equivalent served as a template for amplification. The expression level was calculated from the cycle threshold values (CT) based on the fluorescence detected within the geometric region of the semi-log plot, and the mRNA ratio (arbitrary units) was calculated with respect to that of GV oocytes. Experiments were repeated at least three times, and data are expressed as the mean ± SEM. PN, pronucleus one-cell zygote; 2C, two-cell; 4C, four-cell; 8C, eight-cell; MO, morula; BL, blastocyst-stage embryo. C, D) *In situ* hybridization of Bcl2l10 mRNA. Mouse ovaries of 2-week-old (C) and 4-week-old (D) were used, and we observed the oocyte-specific Bcl2l10 mRNA expression. Scale bars represent 25 □m and 50□m for C and D, respectively (C, D cited in Yoon et al., 2006, *Korean J Fertil Steril*).

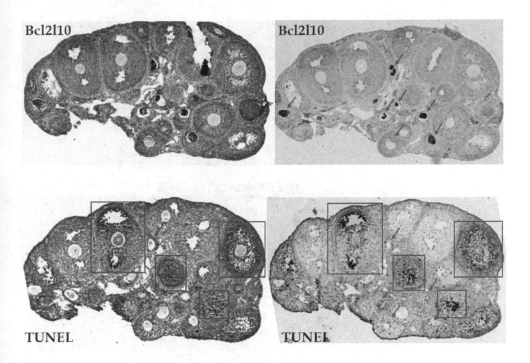

Fig. 2. BCL2L10 immunohistochemical labeling and TUNEL assay in serial sections. Tissue sections were obtained from 4-week mouse ovaries. Arrows indicate peculiar BCL2L10 expression in oocytes of preantral follicles, while boxes indicate apoptotic follicles with TUNEL-positive granulosa cells.

3. Bcl2l10 binding proteins

Next, we tried to identify the working partner(s) of Bcl2l10 using immunoprecipitation followed by mass spectrometry (Yoon et al., 2006). We transiently expressed FLAG-Bcl2l10 in NIH/3T3 cells and purified Bcl2l10-associated proteins (Figure 3). Specifically detected 14 bands were excised in the Coomassie Blue-stained 12% SDS-PAGE gels and identified using LC/MS/MS analysis. The protein band specific for Diva was identified as Bcl2-like 10 in the size of 21 kDa (Fig. 3, gel slice number 2) suggesting a successful IP analysis using monoclonal anti-FLAG antibody. List of potential Bcl2l10-binding partners is summarized in Table 1. Interestingly, many of the identified Bcl2l10-associated proteins are known to be associated with the cytoskeletal system. The identification of actin-related proteins such as actin, α-actinin, gelsolin, myosin, tropomyosin, and tropomodulin 3 as Diva-binding proteins suggests that Diva protein associates with the components of actin filaments. Actin was the most abundant Diva-associated protein as shown in Figure 3 (gel slice number 6). All these proteins, with the exception of actin, are known to bind actin to form, stabilize, and cross-linking the microfilaments.

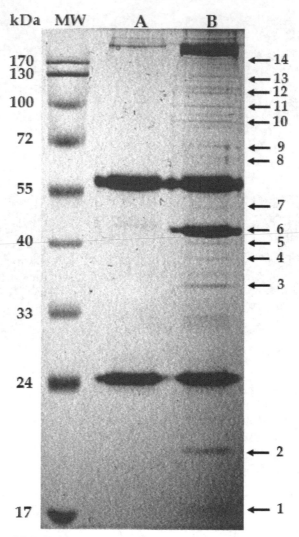

Fig. 3. Identification of BCL2L10-associated proteins. NIH/3T3 cells were transiently transfected for 24 h with empty vector or FLAG-BCL2L10-expressing vector. Total protein extracts were incubated with anti-FLAG-agarose beads. Anti-FLAG-agarose elutes from an equivalent amount of transfected cells (500 µg) were analyzed by 12% SDS polyacrylamide gel electrophoresis, and proteins were visualized by Coomassie Blue staining. The positions and relative molecular masses in kilodaltons (kDa) of protein size markers are indicated on the left. Protein-containing bands, as indicated by numbers on the right, were excised, and proteins present in gel slices were identified by mass spectrometry. MW, protein size markers; A) NIH/3T3, anti-FLAG-agarose elutes from NIH/3T3 cells transfected with pCMV-FLAG empty vector; B) FLAG-BCL2L10, NIH/3T3 cells transfected with pCMV-FLAG-BCL2L10 vector (Yoon et al., 2006).

Gel slice number	Accession No.	Score	Description	Unigene
1	gi_16307437	95	RIKEN cDNA 2900073G15	2900073G15Rik
2	gi_30851239	69	Bcl2l10	Bcl2l10
3	gi_1351289	69	Tropomyosin 1 alpha chain (alpha-tropomyosin)	Tpm1
4	gi_19353393	79	Tpm2 protein; Tropomyosin 2, beta	Tpm2
5	gi_52139168	90	Tropomodulin 3	Tmod3
6	gi_49868	109	Beta-actin (aa 27-375); Actin, beta, cytoplasmic	Actb
7	gi_80478706	45	ARP3 actin-related protein 3 homolog (yeast)	Actr3
8	gi_194362	123	Igh-4 protein	Igh-4
9	gi_27369615	29	PIF1 homolo: Expressed sequence AI449441	AI449441
10	gi_34871482	34	PREDICTED: similar to Transcription initiation factor	Taf6
11	gi_90508	39	Gelsolin, cytosolic	Gsn
12	gi_61097906	35	Actinin, alpha 1	Actn1
13	gi_47847434	42	mFLJ00150 protein: Centrosomal protein 110	Cep110
14	gi_17978023	80	Myosin, heavy polypeptide 9, non-muscle	Myh9

Table 1. Bcl2l10-associated proteins identified by immunoprecipitation and mass spectrometry.

Immunoprecipitation/Western blot analysis of ovarian tissue homogenates confirmed the association of actin and tropomyosin with Bcl2l10 (Figure 4). These findings suggest that, in the ovary, Bcl2l10 plays roles unrelated to apoptosis and instead participates in the regulation of cytoskeletal systems. During meiosis, actin filaments have roles in migration of chromosomes, segregation of homologous chromosomes, development and maintenance of the cortex, formation of polarity, movement of peripheral cortical granules, and extrusion of the first polar body (Sun and Schatten, 2006).

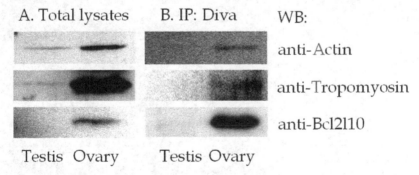

Fig. 4. Identification of BCL2L10-associated proteins using immunoprecipitation (IP)/Western blot analysis (WB). A) WB analysis. Total protein lysates (20 µg) from mouse testes and ovaries were electrophoresed and probed with anti-actin, anti-tropomyosin, or anti-BCL2L10 antibody. Testes lysates served as a negative control for BCL2L10 protein expression. B) IP/WB analysis. Total protein lysates from testes and ovaries (500 µg) were subjected to IP using anti-Bcl2l10 antibody. Anti-BCL2L10-agarose elutes were then electrophoresed and probed with anti-actin, anti-tropomyosin, or anti-BCL2L10 antibody to confirm of the association of actin and tropomyosin with BCL2L10 (as cited in Yoon et al., 2006, *Korean J Fertil Steril*).

A non-apoptotic function of Bcl2l10 has also been demonstrated. Specifically, Bcl2l10 plays a role in Huntington-interacting protein 1-related (HIP1R) protein-mediated endocytosis as well as in the regulation of actin machinery in 293T cells. HIP1R regulates clathrin-mediated endocytotic apparatus and actin assembly. An interaction between endogenous BCL2L10 and HIP1R has been shown by immunoprecipitation and Far-Western analysis (Kim et al., 2009).

4. Function of Bcl2l10 in oocytes

We have investigated the role of Bcl2l10 not only in the ovary, but also in oocytes. RNA interference (RNAi) was used for this purpose. To determine the role of Bcl2l10 during oocyte maturation, we microinjected *in vitro* transcribed dsRNA for Bcl2l10 into the cytoplasm of germinal vesicle-stage (GV) oocytes and monitored *in vitro* oocyte maturation. Bcl2l10 RNAi selectively reduced levels of endogenous Bcl2l10 and resulted in incomplete meiosis that was arrested in meiosis I (MI) (Yoon et al., 2009). MI-arrested oocytes had abnormalities in the spindles and chromosomes. The most prominent changes were the disappearance of spindles and aggregation of chromosomes (Figure 5).

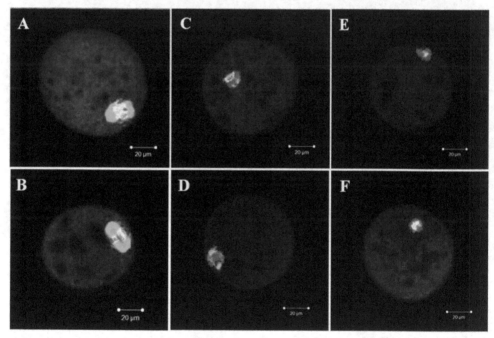

Fig. 5. Bcl2l10 RNAi-induced abnormalities in oocytes, as seen by α-tubulin immunofluorescent staining and chromosomal staining. Germinal vesicle oocytes were injected with Bcl2l10 dsRNA and cultured for 16 h. Oocytes were then fixed in 4% paraformaldehyde and stained with an α-tubulin antibody (green). Chromosome material was counterstained with propidium iodide (red). A) Control, uninjected oocyte. B) Buffer-injected control oocyte. C-F) Bcl2l10 dsRNA-injected oocytes arrested in meiosis I (as cited in Yoon et al., 2010, From Biol Reprod). Bars represent 20 μm.

5. Genes downstream of Bcl2l10

After discovering that Bcl2l10 RNAi induced changes in spindles and chromosomes, we set out to identify the network of factors acting downstream of Bcl2l10 in oocytes so that we could better understand the regulatory mechanisms underlying meiosis in oocytes. To identify downstream genes, we conducted microarray analysis of Bcl2l10 RNAi-induced changes in gene expression in mouse oocytes (Kim et al., 2011).

Due to the small amounts of initial total RNA obtained from the 350 oocytes, we performed an amplifying two-cycle target labeling assay so that we could obtain sufficient amounts of labeled cRNA target for microarray analysis. The labeled cRNA was hybridized to the Affymetrix GeneChip Mouse Genome 430 2.0 Array, which covers transcripts and variants from 34,000 well-characterized mouse genes.

Bcl2l10 RNAi induced a more than 2-fold up-regulation of 644 genes and down-regulation of 1,166 genes. Notably, the top 20 up-regulated genes included five enzymes (*i.e.*, pyrophosphate synthase, N-methyltransferase, and three kinases). The top 20 down-regulated genes were related to cytoskeletal organization (Table 2). These genes encoded proteins such as Tpx2, Cep192, Kir20b, Myo6, and Cd2ap. Tpx2 functions in spindle assembly, Cep192 in microtubule nucleation, Kir20b in microtubule-based movement, Myo6 in physical interactions, and Cd2ap in actin polymerization.

Genes	Gene title	Fold change
Tpx2	TPX2, microtubule-associated protein homolog (*Xenopus laevis*)	- 16.1
Rbm12b	RNA binding motif protein 12B	- 15.3
Ptp4a1	protein tyrosine phosphatase 4a1-like	- 14.6
Ranbp2	RAN binding protein 2	- 10.1
Eea1	early endosome antigen 1	- 9.3
Arid4a	AT rich interactive domain 4A (RBP1-like)	- 9.1
Cep192	centrosomal protein 192	- 8.2
Kif20b	kinesin family member 20B	- 7.9
Psip1	PC4 and SFRS1 interacting protein 1	- 7.8
Atad2b	ATPase family, AAA domain containing 2B	- 7.5
Mki67	antigen identified by monoclonal antibody Ki 67	- 7.4
Nexn	nexilin	- 7.2
Eif4g3	eukaryotic translation initiation factor 4 gamma, 3	- 6.9
Ccnb3	cyclin B3	- 6.9
C430048L16Rik	RIKEN cDNA C430048L16 gene	- 6.8
Cenpm	centromere protein M	- 6.8
Leo1	Leo1, Paf1/RNA polymerase II complex component	- 6.7
Myo6	myosin VI	- 6.6
Tnfaip8	tumor necrosis factor, alpha-induced protein 8	- 6.6
Cd2ap	CD2-associated protein	- 6.5

Table 2. Top 20 genes down-regulated more than 2-fold by Bcl2l10 RNAi.

Tpx2 was found to be down-regulated by 16.1-fold and Cep192 by 8.2-fold. The functions of these proteins are intimately related. Interfering with TPX2 function in HeLa cells causes defects in microtubule organization during mitosis, and Tpx2 RNAi produces abnormalities in spindle formation (Gruss et al., 2002). Tpx2, a microtubule-binding protein, and Cep192, a centromere protein, are well-known cofactors of Aurora A kinase. Both proteins act to control the activity and localization of this kinase (Joukov et al., 2011). Eukaryotes have one to three members of the Aurora family of serine-threonine kinases. Aurora A is an important oncogenic kinase that has well-established roles in spindle assembly (Xu et al., 2011). During mitosis, a fraction of Aurora A binds Tpx2, activates the kinase, and targets it to spindle microtubules (Eyers et al., 2003, Kufer et al., 2002, Ozlu et al., 2005, Tsai et al., 2003). Tpx2 controls localization of Aurora A at centrosomes, whereas Cep192 controls its activity in microtubules (Joukov et al., 2010).

Our finding that Bcl2l10 RNAi induced concurrent down-regulation of Tpx2 and Cep192 leads us to conclude that Bcl2l10 may have important roles in regulating oocyte meiosis through its ability to act as an upstream regulator of Tpx2 and Cep192. Association between Bcl2l10 and Aurora kinase A is an interesting new area that warrants further investigation.

6. Conclusion

We have identified Bcl2l10 expression in oocytes and uncovered a role for this factor in regulating meiosis. Our findings point to new non-apoptotic function for this Bcl2 family member and open a challenging new area of research on Bcl2l10 regulation of meiosis through Bcl2l10 involvement in spindle assembly. We propose that Bcl2l10 is an important regulator of meiotic spindle formation and works closely with Tpx2, Cep192, and Aurora A kinase. The molecular mechanisms underlying meiotic regulation by Bcl2l10 and its downstream genes (Tpx2 and Cep192) as well as the relationship between Bcl2l10 and Aurora A kinase are currently under careful investigation.

7. Acknowledgment

This research was supported by the Basic Science Research Program through the National Research Foundation of Korea (NRF) funded by the Ministry of Education, Science, and Technology (20100022600 and 2009-0068363), by the Priority Research Centers Program through the NRF of Korea funded by the Ministry of Education, Science and Technology (2009-0093821), and by a Korea Research Foundation Grant funded by the Korean Government (KRF-2008-314-E00120).

8. References

Eyers PA, Erikson E, Chen LG & Maller JL. (2003). A novel mechanism for activation of the protein kinase Aurora A. *Current Biology*, 13, 8, pp.691-697, ISSN 0960-9822

Gruss OJ, Wittmann M, Yokoyama H, Pepperkok R, Kufer T, Sillje H, Karsenti E, Mattaj IW & Vernos I. (2002). Chromosome-induced microtubule assembly mediated by TPX2 is required for spindle formation in HeLa cells. *Nature Cell Biology*, 4, 11, pp.871-879, ISSN 1465-7392

Hsu SY & Hsueh AJ. (2000). Tissue-specific Bcl-2 protein partners in apoptosis: an ovarian paradigm. *Physiological Reviews*, 80, 2, pp.593-614, ISSN 0031-9333

Hsu SY, Kaipia A, McGee E, Lomeli M & Hsueh AJ. (1997). Bok is a pro-apoptotic Bcl-2 protein with restricted expression in reproductive tissues and heterodimerizes with selective anti-apoptotic Bcl-2 family members. *Proceedings of the National Academy of Sciences of the United States of America*, 94, 23, pp.12401-12406, ISSN 0027-8424

Inohara N, Gourley TS, Carrio R, Muniz M, Merino J, Garcia I, Koseki T, Hu Y, Chen S & Nunez G. (1998). Diva, a Bcl-2 homologue that binds directly to Apaf-1 and induces BH3-independent cell death. *Journal of Biological Chemistry*, 273, 49, pp.32479-32486, ISSN 0021-9258

Joukov V, De Nicolo A, Rodriguez A & Walter JC. Livingston DM. (2010). Centrosomal protein of 192 kDa (Cep192) promotes centrosome-driven spindle assembly by engaging in organelle-specific Aurora A activation. *Proceedings of the National Academy of Sciences of the United States of America*, 107, 49, pp.21022-21027, ISSN 0027-8424

Joukov, V. (2011). Aurora kinases and spindle assembly: Variations on a common theme? *Cell Cycle*, 10, 6, pp.895-903, ISSN 1538-4101

Kaipia A, Hsu SY & Hsueh AJ. (1997). Expression and function of a proapoptotic Bcl-2 family member Bcl-XL/Bcl-2-associated death promoter (BAD) in rat ovary. *Endocrinology*, 138, 12, pp.5497 5504, ISSN 0013-7227

Ke N, Godzik A & Reed JC. (2001). Bcl-B, a novel bcl-2 family member that differentially binds and regulates bax and bak. *Journal of Biological Chemistry*, 276, 16, pp.12481-12484, 0021-9258

Kim EA, Kim KH, Lee HS, Lee SY, Kim EY, Seo YM, Bae J, Lee KA. (2011). Downstream genes regulated by Bcl2l10 RNAi in the mouse oocytes. *Development and Reproduction*, 15, 1, pp.61-69, ISSN 1226-6752

Kim JD, Won M, Yoon S, Ko JJ, Lee KA, Lee K & Bae J. (2009). HIP1R interacts with a member of Bcl-2 family, BCL2L10, and induces moderate apoptosis. *Cellular Physiology and Biochemistry*, 23, 1-3, pp.43-52, ISSN 1015-8987

Kim MR & Tilly JL. (2004). Current concepts in Bcl-2 family member regulation of female germ cell development and survival. *Biochimica and Biophysica Acta*, 1644, 2-3, pp.205–210, ISSN 1672-9145

Kufer TA, Sillje HH, Korner R, Gruss OJ, Meraldi P, Nigg EA. (2002). Human TPX2 is required for targeting Aurora-A kinase to the spindle. *Journal of Cell Biology*, 158, 4, pp.617-623, ISSN 0021-9525

Lee R, Chen J, Matthews CP, McDougall JK & Neiman PE. (2001). Characterization of nr13-related human cell death regulator, boo/diva, in normal and cancer tissues. *Biochimica and Biophysica Acta*, 1520, 3, pp.187-194, ISSN 1672-9145

Ozlu N, Srayko M, Kinoshita K, Habermann B, O'toole ET, Müller-Reichert T, Schmalz N, Desai A & Hyman AA. (2005). An essential function of the C. elegans ortholog of TPX2 is to localize activated aurora A kinase to mitotic spindles. *Developmental Cell*, 9, 2, pp.237-248, ISSN 1534-5807

Sun QY, & Schatten H. (2006). Regulation of dynamic events by microfilaments during oocyte maturation and fertilization. *Reproduction*, 131, 2, pp.193-205, ISSN 1470-1626

Tilly JL, Tilly KI, Kenton ML & Johnson AL. (1995). Expression of members of the Bcl-2 gene family in the immature rat ovary: equine chorionic gonadotropin-mediated inhibition of granulosa cell apoptosis is associated with decreased Bax and constitutive Bcl-2 and Bcl-xlong messenger ribonucleic acid levels. *Endocrinology*, 136, 1, pp.232–241, ISSN 0013-7227

Tsai MY, Wiese C & Cao K. (2003). A Ran signaling pathway mediated by the mitotic kinase Aurora A in spindle assembly. *Nature Cell Biology*, 5, 3, pp.242-248, ISSN 1465-7392

Xu X, Wang X, Xiao Z, Li Y & Wang Y. (2011). Two TPX2-dependent switches control the activity of Aurora A. *PLoS ONE*, 6, 2, e16757, ISSN 1932-6203

Yoon SJ, Chung HM, Cha KY, Kim NH & Lee KA. (2005). Identification of differential gene expression in GV vs. MII mouse oocytes by using annealing control primers. *Fertility and Sterility*, 83, S4, pp.1293-1296, ISSN 0015-0282

Yoon SJ, Kim EY, Kim YS, Lee HS, Kim KH, Bae J, & Lee KA. (2009). Role of Bcl2-like 10 (Bcl2l10) in regulating mouse oocyte maturation. *Biology of Reproduction*, 81, 3, pp.497-506, ISSN 0006-3363

Yoon SJ, Kim JW, Choi KH, Lee SH & Lee KA. (2006). Identification of oocyte-specific Diva-associated proteins using mass spectrometry. *Korean Journal of Fertility and Sterility*, 33, 3, pp.189-198, ISSN 1226-2951

5

Nuage Components and Their Contents in Mammalian Spermatogenic Cells, as Revealed by Immunoelectron Microscopy

Yuko Onohara and Sadaki Yokota
Nagasaki International University,
Japan

1. Introduction

The nuage is a germ cell-specific organelle that has been studied for more than a hundred years. It was first discovered in spermatogenic cells as a perinuclear granule stained by basic dyes and visualized using a light microscope. Morphological studies using electron microscopes demonstrated that "chromatoid body (CB)" is a nuage component in spermatids, and that "inter-mitochondrial cement (IMC)" is another nuage component found in spermatocytes. During meiosis, the IMC disappears, but the CB soon reforms in post-meiotic spermatids. Recent morphological and molecular biological studies have identified many components of nuage, and suggest that they play roles in the silencing, decay, and storage of RNA, and in the aggresome system. In this report, we summarize recent findings related to nuage and discuss their functions in spermatogenic cells.

2. Historical aspect of nuage components in germ line cells

By the 19th century, the basic techniques for histology, including preparation of tissue sections and staining of sections with various dyes, had largely been established, and light microscopic studies of tissues and cells from various animals and plants had begun to increase in terms of both number and detail. The subcellular structures in spermatogenic cells strongly attracting histologists' interest were strongly stained by basic dyes such as safranin, crystal violet, and Heidenhain's iron-hematoxylin. Two types of structures that today correspond to the chromatoid body (CB) were first described by von Brunn, who termed them "Protoplasmaanhäufungen" (protoplasmic depositions) and followed their fate, concluding that one formed the acrosome cap and the other moved to the caudal portion of spermatids to create the flagellum of spermatozoa (von Brunn, 1876). Afterwards, many histologists described similar structures using various names, such as "Nebenkörper" in spermatogenic cells from salamander and mouse (Herman 1889); "chromatoide Nebenkörper," derived from "intranuclearkörper" (the intranuclear body), which is different from the nucleolus (von Ebner, 1888; Benda, 1891); "corps chromatoides," which were stained black by iron-hematoxylin and deep red by safranin and finally degraded to "corps résiduel" (Regaud, 1901); and "chromatoide Körpers," which appeared as "chromatoiden Nuckeolens" in the nucleus and were dispersed in the cytoplasm (Schreiner & Schreiner, 1905). In the first half of the 20th century, CBs from various animals including insects (Wilson, 1913; Pollister, 1930), crustaceans (Fasten, 1914), arachnids (von Korff, 1902), fish (Schreiner & Schreiner, 1905), fowl (Zlotnik, 1947), and mammals (Duesberg, 1908; Wodsedalek, 1914; Gatenby & Beams, 1935) were studied extensively by light microscopy. In these classic studies, the potential origins of CBs were discussed. The arguments can be summarized into two ideas: 1) that the CB appears from the beginning in the cytoplasm of spermatocytes or spermatids, and 2) that the CB arises within the nucleoplasm and then moves to the cytoplasm. The fate of CBs can be summarized as follows: the CB exists in a perinuclear area and forms a part of the acrosome or the axis of the flagellum, or enters the bag-like structure that is formed in the cytoplasm at the base of the flagellum, and is degraded there. The studies cited above only involved observing stained tissue sections by light microscopy. Although most of these ideas have been shown to be incorrect, some of them have been accepted.

3. Ultrastructural characteristics of nuage components in mammalian spermatogenic cells

3.1 The CB as a representative nuage structure in mammalian spermatogenic cells

Electron microscopic studies of mammalian spermatogenic cells commenced in the 1950s (Watson, 1952; Challice, 1953). Burgos and Fawcett (1955) were the first to observe the CB by electron microscopy (Figure 1).

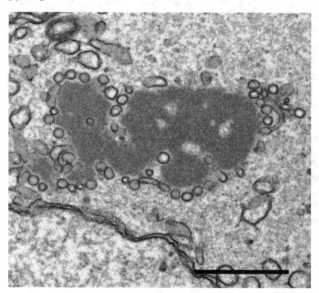

Fig. 1. Electron micrograph of a typical CB of a step 4 spermatid. The CB consists of an electron-dense matrix and is surrounded by small clear vesicles and tubules. The CB has no limiting membrane. Bar = 1 μm.

Burgos and Fawcett described an irregular and osmiophilic component corresponding to the "chromatic body" or "accessory body" detected by light microscopy. This structure was initially found near the Golgi complex, subsequently moving towards the caudal region in late spermatids. In 1955, Minamino described the material as an "idiosome remnant." In 1959, Sasa reported the ultrastructure of the rat testis, and referred to "Chromatoider

Nebenkörper." He suggested that osmiophilic components derived from the cluster of cytoplasmic vesicles might fuse to each other to form the big honeycomb structure.

In 1970, Fawcett et al. studied the origin and fate of CBs and satellite bodies during spermatogenesis in several mammals at the ultrastructural level. They presented evidence that the CB was not derived from the nucleus, as had been reported previously. CBs were not present in spermatocytes. Instead, they were found between clusters of mitochondria, in the dense interstitial material that is today known as inter-mitochondrial cement (IMC) (Figure 2). Because its texture and density were similar to those of CB, the researchers assumed that IMC might be the origin of the CB, and that two CBs might be assembled from IMC and possibly distributed to two daughter cells during meiosis I. In late spermatids, the CB became associated with the base of the flagellum, so they assumed that it might contribute to formation of the "annulus," a ring associated with the plasma membrane in spermatids and subsequently located at the junction of the middle and principal pieces in spermatozoa.

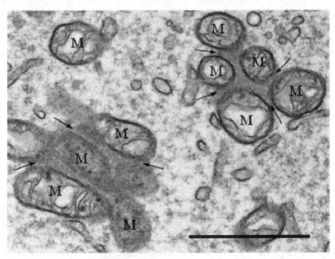

Fig. 2. Electron micrograph of typical IMC in a stage X pachytene spermatocyte. IMC appears as a dense interstitial material (arrows) between the clustered mitochondria (M). Bar = 1 μm.

Susi and Clermont (1970) studied the morphology and fate of CBs by routine electron microscopy. In spermatocytes, the CB was composed of two elements: 1) an electron-dense material with a sponge-like texture; and 2) numerous small vesicles with a diameter of 40–80 nm, located in the cytoplasm near the nuclear envelope. In step 1–7 spermatids, the CB was observed as a single irregular granule with a diameter of 1–2 μm, whose material was more condensed compared with CBs in spermatocytes. It was located proximal to the nuclear envelope, or sometimes near the Golgi apparatus, and was occasionally associated with the "multivesicular body (MB)." The CB and MB were found in the caudal region of the nucleus in step 8 spermatids. In step 9–10 spermatids, the CB was spherical, had a diameter of 1.5 μm, and was surrounded by vesicles positive for glycoproteins (stained by the periodic acid–chromic acid–silver methenamine method). In later spermatids, the CB was fragmented into smaller masses. Susi and Clermont suggested that the CB might be highly active in dramatically changing its form.

Based on the results of indium trichloride staining, Eddy (1970) concluded that no RNA existed in the CB. Similarly, the results of radioautography with isotope-labeled bases suggested that the dense granules observed in oocytes from *Rana* spp. contained neither RNA nor any other nucleic acid (Eddy & Ito, 1971). These results conflicted with previous reports (Daoust & Clermont, 1955).

In 1974, Eddy studied "nuage" (clouds in French) structures in primordial germ cells from rats of both sexes. He identified nuage in both male and female cells, and proposed that the nuage is the characteristic component of mammalian germ cells. He further compared the nuage with the "polar granule" found in insects (Dhainaut, 1970; Mahowald, 1962) and with "germinal plasm" found in amphibians (Nieuwkoop & Faber, 1956). These structures were previously speculated to play a role in germ cell determination. Eddy noted the common morphological profile and distribution of nuage with these components, which led him to suggest that nuage might be involved in the determination of mammalian germ cells, as had been reported for insects and amphibians. In terms of mammals, nuage was identified in guinea pig (Adams & Hertig, 1964), hamster (Weakley, 1969; Fawcett et al., 1970), rabbit (Nicander & Plöen, 1969), mouse (Fawcett et al., 1970), rat (Brökelmann, 1963), and human (Burgos et al., 1970). In germ cells of both sexes, nuage appeared either as a discrete mass or in association with mitochondrial clusters during the different stages of oogenesis and spermatogenesis. Eddy (1974) suggested that that the CB was a component of the nuage in spermatogenic cells.

In the same year, Kerr and Dixon (1974) studied germ plasm, which they regarded as a nuage found in spermatogenic cells, based on its appearance, expression stage, and association with mitochondria. They assumed that germ plasm might be a precursor of the CB. Russell and Frank (1978) classified nuage into six different types based on its form, distribution, and interrelationships with other organelles: 1) 70–90 nm spherical particles: fine fibrous and partly high-dense material found in spermatogonia, secondary spermatocytes, and intermediate forms; 2) sponge bodies: loosely organized masses, not only perinuclear but found throughout the cytoplasm in spermatocytes of all stages, and occasionally in spermatogonia and early spermatids; 3) IMC: 0.1 µm or narrower loosely organized strands, whose density and appearance resemble those of the sponge body, observed mainly in early spermatocytes and occasionally in spermatogonia; 4) 30 nm particles: clusters of dense particles different from ribosomes in the cytoplasm of meiosis I cells and secondary spermatocytes; 5) CBs: located in the cytoplasm proximal to the nucleus in pachytene spermatocytes; and 6) definitive CBs. Temporary CBs had been described previously as large spherical bodies with a diameter of approximately 0.5 µm that were closely associated with the few mitochondria observed in secondary spermatocytes and spermatids. Russell and Frank (1978) first classified the CB into two types that were present during different stages of spermatogenesis, although it was unclear whether the two types were the same in terms of their composition. They further reported that the CBs were not found in diplotene spermatocytes or during meiosis I. During meiosis II, they found numerous CBs in the cytoplasm. These CBs were separated into two components, one of which had a large, round honeycomb structure, and the other of which was a loose network of irregularly shaped dense strands.

The ultrastructural studies of CBs described above can be summarized as follows: 1) the CB is a component of nuage that was specifically detected in spermatogenic cells; 2) CBs displayed irregular shapes, comprising dense fibrous material surrounded by small vesicles, and had no limiting membrane; 3) CBs appeared first in spermatocytes, disappeared during meiosis, and reappeared in spermatids.

3.2 Relationship between the CB and the nuclear envelope

In classical studies, CBs were assumed to be derived from the intranuclear structure, based on the similarity in safranin staining in both structures (Benda, 1891; Regaud, 1901). Sud (1961) first suggested the presence of highly polymerized RNA and arginine-rich basic proteins in CBs based on the results of histochemical studies.

In 1970, Fawcett et al. proposed a new idea: that CB originated not from nucleus, but from the dense materials between clustered mitochondrial, i.e., IMC, which appeared during early spermatocytes. IMC was gradually deposited there as spermatocytes differentiated. Their hypothesis that CB is derived from IMC was supported by the facts that 1) no dense substance was observed on the intranuclear side of nuclear pores when the CB was in contact with the cytoplasmic side, and 2) that CBs already existed in spermatocytes before the CB made contact with nuclear pores. Although the CB is not derived from the nucleus, Fawcett et al. (1970) assumed that some kind of exchange might take place between the nucleus and the CB. On the other hand, in 1972, Comings and Okada observed that a nuclear granule exited to the cytoplasm of primary spermatocytes via the nuclear pore, and therefore concluded that the CB originated from the nucleus. This idea was discussed by Eddy in 1974.

The movement of CBs in the rat testis was studied using time-lapse cinemicrography, the results of which revealed that the CB rapidly moved around the nuclear envelope and Golgi complex, and occasionally seemed to be detected as an intranuclear particle (Parvinen & Jokelainen, 1974; Parvinen & Parvinen, 1979). Based on these observations, they assumed that the CB played a role in the transport of haploid gene products in early spermatids.

Ventelä et al. (2003) reported that the CB moved from one spermatid to another through the cytoplasmic bridge, and that this movement was inhibited by the microtubule-depolymerizing agents, nocodazole and vincristine. These results demonstrated that the microtubular network is involved in CB mobility.

Communication between the CB and nucleus, suggested by the results of classic light microscopic studies, was confirmed by electron microscopic studies. Cinemicrography studies showing real-time movement of CBs provided clear evidence of CB-nucleus communication (Parvinen & Jokelainen, 1974). Consequently, it was concluded that the CB may not be derived from the nucleus, but communicates with the nucleus via the nuclear pore complex during spermatogenesis.

3.3 Clusters of 30–40 nm particles

Russell and Frank (1978) observed clusters of small, 30–40 nm particles in dividing meiosis I cells and pachytene spermatocytes, which were morphologically different from ribosomes and glycogen particles. These clusters were composed of hundreds of aggregated 30–40 nm particles, and contained a filamentous material. Russell and Frank suggested that these particles might stabilize RNAs encoding proteins required for the subsequent development of spermatids.

3.4 IMC as an origin of the CB

Some concluded that IMC was a precursor of mitochondria because its density was similar to that of the mitochondrial matrix (Odor, 1965; André, 1962). However, this hypothesis was disproved by confirmation of the striking difference in the density of the two components after improved fixation (Fawcett, et al., 1970). As mentioned above, Fawcett et al. proposed that the CB was derived from IMC (Fawcett, et al. 1970). According to them, IMC gradually increased in size by coalescence as spermatids developed, subsequently dissociated from the mitochondria, and assembled to form the CB.

Russell and Frank (1978) reported that IMC appeared when mitochondria were noticeably clustered in pachytene spermatocytes, and that there was no evidence of IMC during or after the first meiotic metaphase, when the mitochondria dispersed.

3.5 Satellite body (SB, sponge body) and its origin

Sud (1961) first named the structure the "satellite," and described its localization near to, and sometimes in contact with, the CB. The chemical composition of the satellite differed from that of the CB. Sud proposed that the satellite might form the basal body of the axial filament.

Fawcett et al. (1970) described the characteristics, origin, and fate of the satellite body (SB), which they referred it "chromatoid body satellite", and first described the distinction between the SB and the CB at the ultrastructural level. He reported that the SB was present in zygotene and pachytene spermatocytes prior to the appearance of the CB, and had a more regular shape and contained less filamentous material than the CB. In the report of Fawcett et al. (1970), they hypothesized that clusters of small particles of 40-60 nm in diameter, which had previously been suggested to be precursors of centrioles (Stockinger & Cirelli, 1965; Sorokin, 1968), were in fact the source of the SB, because they were the only conceivable precursor of the SB and had a strikingly resemblance to the SB. The fate of the SB was unknown, but it was found to move from the Golgi region to the caudal region of the nucleus in spermatids. In late spermatids, a spherical mass was found near the developing connecting piece, and was assumed to be derived from the SB on the basis of its size.

Russell and Frank (1978) described the SB as a "sponge body" found in the cytoplasm of secondary spermatocytes and young spermatids, and very occasionally spermatogonia. They indicated that the density and appearance of the SB resembled those of IMC.

3.6 Nuage components in meiotic cells

Nuage was found to dissociate from mitochondrial clusters during the pachytene phase of meiosis, forming the CBs in spermatids (Eddy, 1974). Eddy also observed that after meiosis II, the CB was dispersed in the cytoplasm as small dark masses. However, in step 1 spermatids the CB was reconstructed. Russell and Frank (1978) reported that while the CB was rarely seen in dividing meiosis I cells, there was loose, dense material that was irregularly shaped and contained granular materials on its inner surface. As mentioned above, the clusters of 30 nm particles were found in dividing meiosis I cells and were thought to possibly contain mRNAs. Russell and Frank assumed that these clusters might contain RNAs that had been transferred from the nucleus to cytoplasm, and that these RNAs might encode proteins involved in the later stages of spermatogenesis. They also reported that the definitive CB was observed in secondary spermatocytes proximal to meiosis II cells. CB was suggested to be a storage organelle for haploid gene products during meiosis, and to possibly participate in the transport of these products from the nucleus to the cytoplasm (Söderström & Parvinen, 1976; Parvinen & Parvinen, 1979). Söderström (1978) reported that the condensed CB was present during meiotic prophase, and that new materials were formed in the nucleus in cells of meiotic prophase and spermatids, and proposed that the materials might be transported to the cytoplasm and be added to the CB. We ourselves have shown that the small particles partly associated with mitochondria during meiosis stain positive for the nuage marker protein DDX4. This suggests that nuage components might dissociate into small particles during meiosis, allowing them to be apportioned to the daughter cells evenly (Figure 3).

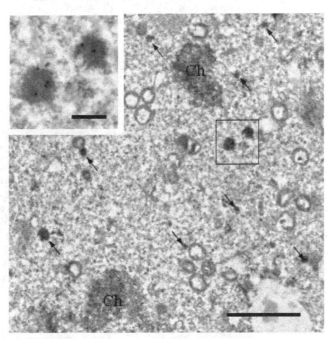

Fig. 3. Rat spermatocyte during meiosis. A dividing chromosome (Ch) is present. No aggregates of particles are seen, but small particles are present (arrows). These particles are positive for DDX4 (inset). Bar = 2 μm (main image) and 0.5 μm (inset).

4. Immunoelectron microscopic localization of nuage proteins in spermatogenic cells

4.1 Argonaute/Piwi and Tudor families in nuage components

The Argonaute/Piwi family of proteins can be split into CB and nuage components. Members of the Argonaute/Piwi family commonly have PAZ and PIWI domains, and are classified into the Ago1 and PIWI subfamilies based on their sequences. Ago 1–4 belong to the Ago1 subfamily, are highly expressed in various tissues, and function in the RNA-induced silencing complex (RISC) pathway (Meister et al., 2004). In contrast, members of the PIWI subfamily are specifically expressed in germ cells, with MIWI and MILI performing particularly important roles in spermatogenesis (Sasaki et al., 2003; Deng & Lin, 2002). MIWI was localized to the CB in spermatids. In a MIWI-null mouse model, the fully compacted CB was not found, with a non-compacted and diffuse CB instead being observed (Kotaja et al., 2006b), which suggests the importance of MIWI for the construction of the compacted CB. The Tudor family proteins Tudor repeat domain containing protein (TDRD)-1, -4, -5, -6, -7, and -9 were reported to be expressed in germ cells, and were shown to co-localize with Piwi family proteins in nuage. TDRD-1 has been detected in nuage in spermatogonia, in IMC in spermatocytes, and in CBs in round spermatids (Chuma et al., 2003). The analysis of a TDRD-1-null mouse model revealed that TDRD-1 was essential for piRNA biogenesis, formation of IMC, and the correct localization of DDX4 in the CB (Hosokawa et al., 2007). In pachytene spermatocytes, TDRD-5 co-localized in IMC and the CB with TDRD-1, MIWI, MILI, and DDX4. In mid-pachytene spermatocytes, TDRD-5 was partly co-localized with TDRD-6 and -7; in diplotene spermatocytes, it was mainly co-localized with TDRD-6. In round spermatids, TDRD-5 co-localized in the CB with DDX4, MIWI, TDRD-1, -6, and -9, and interestingly, at the same time, TDRD-7 localized to another perinuclear structure distal from the CB (Yabuta et al., 2011), which seemed to be the second nuage component. In a TDRD-6-null mouse model, CB was not condensed, but rather was dispersed (Vasileva et al., 2009).

4.2 RNF17/TDRD-4 in nuage components

RNF17/TDRD-4 was first described as a germ cell-specific gene that contains repeated Tudor domains and a RING finger motif, which is present in ubiquitin E3 ligase. Yin et al. (1999, 2001) reported that RNF17 interacted with Mad family proteins and repressed transcription of the gene encoding the oncoprotein Myc. RNF17 contributes to activation of the transcription of Myc-responsive genes by supplying Mad proteins. In RNF17-null mice, which are sterile, spermatogenesis completely stops at the round spermatid stage (Pan et al., 2005), which indicates the importance of RNF17 in the development of spermatids. Pan et al. (2005) described a new dense granule that is positive for RNF17 and located in the perinuclear region near to the CB throughout spermatogenesis, and named it the "RNF17 granule." The morphological appearance of the RNF17 granule closely resembles that of the SB. If the SB contains RNF17, it would be one of the first SB components to be identified. Our preliminary data show that MAELSTRÖM (MAEL) is also localized to the SB, as well as other nuage components (Fig. 4).

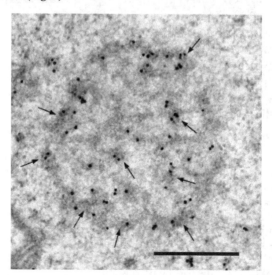

Fig. 4. Typical SB of a stage X pachytene spermatocyte stained for MAEL by immunoelectron microscopy (IEM). Gold particles showing MAEL antigenic sites are present on denser patches (arrows). The SB is 0.5–1.5 μm in diameter, and consists of a network of fibrils overlaid by denser patches of amorphous material. Bar = 0.5 μm.

4.3 DDX4 in nuage components

DDX4, the mouse homolog of the *Drosophila* vasa gene MVH, is an ATP-dependent RNA helicase belonging to the DEAD box family. DDX4 has been studied as a CB marker protein (Noce et al., 2001; Kotaja et al., 2006a; Kotaja & Sassone-Corsi, 2007; Onohara et al., 2010). In DDX4-null mice, spermatogenesis arrests at the prophase of meiosis, and most spermatocytes undergo degeneration. Furthermore, the amount of IMC in the spermatocytes is greatly reduced (Tanaka et al., 2000; Chuma et al., 2006). DDX4 was reported to interact with MIWI and MILI. Moreover, MILI-/- and DDX4-/- mice have similar phenotypes, which suggest that these proteins work cooperatively (Kuramachi-Miyagawa et al., 2004). In MILI-null mice, DDX4 was not detected in the nuage components (Kuramachi-Miyagawa et al., 2004), which suggests that DDX4 requires MILI to localize to the nuage components.

We previously reported the ultrastructual localization of DDX4 in nuage components during rat spermatogenesis (Onohara et al., 2010). Our report showed the wide expression of DDX4 in nuage components and other structures. In pachytene spermatocytes, DDX4 mainly localized to the surface of IMC, while loose aggregates were observed in the juxtanuclear area (Figure 5). The loose aggregate consisted of 70–90 nm particles and the material was dissociated from the IMC. In meiosis, structures positive for DDX4 disappeared, but small particles were identified in the cytoplasm (Figure 3). In spermatids, DDX4 mainly localized to the CB until step 6, after which the signal became gradually weaker, which suggests that the expression of CB proteins was dependent on spermatid step. No DDX4 signal was observed in the SB.

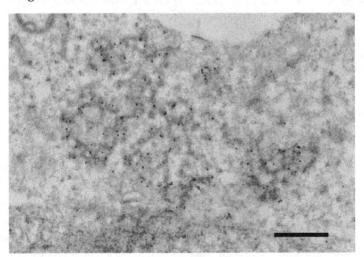

Fig. 5. IEM image of the loose aggregate consisting of 70–90 nm particles. Note that the particles contain gold particles used to label DDX4. Bar = 0.5 μm.

4.4 DDX25/GRTH in nuage components

DDX25 belongs to the DEAD-box family, and is essential for the completion of spermatogenesis. DDX25 acts as a hormone-dependent RNA helicase in Leydig cells and germ cells, and is referred to as a gonadotropin-regulated testicular helicase (GRTH) (Tang et al., 1999). DDX25-null mice, which are sterile, showed arrest of spermatogenesis at the round spermatid stage (Sheng et al., 2006). Sato et al. (2010) treated spermatogenic cells with inhibitors of RNA polymerase and nuclear protein export. As a result, the amount of DDX25 in the nucleus increased, while the amount in the cytoplasm decreased. Furthermore, after treatment, the CB became smaller and condensed, as observed in DDX25-deficient mice. The results of co-immunoprecipitation studies using an anti-DDX25 antibody revealed that DDX25 bound to mRNAs, including DDX25 mRNA, in both the nucleus and cytoplasm. This suggested that the main function of DDX25 is to transport mRNA from the nucleus to cytoplasm, and to transport DDX25-ribonucleoprotein (RNP) complexes essential for controlling CB construction.

We ourselves found that DDX25 was abundant in small particles and loose aggregates of 70–90 nm particles in spermatocytes, but was less abundant in IMC (in press). The loose aggregate is closely associated with the nuclear envelope (Onohara et al., 2010). In spermatids, DDX25 is also localized to CB, but rarely or not at all to the SB.

4.5 Other proteins in nuage components

Haraguchi et al. (2005) reported the localization of the ubiquitin-conjugating enzyme cytochrome oxidase subunit I (COXI), Hsp70, and phospholipid hydroperoxide glutathione peroxidase (PHGPx) to the CB, and the localization of vimentin, the 20S proteasome subunit, and Lamp1 in the region around the CB. The protein composition of the CB overlapped with that of the aggresome, which indicates that the CB has a role not only in protein synthesis, but also in degradation, as an aggresome system.

Dicer, a key protein in the biogenesis of mature microRNAs and small-interfering RNAs (siRNAs), has a highly dynamic expression pattern in the CBs of meiotic spermatocytes and round spermatids, and interacts with DDX4 at its C-terminus, as revealed by immunoprecipitation (Kotaja et al., 2006a). The localization of Ago2, -3, Dcp1a, and GW182 to the CB has been reported. Because these proteins are known to be markers of the processing body (P-body), a dense granule present in the cytoplasm in somatic cells (Kedersha et al., 2005), the CB may have functions in common with the P-body.

The Kinesin motor protein KIF17b is involved in the nucleus-to-cytoplasm transport of RNA, and was reported to co-localize with MIWI in the CB of round spermatids (Kotaja et al., 2006b). It may therefore be involved in the microtubule-mediated transport of CBs.

Brunol2/CUG-BP1/CELF1 is a homolog of the *Drosophila melanogaster* protein Bruno, and is a member of the CELF family that controls the splicing of pre-mRNAs. Bruno was found in nuage in ovaries and embryos (Snee & Macdonald, 2004). Spermatogenesis was arrested in step 7 spermatids in Brunol2-deficient mice (Kress et al., 2007). Recently we showed that Brunol2 localizes to the CB and other nuage components during rat spermatogenesis (in press).

HuR, an RNA-binding protein in somatic cells that controls the stability or translation of AU-rich mRNAs, was found in the CB and co-localized with DDX4 (Nguyen et al., 2009). HuR is involved in the shuttling of mRNAs from the nucleus to the CB for storage and protection against degradation.

MAEL, which is an ortholog of the *Drosophila* protein, MAELSTROM, is localized to the CB and unsynapsed chromosomes in spermatocytes, and has been implicated in the silencing of unsynapsed chromatin in spermatocytes during meiosis (Costa et al., 2006). MAEL co-localizes with DDX4, MIWI2, TDRD-9, and the P-body components Dcp1a, DDX6, GW182, and Xrn1 in the CB, but does not localize to the IMC. In MAEL-knockout mice, Piwi-interacting RNA (piRNA) production and post-transcriptional transposon silencing are perturbed in fetal gonocytes, suggesting that MAEL may be involved in these processes (Aravin et al., 2009).

GASZ, a germ cell protein with ankyrin repeats, a sterile alpha motif, and a leucine zipper, is localized to the CB. In GASZ-null mice, levels of MILI/piRNAs in the CB were dramatically reduced, and spermatogenesis was arrested before meiosis (Ma et al., 2009).

In 2010, MOV10L1, a member of the DEAD-box RNA helicase family essential for retrotransposon silencing in mouse germ cells, was reported to be expressed in spermatogonia, pachytene spermatocytes, and all intermediate forms in the mouse testis, and to interact with MILI, MIWI, and heat shock 70 kDa protein 2 (Frost et al., 2010). MOV10L1 appeared in a complex termed "MIWI-MILI block," which may be identical to IMC.

MitoPLD, a member of the phospholipase D superfamily, is abundant in the outer membrane of mitochondria, and degrades cardiolipin, yielding the lipid signaling molecule phosphatidic acid (PA). In MitoPLD-knockout mice, the nuage was arranged like a donut in the perinuclear region, and γ-tubulin was detected in the center of the aberrant nuage in spermatogonia, suggesting that MitoPLD may be involved in microtubule-dependent nuage transport. Furthermore, in MitoPLD-null mice, spermatogenesis arrests in late or post-meiotic spermatocytes, and the IMC is absent from spermatocytes. In addition, TDRD-1 and mitochondria aggregate to the pericentriolar region instead of the perinuclear region in gonocytes. In mice lacking Lipin1, which metabolizes PA, nuage increases in size and density, and the localization of TDRD-1 is altered. These results suggest that MitoPLD and/or PA may be involved in the generation of IMC and the correct localization of nuage and maintenance of its components (Huang et al., 2011; Watanabe et al., 2011).

NANOS, which is essential for primordial germ cell (PGC) in *Drosophila*, plays a critical role in spermatogenesis, particularly during meiosis. NANOS2 blocks meiosis by suppressing Stra8, an inducer of meiosis (Saga, 2008). NANOS1, which is known to regulate the translation of specific mRNAs, is found in the CB and co-localizes with PUMILIO protein, as well as the microRNA biogenesis factor GEMIN3 (Ginter-Matuszewska et al., 2011).

During the last decade, many nuage constituent proteins have been identified. Above all, it is noteworthy that RNA-binding proteins and RNA-regulating proteins are abundant in nuage, indicating that nuage may contribute to RNA silencing and RNA decay during spermatogenesis. Many nuage proteins co-localize simultaneously with other proteins in the same compartment, suggesting that most nuage components may function cooperatively during germ cell development. Functional nuage proteins are listed in Table 1.

Component	Domain/Function	Where found	References
TDRD-1	Tudor proteins; contain Tudor domains; essential for piRNA biosynthesis	Spermatogonia, spermatocytes, round spermatids	Chuma et al., 2003
TDRD-5		Spermatocytes, spermatids	Yabuta et al., 2011
TDRD-6		Spermatocytes, spermatids	Vasileva et al., 2009; Yabuta et al., 2011
TDRD-7		Spermatids	Tanaka et al., 2000; Yabuta et al., 2011
TDRD-9	Tudor protein; ATPase/DExH-type helicase; essential for piRNA biosynthesis	Spermatocytes	Shoji et al., 2009
TDRD-4/RNF17	Tudor protein; contains a RING finger motif	Spermatogonia, spermatocytes, round spermatids	Pan et al., 2005
Ago2/Ago3	Argonaute family protein; component of the RISC	Spermatids	Kotaja et al., 2006a
MILI	Piwi family proteins; associate with piRNA	Spermatocytes	Sasaki et al., 2003; Kotaja et al., 2006b
MIWI2		Spermatids	Sasaki et al., 2003; Kotaja et al., 2006b
DDX4/MVH	DEAD box RNA helicases	Spermatocytes, spermatids	Noce et al., 2001; Kotaja et al., 2006a
DDX6		Spermatocytes	Aravin et al., 2009
DDX25		Spermatocytes, spermatids	Sheng et al., 2006; Sato et al., 2010
MOV10L1	DEAD box RNA helicase family; silencing retrotransposons in germ cells	Spermatogonia to pachytene spermatocytes	Frost et al., 2010
Dicer	RNase III; major protein in the biogenesis of small RNAs	Spermatocytes, round spermatids	Kotaja et al., 2006a
Dcp1a	P-body markers	Pachytene spermatocytes, spermatids	Kotaja et al., 2006a
GW182		Spermatids	Kotaja et al., 2006a; Aravin et al., 2009
Xnt1		Spermatids	Aravin et al., 2009
Brunol2/CUG-BP1	CELF family protein; controls the splicing of pre-mRNAs	Spermatocytes, spermatids	Kress et al., 2007
GASZ	Essential for fertility in females and spermatogenesis	PGCs, gonocytes, spermatogonia, spermatocytes	Ma et al., 2009
HuR	RNA-binding protein in somatic cells; controls the stability or translation of AU-rich mRNAs	Spermatids	Nguyen et al., 2009
KIF17b	Kinesin motor protein	Round spermatids	Kotaja et al., 2006b; Aravin et al., 2011
MAEL	HMG box protein	Gonocytes, meiosis, spermatocytes, round spermatids	Costa et al., 2006; Aravin et al., 2009
NANOS1	Suppression of meiosis; regulates the translation of specific mRNA	PGCs, gonocytes, meiosis, spermatids	Ginter-Matuszewska et al., 2011
PUMILIO2	RNA-binding protein; translational regulation	Round spermatids	Ginter-Matuszewska et al., 2011
GEMIN3	MicroRNA biogenesis factor	Round spermatids	Ginter-Matuszewska et al., 2011

Table 1. Functional nuage-related proteins that play roles in RNA silencing, RNA decay, and nuage morphogenesis.

5. Functions of nuage proteins and their relationships to morphological features

5.1 A transcription context specific for spermatogenic cells

Proteins such as Dicer and PIWI family proteins that are involved in the posttranscriptional regulation of microRNAs during spermatogenesis are highly concentrated in nuage. Dicer processes the precursors of both siRNAs and microRNAs (miRNAs) into small mature RNAs in the cytoplasm in somatic cells. These small mature RNAs are assembled into an RISC, which contains Argonaute family proteins and causes translational repression or mRNA cleavage in the cytoplasm in somatic cells (Stefani & Slack, 2008). Ago family proteins, Dicer, miRNAs, and miRNA-repressed mRNA are localized to the P-bodies of somatic cells, suggesting that RNA silencing and RNA disruption may occur there (Sen & Blau, 2005). Some P-body markers are also expressed in the CB, suggesting that the CBs of germ cells may be involved in RNA silencing and RNA disruption, as in the P-bodies of somatic cells (Kotaja et al., 2006a). The function of the P-body in somatic cells has been studied extensively (Sen & Blau, 2005). The CB has many features in common with the P-body, strongly suggesting that the CB may play similar roles to the P-body.

5.2 Function of the CB in mRNA storage and transport

Söderström and Parvinen (1976) studied RNA synthesis during spermatogenesis by autoradiography using tritiated uridine. In step 1–8 spermatids, the rate of RNA synthesis was low in the cell as a whole, but very high in the CB, and was arrest in step 8 spermatids in both places. Therefore, it is likely that CBs in spermatids contain RNAs, which may encode proteins involved in the regulation of spermiogenesis in late spermatids.

Kotaja et al. (2006a) performed *in situ* hybridization to determine whether or not the CB contains miRNA, and demonstrated the localization of miRNAs in the CB. HuR is involved in the transport of mRNAs from the nucleus to the CB so that they can be stored and protected against degradation (Nguyen et al., 2009). Kotaja et al. (2006b) detected KIF17b, known to bind to mRNAs, in the CB, and suggested that the CB might be involved in the transport of mRNAs in early spermatids.

These studies strongly suggest that the CB is also a site for the storage and transport of mRNAs that encode proteins with roles in late spermatids.

5.3 The function of nuage components other than the CB

During the differentiation of spermatogonia and spermatocytes, several nuage structures with specific shapes appear, persist for a period of time, and then disappear, as described above. The other organelles, including mitochondria, the Golgi complex, and lysosomes, do not undergo marked changes in morphology. However, the differentiation of spermatids to spermatozoa is dynamic and dramatic. Spermatids transform into spermatozoa, whose role is to enable transport of DNA to the target cell, the oocyte, through an extraordinarily intricate process of cell differentiation termed spermiogenesis, which lasts approximately 20 days in rats (Clermont et al., 1959) and 24 days in humans (Heller & Clermont, 1964). Major events that occur during spermiogenesis include formation of the acrosome, nuclear condensation, formation of the tail, trimming of the cytoplasm, and organelle reorganization (de Kretser & Kerr, 1988; Clermont et al., 1993). Ultrastructural studies have described various specialized structures, including the radial body-annulate lamellae complex; small puffs of a fine, filamentous, fuzzy material; a granulated body composed of fine, dense granular materials; a reticulated body consisting of several dense anastomosed cords; and a large, dense granule surrounded by crescentic mitochondria (Clermont et al., 1993), which we termed the mitochondria-associated granule (MAG).

Our own IEM studies provide evidence that some of these structures contain nuage proteins.

1. The radial body-annulate lamellae complex, a membrane-bound structure that is continuous with the endoplasmic reticulum (ER), is observed in the cytoplasmic lobe of late spermatids, and is believed to be the site at which resorption of the ER membrane occurs (Clermont & Rambourg, 1978). No nuage proteins have yet been detected in this structure. Therefore, it seems that it is not a structural site for the function of nuage proteins.

2. Small puffs have no limiting membrane, are attached to the cytoplasmic surface of the ER, and are observed in step 8–10 spermatids. The only nuage protein to have been detected in the puffs so far is Brunol2 (in press).

3. Granulated bodies appear in the cytoplasmic lobes of step 14 spermatids, are most abundant in step 17 spermatids, and decrease in number during subsequent stages of spermatogenesis. It was reported by Clermont et al. (1990) that they contain outer dense fiber (ODF) proteins. It is assumed that ODF polypeptides are temporarily stored in these bodies (Clermont et al., 1993). However, their function remains unclear. Our preliminary IEM studies detected BRUNOL2, DDX25, NANOS1, and MAEL in these bodies, but not DDX4 (unpublished data). Gold particles that were used to label these nuage proteins were confined to the fine granular matrix.

4. The reticulated body first appears in the cytoplasm of step 14 spermatids and completely disappears in step 18 spermatids (Clermont et al. 1990). The body is characterized by several dense, anastomosed cords with a width of 80–100 nm. We detected BRUNOL2, DDX25, NANOS1, and MAEL in this structure by IEM, but not DDX4. Other than the nuage proteins, only cathepsin H has so far been detected in the structure (Haraguchi et al., 2003). The nature and function of this structure remains unclear.

5. The MAG is 1–2 μm in diameter, is composed of a fine granular material, and appears in the elongated cytoplasmic lobes of step 9–17 spermatids (Figure 6). Its strongest characteristic is its close association with mitochondria (Clermont et al., 1990). We detected the nuage proteins BRUNOL2, DDX4, DDX25, NANOS1, and MAEL in this structure in our preliminary studies (unpublished data). Although dense materials in the MAG were assumed to be a source of ODF proteins, the results of IEM experiments that detected no ODF protein signals in the MAG indicates that this is not the case (Clermont et al., 1990). No proteins other than the above five nuage proteins have so far been detected in this structure. Although the nature and function of the MAG are unknown, the existence of nuage proteins may provide important clues about the function of the MAG.

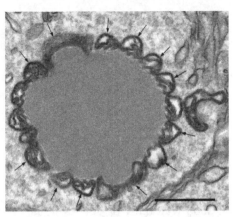

Fig. 6. Electron micrograph of an MAG in a step 16 spermatid. The MAG is a large spherical mass with a diameter of 1–2 μm, composed of granular material and surrounded by mitochondria. Bar = 1 μm.

In the neck cytoplasm of step 11 spermatids, small, discrete, dense masses are frequently observed (Fawcett & Phillips, 1969), closely associated with other structures or free in the cytoplasm. One, which is associated with the plasma membrane and subsequently moves to a position at the junction of the middle and principal regions of the tail, is known as the annulus. It was assumed that the annulus was derived from the CB (Fawcett et al., 1970). However, our study (Onohara et al., 2010) and preliminary results (unpublished data) indicate a lack of signals in the annulus for the nuage proteins listed above, indicating that the CB and the nuage proteins are not involved in the formation of the annulus. Spherical masses are frequently observed at the base of the flagellum of late spermatids (Fawcett & Phillips, 1969). Our IEM studies show that these masses are positive for Dicer1 and MAEL (unpublished data). Interestingly, these Dicer1-positive particles are maintained in epididymal sperm, suggesting that they are carried into the cytoplasm of the ovum. Furthermore, a less electron-dense mass composed of fine filamentous materials surrounds the connecting piece in late spermatids (Clermont et al., 1993). This structure is positive for BRUNOL2 (unpublished data). In the neck region, small dense particles are present. These particles contain small amounts of the nuage proteins BRUNOL2, DDX25, NANOS1, and MAEL. In addition, BRUNOL2 and MAEL have been detected in large aggregates of free ribosomes and other unknown materials in the residual bodies of step 19 spermatids (unpublished data). Thus, nuage proteins are associated with various structures that are unrelated to the nuage. Although it is not clear whether the nuage proteins function there, it is likely that they are active in some of the structures mentioned above.

6. Conclusion - Future nuage research

In classic reports, the CB was detected in either the nucleus or the cytoplasm, but there was no obvious evidence to confirm either location. More recent studies have suggested that the IMC is a source of the CB. However, in TDRD-1-null mice, CBs were present in spite of the fact that the IMC disappeared (Chuma et al., 2006). It therefore appears that IMC is not essential for the formation of the CB. Recently, Meikar et al. (2010) purified the CB by immunoprecipitation and found that it contained more than 100 proteins. MitoPLD is essential for the formation of IMC, while PA, which is produced by MitoPLD, affects the architecture of the CB (Huang et al., 2011; Watanabe et al., 2011), suggesting that various proteins and factors are involved in the construction of nuage. Nuage has various forms and is associated with other subcellular organelles such as the nucleus, mitochondria, and the Golgi apparatus (Parvinen, 2005), suggesting that communication between nuage and the other organelles is important for the completion of spermatogenesis. More detailed studies of the properties of nuage will elucidate its function and origin.

One big question remains to be answered: during meiosis, after chromosome recombination has occurred, where do the nuage components go, how are they divided between the daughter cells equally, and how is the nuage reformed? It is also unclear how nuage or nuage components contribute to meiosis. MAEL suppresses transcription via the small RNA pathway during meiosis (Costa et al., 2006). Defects in several nuage components cause the arrest of spermatogenesis before meiosis, suggesting that the correct function of these components is essential for spermatogenesis. Microtubule-dependent movement of the CB is critical to enable communication with other organelles (Kotaja et al., 2006b; Huang et al., 2011; Watanabe et al., 2011).

Frost et al. (2010) suggested that the piRNA silencing complex may be constructed hierarchically. In other reports, nuage protein immunofluorescence staining patterns in some cases overlap completely, but in others only partially, even in the same stage of spermatogenesis (Figure 7). These observations suggest that nuage is formed hierarchically by the coalescence of small complexes assembled from a few proteins. To confirm whether this is indeed the case, we need to clarify the precise relationships among nuage components.

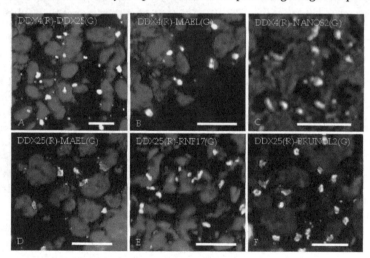

Fig. 7. Immunofluorescence staining of rat seminiferous epithelium with a combination of two different antibodies. Different antigens were stained using Alexa Fluor® 488 and Alexa Fluor® 568. The images were merged using Adobe Photoshop®. Staining for (A) DDX4 and DDX25, (B) DDX4 and MAEL, (C) DDX4 and NANOS2, (D) DDX25 and MAEL, (E) DDX25 and RNF17, and (F) DDX25 and BRUNOL2. All proteins examined were localized to CBs and two staining patterns were seen: 1) the patterns for both proteins completely overlapped within the CBs (A–C); and 2) the patterns for both proteins partially overlapped within the CBs (D–F). This clearly indicates that some nuage proteins distribute homogeneously, while others are segregated within CBs. Bar = 50 μm.

Figure 8 summarizes the possible functions of nuage in molecular biology and morphology. While the function and significance of nuage in spermatogenic cells have begun to be elucidated, they still remain largely unclear. Collecting information on nuage components and clarifying the function of other nuage-like structures may reveal insights into the significance of nuage in the development and maintenance of germ cells.

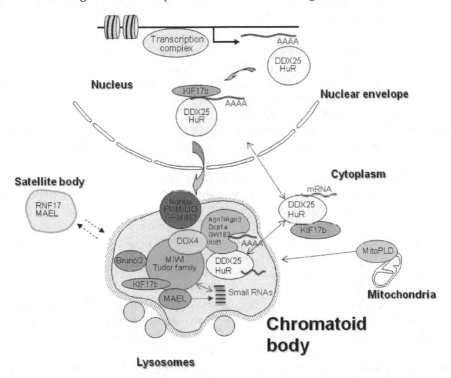

Fig. 8. Possible functions of nuage. Haploid gene products, including RNA binding proteins such as DDX25 and HuR, and microtubule-associated proteins such as KIF17b, bind to RNP. The mRNA-RNP complex is transported from the nucleus to the cytoplasm via the nuclear pore complex, or directly, to the CB. The movement of this complex is known to be supported by the microtubule network. In the cytoplasm, the CB frequently associates with the nuclear pores to interact with mRNAs. The CB contains RNA-binding and RNA-processing proteins such as DDX4, DDX6, and DDX25, and components of the RNA silencing and RNA decay pathways, such as small RNAs, Dicer, MAEL, Piwi family proteins, and Tudor family proteins. Furthermore, the CB contains P-body components such as Ago2, Dcp1a, GW182, and Xnt1, and thus, like the P-body, functions as a post-transcriptional regulator. The CB also contains an RNA-protecting protein, HuR. mRNAs stored in the CB are released into the cytoplasm in response to appropriate stimuli and are translated. Dicer produces small RNAs from their precursors in the CB. The mitochondrial membrane protein MitoPLD is involved in the movement and construction of the CB. The CB also contains the ubiquitin-conjugating enzyme E2. Some CB proteins are polyubiquitinated and degraded by proteasomes located on the surface of the CB (dots). Other components are degraded by lysosomes in close contact with the CB.

7. References

Adams, E.C. & Hertig, A.T. (1964) Studies on guinea pig oocytes. I. Electron microscopic observation on the development of cytoplasmic organelles in oocytes of primordial and primary follicles. *J. Cell Biol.* Vol. 21, pp. 397–427, ISSN 0021-9525

André J. (1962) Contribution of à la conaissance du chondriome. *J. Ultrastruct. Res.* Vol. 6, No. 3, pp. 1-185, ISSN 0022-5320

Aravin, A.A., van der Heijden, G.W., Castañeda, J., Vagin, V.V., Hannon, G.J. & Bortvin, A. (2009) Cytoplasmic compartmentalization of the fetal piRNA pathway in mice. *PLoS Genet.* Vol. 5, No. 12, pp e1000764. ISSN 1553-7404

Benda, C. (1891) Neue Mittheilungen über die Entwickelung der Genitadrüsen und über die Metamorphose der Samenzellen (Histogenese der Spermatozoen) *Arch. Anat. Physiol. Physiol.* Vol. 30, pp. 549–552

Brökelmann, J. (1963) Fine structure of germ cells and Sertoli cells during the cycle of the seminiferous epithelium in the rat. *Z. Zellforsch.* Vol. 59, pp. 820-850, ISSN 0340-0336

Burgos, M.H. & Fawcett, D.W. (1955) Studies on the structure of the mammalian testis. I. Differentiation of the spermatids in the cat (Felis domestica) *J. Biophys. Biochem. Cytol.* Vol. 1, pp. 287–315, ISSN 0095-9901

Burgos, M.H., Vitale-Calpe, R. & Aoki, A. (1970) Fine structure of the testis and its functional significance Vol. 1. In: "The Testis", Johnson, A.D., Gomes, W.R. & Vandemark, N.L. (Ed.) pp. 552–649, Acad. Press, New York.

Challice, C.E. (1953) Electron microscopic studies of spermiogenesis in some rodents. *J. Roy. Micros. Soc.* Vol. 73, No. 3, pp. 115–127, ISSN 0368-3974

Chuma, S., Hiyoshi, M., Yamamoto, A., Hosokawa, M., Takamune, K. & Nakatsuji, N. (2003) Mouse Tudor Repeat-1 (MTR-1) is a novel component of chromatoid bodies/nuages in male germ cells and forms a complex with snRNPs. *Mech Dev.* Vol. 120, No. 9, pp. 979-990, ISSN 1872-6356

Chuma, S., Hosokawa, M., Kitamura, K., Kasai, S., Fujioka, M., Hiyoshi, M., Takamune, K., Noce, T. & Nakatsuji, N. (2006) Tdrd1/Mtr-1, a tudor-related gene, is essential for male germ-cell differentiation and nuage/germinal granule formation in mice. *Proc Natl Acad Sci U S A.* Vol. 103, No. 43, pp. 15894-15899, ISSN 1091-6490

Clermont, Y., Leblond, C.P. & Meissier, B. (1959) Durée du cycle de l'épithélium séminal du rat. *Arch. Anat. Micrscop. Morphol. Exptl.* Vol. 48, pp. 37-56.

Clermont, Y., Oko, R. & Hermo, L. (1990) Immunocytochemical localization of proteins utilized in the formation of outer dense fibers and fibrous sheath in rat spermatids: An electron microscope study. Vol. 227, No.4, pp. 447-457, ISSN 1097-0185

Clermont, Y., Oko, R. & Hermo, L. (1993) Cell Biology of Mammalian Spermiogenesis, In: *Cell and Molecular Biology of the Testis*, Desjardins, C. & Ewing, L.L. (eds.) pp. 332-376

Clermont, Y. & Rambourg, A. (1978) Evolution of the endoplasmic reticulum during rat spermiogenesis. *Am. J. Anat.* Vol. 151, No. 2, pp. 191-212, ISSN 0002-9106

Comings, D.E. & Okada, T A. (1972) The chromatoid body in mouse spermatogenesis: evidence that it may be formed by the extrusion of nucleolar components. J. Ultrastruct. Res. Vol. 39, No.,1 pp. 15-23, ISSN 0022-5320

Costa, Y., Speed, R.M., Gautier, P., Semple, C.A., Maratou, K., Turner, J.M. & Cooke, H.J.
(2006) Mouse MAELSTROM: the link between meiotic silencing of unsynapsed
chromatin and microRNA pathway? Hum Mol Genet, Vol. 15, No.15, pp. 2324-2334,
ISSN 0964-6906

Daoust, R. & Clermont, Y. (1955) Distribution of nucleic acids in germ cells during the cycle
of the seminiferous epithelium in the rat. Am J Anat. Vol. 96, No. 2, pp. 255-83, ISSN
0002-9106

De Kretser, D.M. & Kerr, J.B. (1988) The cytology of the testis, In: The Physiology of
Reproduction, Knobil, E. Neill, J. D. et al. (eds.) pp. 837-932, Raven Press, New York.

Deng, W. & Lin, H. (2002) miwi, a murine homolog of piwi, encodes a cytoplasmic protein
essential for spermatogenesis. Dev Cell. Vol. 2, No. 6, pp. 819-830, ISSN 1878-1551

Dhainaut, A. (1970) Etude en microscopie électronique et par autoradiographie á haute
résolution des extrusions nucléaires au cours de l'ovogenèse de Nereis pelagica
(Annélide polychète) J. Microscopie. Vol. 9, pp. 99–118

Duesberg, J. (1908) Les division des spermatocytes chez le rat (Mus decumanus Pall., variété
albinos) Arch. Zellforsch. Vol.1, pp. 399–449

Eddy, E.M. (1970) Cytochemical observations on the chromatoid body of the male germ
cells. Biol. Reprod. Vol. 2, No. 1, pp. 114–128, ISSN 1529-7268

Eddy, E.M. (1974) Fine structural observations on the form and distribution of nuage in
germ cells of the rat. Anat Rec. Vol. 178, No. 4, pp. 731-757, ISSN 1097-0185

Eddy, E.M. & Ito, S. (1971) Fine structural and radioautographic observations on dense
perinuclear cytoplasmic material in tadpole oocytes. J Cell Biol., Vol. 49, No.1, pp.
90-108, ISSN 1540-8140

Fasten, N. (1914) Spermatogenesis of the American grayfish, Cambarus virilis and Cambrarus
immunis (?), with special reference to synapsis and the chromatoid bodies. J. Morph.
Vol. 25, pp. 587–649, ISSN 1097-4687

Fawcett, D.W., Eddy, E.M. & Phillips, DM. (1970) Observations on the fine structure and
relationships of the chromatoid body in mammalian spermatogenesis. Biol Reprod.
Vol. 2, No. 1, pp. 129-153, ISSN 1529-7268

Fawcett, D.W. & Phillips, D.M. (1969) The fine structure and development of the neck region
of the mammalian spermatozoon. Anat. Rec. Vol. 165, pp. 153-184, ISSN 1097-0185

Frost, R.J., Hamra, F.K., Richardson, J.A., Qi X., Bassel-Duby, R. & Olson, E.N. (2010)
MOV10L1 is necessary for protection of spermatocytes against retrotransposons by
Piwi-interacting RNAs Proc Natl Acad Sci U S A. Vol. 107, No. 26, pp. 11847-11852,
ISSN 1091-6490

Gatenby, J.E. & Beams, H.W. (1935) The cytoplasmic inclusions in the spermatogenesis of
man. Q. J. Microsc. Sci. Vol. 18, pp. 1-33, ISSN 0370-2952

Ginter-Matuszewska, B., Kusz, K., Spik, A., Grzeszkowiak, D., Rembiszewska, A.,
Kupryjanczyk, J. & Jaruzelska, J. (2011) NANOS1 and PUMILIO2 bind microRNA
biogenesis factor GEMIN3, within chromatoid body in human germ cells.
Histochem Cell Biol. Vol.136, No. 3, pp.279-287

Haraguchi, C.M., Ishido, K., Kominami, E. & Yokota, S. (2003) Expression of cathepsin H in
differentiating rat spermatids: Immnuoelectron microscopic study. Histochem. Cell
Biol. Vol. 120, No. 1, pp. 63-71, ISSN 0948-6143

Haraguchi, C. M., Mabuchi, T., Hirata, S., Shoda, T., Hoshi, K., Akasaki, K. & Yokota, S.
    (2005) Chromatoid bodies: Aggresome-like characteristics and degradation sites for
    organelles of spermiogenic cells. *J. Histochem. Cytochem.* Vol. 53, No. 4, pp. 455–465,
    ISSN 0022-1554

Heller, C.G. & Clermont, Y. (1964) Kinetics of the germinal epithelium in man. *Rec. Prog.
    Hormone Res.* Vol. 20, pp. 545-575

Hermann, F. (1889) Beiträge zur Histologie des Hodens. *Arch. f. mikr. Anat.* Vol. 34, pp. 58-
    105

Hosokawa, M., Shoji, M., Kitamura, K., Tanaka, T., Noce, T., Chuma, S. & Nakatsuji, N.
    (2007) Tudor-related proteins TDRD1/MTR-1, TDRD6 and TDRD7/TRAP: domain
    composition, intracellular localization, and function in male germ cells in mice. *Dev
    Biol.* Vol. 301, No. 1, pp. 38-52, ISSN 0012-1606

Huang, H., Gao, Q., Peng, X., Choi, S.Y., Sarma, K., Ren, H., Morris, A.J. & Frohman, MA.
    (2011) piRNA-associated germline nuage formation and spermatogenesis require
    MitoPLD profusogenic mitochondrial-surface lipid signaling. *Dev Cell.* Vol. 20, No.
    3, pp. 376-387, ISSN 1878-1551

Kedersha, N., Stoecklin, G., Ayodele, M., Yacono, P., Lykke-Andersen, J., Fritzler, M.J.,
    Scheuner, D., Kaufman, R.J., Golan, D.E. & Anderson, P. (2005) Stress granules and
    processing bodies are dynamically linked sites of mRNP remodeling. *J Cell Biol.*
    Vol. 169, No. 6, pp. 871-884, ISSN 1540-8140

Kerr, J.B. & Dixon, K.E. (1974) An ultrastructural study of germ plasm in spermatogenesis of
    Xenopus laevis. J Embryol Exp Morphol. Vol. 32, No. 3, pp. 573-592, ISSN 0022-0752

Kirino, Y., Vourekas, A., Kim, N., de Lima Alves F., Rappsilber J., Klein P.S., Jongens T.A. &
    Mourelatos Z. (2010) Arginine methylation of vasa protein is conserved across
    phyla. *J Biol Chem.* Vol. 285, No. 11,pp. 8148-8154, ISSN 0021-9258

Kotaja, N., Bhattacharyya, S.N., Jaskiewicz, L., Kimmins, S., Parvinen, M., Filipowicz, W. &
    Sassone-Corsi P. (2006a) The chromatoid body of male germ cells: similarity with
    processing bodies and presence of Dicer and microRNA pathway components. *Proc
    Natl Acad Sci U S A.* Vol. 103, No. 8, pp. 2647-52, ISSN 1091-6490.

Kotaja, N., Lin, H., Parvinen, M. & Sassone-Corsi, P. (2006b) Interplay of PIWI/argonaute
    protein MIWI and kinesin KIF17b in chromatoid bodies of male germ cells. *J. Cell
    Sci.* Vol. 119, No. 13, pp. 2819-2825, ISSN 0021-9533

Kotaja, N. & Sassone-Corsi, P. (2007) The chromatoid body: a germ-cell-specific RNA-
    processing centre. *Nat Rev Mol Cell Biol.* Vol. 8, pp. 85-90, ISSN 1471-0072

Kress, C., Gautier-Courteille, C., Osborne, H.B., Babinet, C. & Paillard, L. (2007) Inactivation
    of CUG-BP1/CELF1 causes growth, viability, and spermatogenesis defects in mice.
    Mol Cell Biol. Vol. 27, No. 3, pp. 1146-1157, ISSN 0270-7360

Kuramochi-Miyagawa, S., Kimura, T., Ijiri, T.W., Isobe, T., Asada, N., Fujita, Y., Ikawa, M.,
    Iwai, N., Okabe, M., Deng, W., Lin, H., Matsuda, Y. & Nakano, T. (2004) Mili, a
    mammalian member of piwi family gene, is essential for spermatogenesis.
    Development. Vol. 131, No. 4, pp. 839-849, ISSN 0959-1991

Ma, L., Buchold, G.M., Greenbaum, M.P., Roy, A., Burns, K.H., Zhu, H., Han, D.Y., Harris,
    R.A., Coarfa, C., Gunaratne, P.H., Yan, W. & Matzuk, M.M. (2009) GASZ is

essential for male meiosis and suppression of retrotransposon expression in the male germline. *PLoS Genet.* Vol. 5, No. 9, pp. e1000635, ISSN 1553-7390

Mahowald, A.P. (1962) Fine structure of pole cells and polar granules in *Drosophila melanogaster. J. Exp. Zool.* Vol. 151, No. 3, pp. 201-215, ISSN 0022-104X

Meikar, O., Da Ros, M., Liljenbäck, H., Toppari, J. & Kotaja, N. (2010) Accumulation of piRNAs in the chromatoid bodies purified by a novel isolation protocol. *Exp Cell Res.* Vol. 316, No. 9, pp. 1567-1575, ISSN 0014-4827

Meister, G., Landthaler, M., Patkaniowska, A., Dorsett, Y., Teng, G. & Tuschl, T. (2004) Human Argonaute2 mediates RNA cleavage targeted by miRNAs and siRNAs. *Mol Cell.* Vol. 15, No. 2, pp. 185-197, ISSN 1097-2765

Minamino, T. (1955) Spermiogenesis in the albino rat as revealed by electron microscopy. *Electron Microsc.* Vol. 4, pp. 249-253

Nguyen, C.M., Chalmel, .F, Agius, E., Vanz,o N., Khabar, K.S., Jégou, B. & Morello, D.. (2009) Temporally regulated traffic of HuR and its associated ARE-containing mRNAs from the chromatoid body to polysomes during mouse spermatogenesis. PLoS One. Vol. 4, No. 3, pp. e4900 (1-13), ISSN 1932-6203

Nicander, L. & Plöen, L. (1969) Fine structure of spermatogonia and primary spermatocytes in rabbits. *Z. Zellforsch.* Vol. 99, pp. 221-234

Nieuwkoop, P.D. & Faber, J. (1956) *Normal table of Xenopus laevis (Daudin).* North Holland Publ. Co., Amsterdam.

Noce, T., Okamoto-Ito, S. & Tsunekawa, N. (2001) Vasa homolog genes in mammalian germ cell development. *Cell Struct Funct.* Vol. 26, No. 3, pp. 131-136, ISSN 0386-7196

Odor, D.L. (1965) The ultrastructure of unilaminar follicles of the hamster ovary. *Am. J. Anat.* Vol. 116, pp. 493–522, 0002-9106

Onohara, Y., Fujiwara, T., Yasukochi, T., Himeno, M. & Yokota, S. (2010) Localization of mouse vasa homolog protein in chromatoid body and related nuage structures of mammalian spermatogenic cells during spermatogenesis. *Histochem Cell Biol.* Vol. 133, No. 6, pp. 627-639, ISSN 0948-6143

Pan, J., Goodheart, M., Chuma, S., Nakatsuji, N., Page, D.C. & Wang, P.J. (2005) RNF17, a component of the mammalian germ cell nuage, is essential for spermatogenesis. *Development.* 132:4029–4039.

Parvinen, M. (2005) The chromatoid body in spermatogenesis. *Int. J. Androl.* Vol. 28, No. 4, pp. 189-201

Parvinen, M. & Jokelainen, PT. (1974) Rapid movements of the chromatoid body in living early spermatids of the rat. *Biol Reprod.* Vol. 11, No. 1, pp. 85-92, ISSN 1529-7268

Parvinen, M. & Parvinen, L.M. (1979) Active movements of the chromatoid body. A possible transport mechanism for haploid gene products. *J. Cell Biol.* Vol. 80, No. 3, pp. 621-628, ISSN 1540-8140

Pollister, A.W. (1930) Cytoplasmic phenomena in the Gerris. *J. Morph.* Vol. 49, pp. 455-507, ISSN 1097-4687

Regaud, C.L. (1901) Études sur la structure des tubes séminifères et sur la spermatogénèse chez les Mammifères. *Arch. Anat. Micr. Morphol. Exp.* Vol. 4, pp. 231–380

Russell, L. & Frank, B. (1978) Ultrastructural characterization of nuage in spermatocytes of the rat testis. *Anat Rec.* Vol. 190, No. 1, pp. 79-97, ISSN 1097-0185

Saga, Y. (2008) Sexual development of mouse germ cells; Nanos2 promotes the male germ cell fate by suppressing the female pathway. *Develop. Growth Differ.* Vol. 50, Supple. 1, pp. S141-S147, ISSN 0012-1592

Sasa, S. (1959) On the ultrastructure of the spermatogenic cells of the albino rat. *J. Chiba Med. Soc.* Vol. 34, pp. 1698-1721, ISSN 0009-3459

Sasaki, T., Shiohama, A., Minoshima, .S. & Shimizu, N. (2003) Identification of eight members of the Argonaute family in the human genome small star, filled. *Genomics.* Vol. 82, No. 3, pp. 323-330, ISSN 0888-7543

Sato, H., Tsai-Morris, C.H. & Dufau, M.L. (2010) Relevance of gonadotropin-regulated testicular RNA helicase (GRTH/DDX25) in the structural integrity of the chromatoid body during spermatogenesis. *Biochim Biophys Acta.* Vol. 1803, No. 5, pp. 534-543, ISSN 0006-3002

Schreiner, A. & Schreiner, K.E. (1905) Über die Entwickelung der männlichen Geschlechtszellen von Myxine glutinosa. *Arch. Biol.* Vol. 21, pp. 183-355

Sen, G.L. & Blau, H.M. (2005) Argonaute 2/RISC resides in sites of mammalian mRNA decay known as cytoplasmic bodies. *Nat Cell Biol.* Vol. 7, No.6, pp. 633-636, ISSN 1465-7392

Sheng, Y., Tsai-Morris, C.H., Gutti, R., Maeda, Y. & Dufau, M.L. (2006) Gonadotropin-regulated testicular RNA helicase (GRTH/Ddx25) is a transport protein involved in gene-specific mRNA export and protein translation during spermatogenesis. *J Biol Chem.* Vol. 281, No. 46, pp. 35048-35056, ISSN 0021-9258

Shoji, M., Tanaka, T., Hosokawa, M., Reuter, M., Stark, A., Kato, Y., Kondoh, G., Okawa, K., Chujo, T., Suzuki, T., Hata, K., Martin, S.L., Noce, T., Kuramochi-Miyagawa, S., Nakano, T., Sasaki, H., Pillai, R.S., Nakatsuji, N. & Chuma, S. (2009) The TDRD9-MIWI2 complex is essential for piRNA-mediated retrotransposon silencing in the mouse male germline. *Dev Cell.* Vol. 17, No. 6, pp. 775-787, ISSN 1878-1551

Snee, M.J. & Macdonald, P.M. (2004) Live imaging of nuage and polar granules: evidence against a precursor-product relationship and a novel role for Oskar in stabilization of polar granule components. *J Cell Sci.* Vol. 117, Pt. 10, pp. 2109-2020, ISSN 0021-9533

Söderström, K.O. (1978) Formation of chromatoid body during rat spermatogenesis. *Z Mikrosk Anat Forsch.* Vol. 92, No. 3, pp. 417-430, ISSN 0044-3107

Söderström, K.O. & Parvinen, M. (1976) In corporation of [3H]uridine by the chromatoid body during rat spermatogenesis. *J. Cell. Biol.* Vol. 70, No. 1, pp. 239-246, ISSN1540-8140

Sorokin, S.P. (1968) Reconstruction of centriole formation and ciliogenesis in mammalian lungs. *J. Cell Sci.* Vol. 3, No. 2, pp. 207-230, ISSN 0021-9533

Stefani, G. & Slack, G. (2008) Small non-coding RNAs in animal development. *Nature review,* Vol. 9, No. 3, pp. 219-230, ISSN9;219-230

Stockinger, .L. & Cirelli, E. (1965) Eine bisher unbekannte Art der Zentiolenvermehrung. *Z. Zellforsh. Mikroscop. Anat. Abt. Histochem.* Vol. 68, pp. 233-740

Sud, B.N. (1961). Morphological and cytochemical studies of the chromatoid body and related elements in the spermatogenesis of the rat. *Q J Microsc Sci* Vol. 102, pp. 495-505, ISSN 0370-2952

Susi, F.R. & Clermont, Y. (1970) Fine structural modifications of the rat chromatoid body during spermiogenesis. *Am J Anat.* Vol. 129, No. 2, pp. 177-191, ISSN 0002-9106

Tanaka, S.S., Toyooka, Y., Akasu, R., Katoh-Fukui, Y., Nakahara, Y., Suzuki, R., Yokoyama, M. & Noce, T. (2000) The mouse homolog of Drosophila Vasa is required for the development of male germ cells. Genes Dev. Vol. 14, No. 7, pp. 841-853, ISSN 0890-9369

Tang, P.Z., Tsai-Morris, C.H. & Dufau, M.L. (1999) A novel gonadotropin-regulated testicular RNA helicase. A new member of the dead-box family. *J Biol Chem.* Vol. 274, No. 53, pp. 37932-37940, ISSN 0021-9258

Vasileva, A., Tiedau, D., Firooznia, A., Müller-Reichert, T. & Jessberger, R. (2009) Tdrd6 is required for spermiogenesis, chromatoid body architecture, and regulation of miRNA expression. *Curr Biol.* Vol. 19, No. 8, pp. 630-639, ISSN 0960-9822

Ventelä, S., Toppari, J. & Parvinen, M. (2003) Intracellular organelle traffic through cytoplasmic bridges in early spermatids of the rat; mechanism of haploid gene product sharing. *Mol. Biol. Cell* Vol. 14, No. 7, pp. 2768-2780, ISSN 1059-1524

von Brunn, A.V. Beiträge zur Entwicklungsgeschiche der Samenkörper. (1876) Arch. f. Mikr. Anat. Vol. 12, pp. 528-536

von Ebner, V. (1888) Zur Spermatogenese bei den Säugethieren. *Arch. f. Mikr. Anat.* Vol. 31, pp. 236-292

von Korff, K. (1902) Zur histogenese der spermien von Phalangista vulpine. *Arch. f. Mikr. Anat.* Vol. 60, pp. 232-260

Watanabe, T., Chuma, S., Yamamoto, Y., Kuramochi-Miyagawa, S., Totoki, Y., Toyoda, A., Hoki, Y., Fujiyama, A., Shibata, T., Sado, T., Noce, T., Nakano, T., Nakatsuji, N., Lin, H. & Sasaki, H. (2011) MITOPLD is a mitochondrial protein essential for nuage formation and piRNA biogenesis in the mouse germline. *Dev Cell.* Vol. 20, No. 3, pp. 364-375, ISSN 1878-1551

Watson, M.L. (1952) Spermatogenesis in the albino rat as revealed by electron microscopy. *Biochim. Biophys. Acta.* Vol. 8, No. 4, pp. 369-374, ISSN 0006-3002

Weakley B.S. (1969) Granular cytoplasmic bodies in oocytes of the golden hamster during the post-natal period. *Z. Zellforsch. Mikrosk. Anat.* Vol. 101, No. 3, pp. 394-400, ISSN 0340-0336

Wilson, E.B. (1913) A chromatoid body simulating an accessory chromosome in pentatoma. *Biol. Bull.* Vol. 24, No. 6, pp. 392-411, ISSN 0006-3185

Wodesdalek, J.E. (1914) Spermatogenesis of the horse with special reference to the accessory chromosome and the chromatoid body. *Biol. Bull.* Vol. 27, No. 6, pp. 295-325, ISSN 0006-3185

Yabuta, Y., Ohta, H., Abe, T., Kurimoto, K., Chuma, S. & Saitou, M. (2011) TDRD5 is required for retrotransposon silencing, chromatoid body assembly, and spermiogenesis in mice. *J Cell Biol.* Vol. 192, No. 5, pp. 781-795, ISSN 1540-8140

Yin, X.Y., Gupta, K., Han, W.P., Levitan, E.S. & Prochownik, E.V. (1999) Mmip-2, a novel RING finger protein that interacts with mad members of the Myc oncoprotein network. *Oncogene.* Vol. 18, No. 48, pp. 6621-6634, ISSN 0950-9232

Yin, X.Y, Grove, L.E. & Prochownik, E.V. (2001) Mmip-2/Rnf-17 enhances c-Myc function and regulates some target genes in common with glucocorticoid hormones. *Oncogene*. Vol. 20, No. 23, pp. 2908-2917, ISSN 0950-9232

Zlotnik, I. (1947) The cytplasmic components of germ-cells during spermatogenesis in the domestic fowl. *Q. J. Microsc. Sci*. Vol. 88, No. 3, pp. 353-365, ISSN 0370-2952

6

OMICS for the Identification of Biomarkers for Oocyte Competence, with Special Reference to the Mare as a Prospective Model for Human Reproductive Medicine

Maria Elena Dell'Aquila[1], Yoon Sung Cho[2], Nicola Antonio Martino[1], Manuel Filioli Uranio[1], Lucia Rutigliano[1] and Katrin Hinrichs[3]

[1]*Dept. Animal Production, University of Bari "Aldo Moro", Bari,*
[2]*Assisted Procreation Unit - Ospedale Accreditato Clinica Santa Maria, Bari,*
[3]*Dept. Physiology and Pharmacology, Texas A&M University, College Station,*
[1,2]*Italy,*
[3]*USA*

1. Introduction

A crucial component of Assisted Reproductive Technologies (ART) is the assessment of oocyte developmental potential, to allow selection of those oocytes most likely to result in fertilization and pregnancy. Currently, oocyte quality assessment is largely based on the morphological appearance of the cumulus-oocyte complex, however the accuracy of morphological methods, as predictive of oocyte competence, is still suboptimal. Therefore, the development of objective, accurate, fast and reliable tests for assessing oocyte developmental potential remains an important aim of human and veterinary reproductive medicine. The process of oocyte meiotic maturation, which is central to the developmental competence of the oocyte, is regulated by numerous genes (Matzuk & Lamb, 2008; Fauser et al., 2011) and protein pathways (Kubiak, 2011) and is accompanied by significant changes within the oocyte at many levels. Better understanding of oocyte meiotic maturation would allow better support of this process to increase the success of reproductive biotechnologies, and thus overcome some forms of infertility. Recently, global assessment strategies, namely OMICS, investigating genomic, transcriptomic, proteomic, lipidomic, and glycomic profiles of oocytes, cumulus or granulosa cells have become increasingly applied to the study of oocyte physiology and pathology. Also being investigated is the oocyte-cumulus metabolome, via measurements of metabolites in biological fluids, such as follicular or tubal fluid, or in culture media. The establishment of these technologies, which are in their initial stages of application to reproductive biology, can require large sample numbers; only animal models can meet this requirement. Because of their wide availability and the body of existing knowledge regarding their biology, oocytes of large animals provide useful models for investigating the relationship between oocyte developmental competence and OMICS biomarkers. This review summarizes recent literature on the application of OMICS strategies to evaluating developmental competence of human oocytes and oocytes of large animals. Among the available animal models, the mare is uniquely applicable to investigation of oocyte developmental competence. Horses represent the most economically valuable domestic animal, with progeny from specific mares worth hundreds of thousands of euros. Thus, there is obvious practical interest in the use of assisted reproduction in this species. In addition, similarities between equine and human follicle growth and oocyte maturation make the mare a particularly valuable model for topics at the interface between animal breeding and biomedical research, such as age-related and obesity-related oocyte dysfunction and the effects of exposure to environmental toxicants, as well as for fundamental research on factors involved in meiotic maturation. For these reasons, particular attention will be dedicated in this review to recent OMICS results obtained in the equine species and to discussion of the potential application of this animal model in future investigations.

2. OMICS as innovative strategies for evaluating oocyte quality

Systems biology is a new and rapidly developing research area in which, by quantitatively describing the interactions among the components of a cell, a systems-level understanding of a biological response can be achieved. Therefore, it requires high-throughput measurement technologies, that is, technologies that can investigate a large number of biological molecules at once. OMICS technologies -- in which aspects of cellular structure or function, such as proteins or RNA transcripts, are studied in their totality (global assessment strategies) -- are opening wider and wider doors into the understanding of all branches of the biology, physiology, and pathology of living organisms. It is likely that information obtained using OMICS will change our concept of "normal" and "pathological," and will enable the efficient evaluation of the effects of extrinsic factors on the status of living systems. Initial studies on the application of OMICS strategies to the oocyte have appeared in the past decade, starting with genomics and transcriptomics, and progressing to the newer fields of glycomics and metabolomics. As noted above, a major concern in the production of viable and competent embryos in vitro is the evaluation of initial oocyte quality and the support of optimal nuclear and cytoplasmic maturation. OMICS approaches to the oocyte will significantly contribute not only to accurate assessment of oocyte quality, but also to the clarification of the mechanisms involved in cell cycle regulation and cell differentiation, thus contributing to the effective utilization of recovered oocytes. Because meiotic maturation and early embryo development involve regulation of the cell cycle and evolution from differentiated, to pluripotent, back to differentiated cells, data generated from study of these processes may also relate to the establishment of innovative targeted cancer treatments and stem cell-based therapies. To introduce the sequential phases of the meiotic process, changes occurring within the oocyte and some of their fundamental regulating factors are briefly described. Upon the luteinizing hormone (LH) surge, M-phase promoting factor (MPF) and the mitogen-activated protein (MAP) kinase ERK 2 (extracellularly-regulated kinase 2) are activated within the oocyte. The oocyte, which at this time is in prophase of meiosis, has replicated chomosomes contained within a nucleus (termed the germinal vesicle, GV). Activation of MPF and ERK 2 trigger nuclear envelope breakdown and chromatin condensation. The condensed chromosomes are subsequently aligned on the spindle of the first meiotic division, forming the metaphase I plate. At this time MPF levels decrease, while ERK 2 levels remain high. The homologous chromosomes

separate: one set of sister chromatids is discarded as the polar body; the other set, in
response to a recrudescence of MPF, lines up on a spindle, forming the second metaphase
plate (MII). Fertilization causes inactivation of MPF and ERK 2, and the second meiotic
division occurs, separating the sister chromatids. One set of chromatids is discarded, as the
second polar body; the other set becomes the female pronucleus. For a comprehensive
description of major pathways involved in oocyte M-phase entry, see Kubiak et al., 2011 and
Tosti & Boni, 2011. Detailed descriptions of OMICS techniques go beyond the intent of this
article; corresponding references are provided in the text. We instead focus on the most
significant results obtained using these techniques, the role of large animal models in
experimental designs that cannot be performed in humans, and on actual and potential
contributions of different animal models to understanding of oocyte biology, with particular
interest in the equine species.

2.1 The mare as a model for human oocyte biology

Large animal models allow the establishment of a wide variety of experimental designs that
can not be applied in humans for obvious ethical reasons, or due to the limited and highly
regulated availability of human biological samples. Among large animals, the mare has
many attributes that make her a good model for reproduction in women (Carnevale, 2008).
These include a long follicular phase, a long interovulatory interval (22 days), presence of a
single dominant follicle, formation a large diameter follicle (~40 mm) with a large volume of
follicular fluid -- the same volume:body weight ratio as in women; a relatively long time
from LH stimulation to ovulation (36 h for both human and horse) and, like the human
oocyte, formation of a markedly dense chromatin mass within the germinal vesicle as the
oocyte gains meiotic competence or undergoes atresia (Parfenov et al, 1989; Hinrichs et al.,
1993). Although seasonality does not occur in women, the equine characteristic of seasonal
reproductive activity provides the potential to examine the influence of applied factors
when cyclic hormonal patterns are not occurring (Carnevale, 2008). Horses are the best
animal model for studies on age-related infertility. Because mares can be of great value,
many mares continue to be bred until they experience subfertility, thus animals with
naturally-occuring age-related subfertility are available for study. Horses have a long
lifespan, thus age-related subfertility occurs at an age (~20-25 years) much closer to that
observed in women than is seen in other animal models. In addition, horses, unlike other
large domestic species, are selected for attributes other than fertility, such as conformation,
athletic prowess, or behavior. Individuals showing subfertility may be worked with
intensively to try to obtain foals, thus, they provide an excellent naturally-occurring model
for many intrinsic causes of subfertility. Horses have similar metabolic responses to nutrient
intake to that in humans, and are used for a wide variety of athletic purposes, thus they can
serve to model important physiological or pathological situations affecting reproduction in
humans (such as stress, life-style, sports activity, obesity or metabolic syndrome) as well as
to examine the effects of external factors such as acute or long-term exposure to drugs or
environmental toxicants. In addition to mimicking the situation in humans, the
development of particularly large follicles allows the possibility of collecting large amounts
of mural granulosa cells (GC) issuing from the follicular wall as well as large amounts of
follicular fluid (30 to 50 ml/follicle) that may be used for OMICS studies in a 1:1 comparison
with the developmental status of the enclosed oocyte. The cumulus-oocyte complex (COC)

of the mare is particularly large, thus allowing a 1:1 evaluation of biological parameters of cumulus cells (CCs) predictive of oocyte meiotic and developmental competence. Moreover, follicular COCs in the mare can be recovered from immature follicles, with initial COC morphological features indicating viability (compact cumulus) or atresia (apoptosis, expanded cumulus) of their surrounding follicle, thus supporting study of the effects of follicle immaturity and early or late atresia on oocyte competence (Hinrichs and Williams, 1997; Dell'Aquila et al., 2003). The equine oocyte is approximately 200 microns in diameter, with good visibility of the perivitelline space and the first polar body (PB), so that mature oocytes are easily identified on morphological examination. The horse oocyte possesses a unique distribution of cytoplasmic lipid droplets, which assume polar aggregation in metaphase II (MII) oocytes, and whose biological meaning is under investigation (Ambruosi et al., 2009). Horses make a valuable model for oocyte assessments associated with penetration of the zona pellucida (e.g. PB biopsy) because, in contrast to species such as cattle and sheep, methods for fertilization via intracytoplasmic sperm injection (ICSI) are well established in the horse (Hinrichs et al., 2005; Choi et al., 2006). Sperm injection is necessary to achieve fertilization after penetration of the zona pellucida for investigative purposes, as the defect in the zona would lead to polyspermy if standard in vitro fertilization (IVF) were to be performed. These features make the equine oocyte a particularly useful model for the establishment of OMICS strategies that could be not only applied to better understanding of human assisted reproductive medicine, but also directly applicable to the horse industry.

2.2 Oocyte genomics – The polar body biopsy and genomic analysis for predicting half of the DNA constitution of an embryo: from FISH to CGH/CNV/SNP-based arrays

The genomic DNA constitution of the oocyte determines the sequence of produced transcripts and proteins, and constitutes half of the early embryo phenotype. The most widely used diagnostic tool for oocyte and embryo genomic investigation to date has been Fluorescent In Situ Hybridization (FISH), a cytogenetic technique which identifies specific DNA sequences on chromosomes by means of fluorescent probes that bind to those parts of the chromosome with which they show a high degree of sequence similarity.

Since the first report in humans (Griffin et al., 1992), several studies have been published reporting the evaluation of human day 3 embryos (4-8 cell stage) for up to 8 pairs of chromosomes (chromosomes 13, 15, 16, 17, 18, 21, 22, and X/Y; review by Seli et al., 2010). However, in recent meta-analysis studies, reported from 2008 to 2010, it has become clear that preimplantation genetic screening by using FISH is not justified. This is because it causes damage to the embryo, it requires embryo cryopreservation and transfer in a subsequent cycle, and it does not significantly contribute to the identification and exclusion of aneuploid embryos. Use of FISH has been reported to be associated with lower implantation rates and it shows errors such as false positives due to mosaicism or false negatives due to the limited number of chromosomes analyzed and the limited targeted regions. Therefore, authoritative scientific committees, including the European Society of Human Reproduction and Embryology (ESHRE) and the American Society of Reproductive Medicine (ASRM), decided to conduct a study to determine whether biopsy of the first and second polar bodies (PBs) of the oocyte would enable the timely identification of the chromosomal status of an oocyte. This aim could be reached by analyzing the complete chromosome complement of the two PBs by Comparative Genomic Hybridization (CGH; reviews by Seli et al., 2010; Geraedts et al., 2010).

Conventional CGH, initially developed by cancer biologists, was applied to human early embryos around 10 years ago (Voullaire et al., 2000 reviewed by Geraedts et al., 2010). These authors performed CGH in association with whole genome amplification by degenerate oligonucleotide-primed polymerase chain reaction. CGH is a competitive hybridization of two fragmented genomes (test and reference genomes) to the chromosomes of a metaphase plate of a normal subject. The tested and subject genomes are labeled with different (red and green) fluorescent dyes so that an increase of red staining will indicate the presence of duplicated regions, an increase of green staining will indicate the presence of deleted regions, whereas the lack of predominance of one of the two colors will indicate normal chromosome structure. This technique allows examination of the whole chromosomal complement, but requires extensive time to get the results. In recent years, CGH-microarray tools have been developed in which the labeled DNAs are affixed to DNA on a microscope slide rather than to metaphase chromosomes. A variety of microarray-CGH platforms are available. As an example, the Cambridge-based company BlueGnome offers an array-based CGH protocol which allows analysis of biopsied PBs within 11 hrs (SurePlex amplification protocol; 24sure analysis, BlueGnome; "http://www.bluegnome.co.uk/"; Geraedts et al., 2010). As regard, Geraedts et al., (2011) and Magli et al., (2011) reported clinical results and technical aspects of a proof-of-principle study performed in associated ART centers in which all mature metaphase II oocytes from patients who consented to the study, fertilized by ICSI, were analyzed. The first and second PBs were biopsied and analysed separately for chromosome copy number by array CGH. If either or both of the PBs were found to be aneuploid, the corresponding zygote was then also processed by array CGH for concordance analysis. It was concluded that the ploidy of a zygote can be predicted with acceptable accuracy by array CGH analysis on both PBs. Interestingly, on the male side, the application of CGH arrays to single human sperm cells has been recently reported (Antonello et al., 2011).

In the aim to move from chromosomal structure to single mutation analysis, SNP (Single Nucleotide Polymorphism) arrays have been developed. A single-nucleotide polymorphism (SNP, pronounced snip) is a DNA sequence variation occurring when a single nucleotide — Adenine (A), Thymine (T), Cytosine (C), or Guanine (G) — differs in a sequence between members of a species or paired chromosomes in an individual. Many common SNPs have only two alleles. Within a population, SNPs can be assigned a minor allele frequency — the lesser of the two allele frequencies for a population. There are variations among human populations, so a SNP allele that is common in one geographical or ethnic group may be rare in another. Unlike the CGH-microarray platforms which involve simultaneous hybridization of differentially labeled DNAs to the same microarray, SNP-microarrays assess test and reference samples, separately, in parallel. From 10.000 to 500.000 SNPs may be evaluated simultaneously. For example, using the Affymetrix platform, analysis of 250.000 SNPs in first PB biopsies (Treff et al., 2010a) and in Day-3 embryos (Treff et al., 2010b) have been reported (review by Seli et al., 2010). The Illumina platform allows the analysis of 370.000 human SNPs.

Because of their utility in recognizing variations associated with disease, recent genetic epidemiology studies have been dominated by genome-wide association studies using

SNPs. However, another form of structural genomic variation, termed copy number variation (CNV), is also widespread throughout the genome. These genomic structural variations range from 1 to 5 Mb and can be highly polymorphic between individuals, and thus can be used for epidemiological study. CNVs in the form of large-scale insertions and deletions, as well as inversions and translocations, may have important roles in meiotic recombination, human genome evolution and gene expression. Many genetic diseases are based on CNVs. However, because they consist of quantitative rather than qualitative changes, show variability in copy numbers and are confounded by the diploidy of the human genome, the detailed genetic structure of CNVs cannot be readily studied by available techniques. Thus, the establishment CNV-microarrays is currently under investigation. New microarray-based technologies will enable more accurate mapping of CNVs, and CNV maps of the human genome are being refined with increasing resolution. The study of CNVs and their effects on human health and disease therefore present a dynamic and exciting challenge for researchers in the field of genetic epidemiology (Wain and Tobin, 2011). The importance of CNVs in human preimplantation genetic screening, or to animal oocyte and embryo testing, has not been reported to date.

Although there is potential economic interest for the application of PB biopsy and subsequent analysis of the chromosomal complement or genome by CGH-, CNV- or SNP arrays in animal husbandry and breeding, to the best of our knowledge no studies have been published to date in large animals. The field is therefore open to future investigations, pending the establishment of the different genomic arrays in these species. A recent report (Le Bourhis et al., 2011) presented for the first time bovine embryo biopsy and genotyping using a 50K SNP Illumina chip. In this study, biopsies of 5 to 10 cells were obtained from *in vitro*-cultured morulae and blastocysts and kept frozen or at room temperature. The genomic DNA of each biopsy was amplified by using a whole-genome amplification kit and was genotyped using a custom CRV 50K Illumina chip. Call rates were calculated from 50.905 SNPs. Percentage of allele drop-out was estimated from the number of heterozygous markers present [% allele drop-out = (calculated heterozygous–observed heterozygous) /calculated heterozygous]. Parentage error was estimated by using the genotypes of the parents of the embryos. A greater quantity of DNA was obtained after amplification of biopsies that were sent frozen to the laboratory than from those at room temperature ($P<0.05$). However, the SNP call rate, % allele drop-out, and parentage error did not differ between groups. These results indicate that genotyping from embryo biopsies following whole genome amplification can be achieved with good efficiency when using high-density marker chips. To the best of our knowledge no studies have been reported to date on genomic analysis in the equine oocyte. A recent paper in the horse by Choi et al., (2010) reports the identification of disease-causing mutations in trophoblastic biopsies from equine in vivo-recovered pre-implantation embryos. These authors demonstrated for the first time the correct identification, by embryo biopsy and whole genome amplification, of sex and genotype at the causative mutation sites for two disease-linked genes (SCN4A and PPIB). The biopsies were performed on Day-6 and Day-7 equine embryos, and after biopsy these embryos were able to produce pregnancies leading to term delivered, normal foals. These two recent studies demonstrate that OMICS technologies have the potential in animal breeding for both marker-assisted selection and for preimplantation diagnosis of genetic diseases.

Another promising genomic investigation area is Epigenetics, the study of changes in gene expression and thus cellular phenotype caused by mechanisms other than changes in the underlying DNA sequence – hence the name *epi-* (Greek: επι- over, above, outer) *genetics*. Examples of such changes are DNA methylation and deacetylation of the histones, the proteins around which DNA are wrapped. Both of these changes serve to suppress gene expression without altering the sequence of the silenced genes. These changes may remain through cell divisions for the remainder of the cell's life, and some epigenetic changes in germ cells may potentially last for multiple generations. Epigenetic changes in eukaryotic biology are the basis of the process of cell differentiation. During embryonic morphogenesis, the totipotent cells of the zygote become the various pluripotent cell lines of the embryo, which in turn become fully differentiated cells. This is accomplished by activating some genes while inhibiting others. Current epigenetic research focuses on chromatin modifications occurring during sequential phases of fertilization (sperm chromatin decondensation, pronuclear formation with DNA duplication and syngamy) and early development (chromosome condensation and assembly in the first metaphase plate of the first mitotic division and the subsequent series of mitotic divisions to the blastocyst stage; Burton & Torres Padilla, 2011). These events may be studied by comparing embryos produced in vivo with those obtained using different technologies, such as IVF, ICSI, parthenogenesis or somatic cell nuclear transfer (Cremer & Zakhartchenko, 2011). Soon, the emergence of quantitative high-throughput microarray technology should allow the development of epigenomic arrays for the evaluation of embryo whole-genome epigenetic status, thus opening the new field of epigenomics (Callinan & Feinberg, 2006) to the study of oocyte and embryo competence.

The methylation pattern of DNA in oocytes may be a key factor for the improvement of efficiency of in vitro embryo production, because it is related to oocyte competence. A recent study (Simarro Fagundes et al., 2011) reported on a differentially-methylated region located in exon 10 of the imprinted gene *IGF2* This study evaluated immature vs *in vitro*-matured bovine oocytes from small (1–3 mm in diameter) and large follicles (≥ 8.1 mm in diameter). It was observed that after IVM, oocytes from ≥ 8.1 mm follicles were less methylated (18.51%) than were those from 1- to 3-mm follicles (49.62%). As oocytes from the larger follicles are more developmentally competent, the less methylated pattern appears to be associated with higher oocyte quality. It was concluded that the methylation pattern of specific genes could be used as a molecular marker for epigenetic reprogramming status in oocytes, helping the development of new in vitro embryo production protocols. A broader study on this wave (Smallwood et al., 2011) reported the first integrated epigenomic analysis of mammalian oocytes (GV vs MII oocytes) and preimplantation embryos (blastocyst stage) identifying over a thousand CG islands methylated in matured oocytes. The authors observed that CG islands were preferentially located within active transcription units, supporting a general transcription-dependent mechanism of methylation, and that very few CG islands were protected from post-fertilization reprogramming, the majority showing incomplete demethylation in Day-3 blastocysts. This study revealed the extent and dynamics of CG island methylation in oocytes, which is a prerequisite for defining the full repertoire of imprinted genes and the mechanistic basis of parent-of-origin expression effects in somatic tissues.

Epigenomic studies have not been reported to date in equine oocytes; however, in the promising field of genomic investigations, the equine oocyte would serve as an excellent model for the comparison of oocyte (metaphase plate) and polar body genomes and epigenomic modifications, due to the ability to investigate the developmental competence of biopsied equine oocytes after fertilization via ICSI.

2.3 Oocyte transcriptomics – The global analysis of oocyte mRNA transcripts

In addition to the genomic constitution of the embryo, it is necessary to know its phenotype: which genes are being utilized at this particular stage of development? Initially, embryonic phenotype is determined by those mRNAs already transcribed and present in the oocyte cytoplasm at the time of fertilization (maternal mRNAs). Evaluation of maternal mRNA content is particularly attractive in the study of developmental biology and for diagnostic and applied purposes in ART (e.g. nuclear reprogramming in cloning, and stem cell research). However, transcriptome analysis of mammalian oocytes and embryos faces three main challenges: 1) the small amount of material available; 2) differing total RNA content in the subsequently-occurring developmental stages, making comparison among stages difficult; 3) existence of oocyte-specific genes often absent from commercially available microarrays (Dalbies-Tran and Mermillod, 2003; Thelie et al., 2009).

Via transcriptomics, it is possible to thoroughly investigate the functional status of a cell line or tissue. Rapidly developing methods consist of RNA extraction, reverse transcription (RT), amplification and labeling, array hybridization, chip scanning, and data interpretation by bioinformatic analysis with subsequent validation by Real Time RT-PCR. Detailed reviews of microarray analysis strategies and interpretation of transcriptomic profiles have been presented by White and Salamonsen, 2005 and Rodriguez-Zas et al., 2008. Some of the public or commercially available software commonly used for trascriptome analysis are:

Public:

- GENE ONTOLOGY: http://www.geneontology.org;
- NCBI Entrez Gene: http://www.ncbi.nlm.nih.gov.sites/entrez?db=gene);
- NCBI Gene Expression Omnibus GEO: http://www.ncbi.nlm.nih.gov/geo;
- KEGG pathway database: http://www.genome.jp/kegg/pathway.html;

Commercially available:

- INGENUITY pathway analysis http://www.ingenuity.com;
- PANTHER Applied Biosystem http://www.pantherdb.org);
- AFFYMETRIX (http://www.affymetrix.com/products/arrays/specific/bovine.affx).

A major aim in oocyte transcriptomics is the analysis of differences among maturation stages, especially between the germinal vesicle (GV) and the metaphase II (MII) stage, as well as differential expression between in vivo- (in vivo-MII) and in vitro- (IVM-MII) matured oocytes. The correct molecular control of meiotic maturation is a fundamental prerequisite for successful development of an early embryo (Tosti & Boni, 2011). Transcriptome microarray technologies have been developed, first in the mouse and more recently in large animals (review by Thelie et al., 2009). At the moment, cattle take center stage in the cast of large animals used as models for human reproductive medicine.

2.3.1 Studies in bovine oocytes

GV vs MII oocyte A pioneering study on oocyte gene expression was conducted at INRA (France) by Dalbies-Tran and Mermillod, in 2003. These authors analyzed gene expression in bovine oocytes before and after IVM, using heterologous hybridization onto a cDNA array. Total RNA was purified from pools of over 200 oocytes either immediately after aspiration from follicles of slaughterhouse cow ovaries, or following IVM. Radiolabeled cDNA probes were generated by RT followed by linear PCR amplification and were hybridized to Atlas human cDNA arrays. To the best of our knowledge, this was the first report of gene expression profiling by this technology in the bovine oocyte. The results demonstrated that cDNA array screening is a suitable method for analyzing the transcription pattern in oocytes, as about 300 identified genes were reproducibly shown to be expressed in the bovine oocyte. The relative abundance of most messenger RNAs appeared stable during IVM; however, it was observed that 70 transcripts underwent a significant differential regulation between meiotic stages (by a factor of at least two). Information obtained in this study constituted the first molecular signature of oocyte cytoplasmic maturation.

GV oocyte vs embryo In a subsequent study at INRA (Thelie et al., 2009) results of an RNA-amplification protocol for bovine oocytes and blastocysts was reported. Using RT-PCR, these authors confirmed that the profiles of both abundant and scarce polyadenylated transcripts were conserved after RNA amplification. Next, amplified probes generated from immature oocytes, in vitro-matured oocytes, and in vitro-produced hatched blastocysts were hybridized onto an in-house cDNA macroarray that included oocyte-specific genes (934 expressed sequence tags of interest including markers of oocyte maturation; Thelie et al., 2009). Following an original approach, two normalization procedures, based on either the median signal or an exogenous standard, were compared and the expected difference in sets of differential genes, depending on the normalization procedure, were calculated. Using a 1.5-fold threshold, no transcript was found to be up-regulated when data were normalized to an exogenous standard, which reflects the absence of transcription during oocyte IVM. In blastocysts, the majority of genes found to be preferentially expressed in oocytes (after normalization) were not activated. This study shed new light on and complemented previous transcriptomic analyses of the bovine oocyte-to-embryo transition using commercial platforms (i.e.: Misirlioglu et al., 2006; Fair et al., 2007; reviewed by Thelie et al., 2009).

In vivo-MII vs IVM-MII oocyte The differences in the MII oocyte transcriptome between oocytes matured in vivo and in vitro were investigated in cattle by Katz-Jaffe et al., (2009). In this study, the Affymetrix Gene Chip Bovine Genome Array, a platform containing over 23.000 bovine transcripts, was used. Transcripts identified as being differentially expressed between the two groups were classified according to gene ontology. Statistical analysis of microarray data identified several processes affected by IVM, including metabolism, energy pathways, cell biogenesis and organization, and cell growth and maintenance. In particular, it was found that 4 genes of the tricarboxylic acid cycle and 14 genes of oxidative phosphorylation were down-regulated in IVM-MII compared with in vivo-MII.

GV vs MII oocyte Mamo et al., (2011) used the Affymetrix GeneChip Bovine Genome Array to perform global mRNA expression analysis of immature (GV) and in-vitro matured (IVM) bovine oocytes. They then used a variety of approaches, including the analysis of transcript abundance in oocytes matured in the presence of alpha-amanitin (a transcription inhibitor), to determine whether the transcriptional changes observed during IVM were real or were artifacts of the techniques used during analysis. It was found that 8489 transcripts were detected across the two oocyte groups, of which ~25.0% (2117 transcripts) were differentially expressed (p<0.001); corresponding to 589 over-expressed and 1528 under-expressed transcripts in the IVM oocytes compared to their immature counterparts. Subsets of the differentially expressed genes were validated by quantitative RT-PCR and the gene expression data was classified according to gene ontology and pathway enrichment. Numerous cell-cycle linked (CDC2, CDK5, CDK8, HSPA2, MAPK14, TXNL4B), molecular transport (STX5, STX17, SEC22A, SEC22B), and differentiation (NACA) related genes were found to be among the over-expressed transcripts in GV oocytes compared to their mature counterparts, while other genes (ANXA1, PLAU, STC1and LUM) were among the over-expressed genes after maturation. This data set provided a unique reference resource for studies of the molecular mechanisms controlling oocyte meiotic maturation in cattle, and by extension to other species, and through use of the alpha-amanitin, addressed the existing conflicting issue of transcription during meiotic maturation.

Adult vs prepubertal oocyte (Romar et al., 2011) This study, rather than applying global OMICS strategies, analyzed the differential expression profile of adult vs prepubertal bovine oocytes by using a specialized panel for genes involved in the maturation process, such as genes known to specifically affect early development after fertilization (maternal-effect genes, shown via mouse knock-outs), biomarkers of oocyte competence or redox metabolism, or genes involved in the regulation of meiotic progression. It was found that some genes (particularly redox genes) are significantly underexpressed in oocytes from prepubertal subjects. This kind of comparison would benefit greatly by using OMICS technologies and underscores the value of animal models, as it would be difficult to perform in humans due to low availability of oocytes from young girls and women.

2.3.2 Studies in human oocytes

Transcriptomic studies have been reported in human oocytes. Kocabas et al., 2006 reported the use of a comprehensive human microarray platform (Affymetrix Human Genome U133 Plus 2.0 GeneChips) to identify the gene transcripts present in **early MII oocytes**, tested within minutes after isolation from the ovary.

In the study by Wells and Patrizio (2008), unfertilized **GV, in vivo-MII** and **IVM-MII** oocytes were analyzed. The study used the Applied Biosystem Human Genome Survey Microarray with 32.878 60mer oligonucleotide probes for the interrogation of 29.098 genes, including 8000 genes not previously included in any commercial array. By bioinformatic analysis, a Venn diagram can be obtained in which each circle represents the transcriptome of a specific cell type, and overlapping areas indicate commonly-expressed genes. The three oocyte categories expressed 12.219, 9.735 and 8.510 genes, respectively. There were extensive overlaps among the three groups, but also some significant differences. In particular, in vivo-MII and IVM-MII oocytes shared similar patterns of gene expression. However, some immature patterns of expression, reminiscent of GVs, persisted in IVM-MIIs. In humans, in-vitro maturation is an attractive strategy for IVF treatment; however, currently IVM oocytes perform poorly after IVF. Data from the this study indicates that although IVM-MII oocytes closely resemble in vivo-MII oocytes in cellular pathways related to nuclear maturity, several pathways associated with cytoplasmic functions continue to be expressed in an

OMICS for the Identification of Biomarkers for Oocyte Competence, with Special Reference to the Mare as a Prospective Model for Human Reproductive Medicine

103

immature manner. Additionally, it was shown that IVM-MII oocytes differ in the expression of genes related to cellular storage and homeostasis. Such differentially expressed genes and their pathways provide clues for the optimization of IVM techniques, and, importantly, a method to assess the effects of those techniques on oocyte competence without having to evaluate development after fertilization, which could raise ethical issues in humans.

In vivo-MII vs IVM-MII In the study by Jones et al., (2008), more than 2000 genes were identified as expressed at more than 2-fold higher levels in oocytes recovered from gonadotropin-stimulated cycles and matured in vitro than those matured in vivo, and 162 of these were expressed at 10-fold or greater levels (this study used Codelink Whole Human Genome Bioarrays, GE Healthcare Biosciences). It was concluded that the overabundance of transcripts in immature oocytes recovered from gonadotropin stimulated cycles, then matured in vitro could be due to dysregulation of either gene transcription or post-transcriptional modifications, resulting in incorrect temporal utilization of genes, culminating in developmental oocyte incompetence.

GV vs IVM-MII vs embryo Zhang et al., 2009a followed the transcriptome changes occurring in human preimplantation development by applying microarray analysis (Affymetrix two-cycle GeneChip Eukaryotic small sample target labeling assay version II; HG-U133 Plus 2.0 array) to human oocytes and embryos at six developmental stages. They observed a dramatic reprogramming of transcription and translation during preimplantation development in a stage-specific manner, with two main transitions (MII to Day 2 and Day 3 to Day 5). Over 47.000 transcripts expressed in oocytes and early embryos were reported, thus providing a fundamental resource for understanding the genetic control of human early development. There was a significant underrepresentation of transcripts responsible for cell signaling and communication (genes associated with the G protein coupled receptor - GPCR - protein signaling pathway, cell communication, immune response, response to external stimuli, cell adhesion, sensory perception and cell-surface receptor-linked signal transduction pathways) in both oocytes and embryos, when compared to adult tissue; the authors concluded that human preimplantation development is almost self-directed -- i.e., oocytes and embryos apparently do not need to communicate with the "external world" to the same degree as adult tissues do. This paper also performed evolutionary comparisons between humans and mice, dogs and chimpanzees. Genes that were highly expressed in human oocytes and embryos varied less from those of other species than did genes of adult tissues: the conclusion was that these "pre-implantation genes" are highly conserved.

Microarray analysis of human oocytes has been subsequently applied to a variety of reproductive issues. Wood et al., (2007) found differences in gene expression between **normal** and **PCOS** (polycystic ovarian syndrome) oocytes for 8123 transcripts, 374 of which were genes related to meiotic spindle dynamics. Grondahl et al., (2010), evaulated 15 independent replicates of single **in vivo-MII** oocytes using the Affymetrix HG-U133 Plus 2.0 gene chip array, which tests around 48.000 well identified genes by using around 56.000 probe sets, and the Affymetrix gene array 2500 scanner. These authors identified 7.470 genes (10.428 transcripts) as present in human in vivo-MII oocytes. Of these, 342 genes showed a significantly different expression level between **young** and **aged** women; notably, genes annotated to be involved in cell cycle regulation, chromosome alignment (e.g. MAD2L1 binding protein), sister chromatid separation (e.g. separase), oxidative stress and ubiquitination. The top signaling network affected by age was 'cell cycle and organism development' (e.g. SMAD2 and activin B1 receptor). Thus, this study provided information on processes that may be associated with lowered oocyte developmental competence due to ageing.

2.3.3 The common gene expression signatures of oocytes and embryonic stem cells

Another interesting research area for transcriptomic investigations is the comparison of the transcriptomic profile between MII oocytes and stem cells. Data have been published on the comparison between human **MII oocytes** and human **embryonic stem cells (ESC)**. Both of these cells types are able to reprogram differentiated nuclei towards pluripotency, either by somatic cell nuclear transfer or by cell fusion, respectively. Comparison of the transcriptome of these two cell types may highlight genes that are involved in induction of pluripotency. Based on a microarray compendium of 205 samples, Assou et al., 2009 compared the gene expression profile of MII oocytes and human ESC to that of somatic tissues. A common oocyte/hESC gene expression profile was identified, which included a strong cell cycle signature, genes associated with pluripotency such as *LIN28* and *TDGF1*, a large chromatin remodelling network (*TOP2A, DNMT3B, JARID2, SMARCA5, CBX1, CBX5*), 18 different zinc finger transcription factors, including *ZNF84*, and several still-poorly annotated genes such as *KLHL7, MRS2*, and Selenophosphate synthetase 1 (*SEPHS1*). Interestingly, a large set of genes in both cell types was found to code for proteins involved in the ubiquitination-proteasome pathway. Upon ESC differentiation into embryoid bodies, the transcription of genes in this pathway declined. In vitro, a selective sensitivity of human ESC to inhibition of proteasome activity was observed. These results shed light on the gene networks that are concurrently overexpressed by the two cell types with somatic cell reprogramming properties.

2.3.4 Prediction of oocyte competence based on analysis of accessory cells (polar bodies, cumulus cells or granulosa cells)

A major problem of reproductive biotechnologies is predicting which oocytes are destined to develop into viable embryos. Analysis of accessory cells, such as PBs, CCs and GCs, allows oocyte quality assessment without interfering with use of the oocyte in ART.

Polar Body Klatsky et al., (2010) reported detection and quantification of mRNA from single human polar bodies, a minimally invasive test of the oocyte gene-expression profile. Gene expression of 12 candidate genes was investigated in PB biopsies and the oocytes from which they originated, and polar-body mRNA was detected for 11 out of 12 genes. This method would allow detection and comparison of individual differences in oocyte gene expression without harming the oocyte.

Granulosa cells The comparative evaluation of the effects of FSH vs human menopausal gonadotrophin on GCs has been reported (Grondahl et al., 2009). These authors found that the drugs used for controlled ovarian hyperstimulation have a significant impact on the gene expression profile of human granulosa cells. Interesting differences were observed for genes involved in the regulation of preovulatory events. For GC in the mare, see the work by Fahiminiya et al., 2010 in section 2.3.5

OMICS for the Identification of Biomarkers for Oocyte Competence, with Special Reference to the Mare as a
Prospective Model for Human Reproductive Medicine

105

Cumulus cells The bi-directional communication between the oocyte and its companion CCs is crucial for the development and function of both cell types. Regassa et al. (2001) investigated the transcripts that are exclusively expressed either in oocytes or in CCs, and the molecular mechanisms affected when communication between the two cell types is removed. The transcriptomic profile of different oocyte and CC samples was analyzed by using Affymetrix GeneChip Bovine Genome array containing 23000 transcripts. Out of 13162 genes detected in GV oocytes and their companion CCs, 1516 and 2727 were exclusively expressed in oocytes and in CCs, respectively, while 8919 were expressed in both. Similarly, of 13602 genes detected in MII oocytes and CCs, 1423 and 3100 were exclusively expressed in oocytes and in CCs, respectively, while 9079 were expressed in both. A total of 265 transcripts were differentially expressed between oocytes cultured with (OO+CCs) and without (OO-CCs) CCs, of which 217 and 48 were over-expressed in the former and the latter groups, respectively. Similarly, 566 transcripts were differentially expressed when CCs were cultured with (CCs+OO) or without (CCs-OO) their enclosed oocytes. Of these, 320 and 246 were over-expressed in CCs+OO and CCs-OO, respectively. While oocyte-specific transcripts include those involved in transcription (IRF6, POU5F1, MYF5, MED18) and translation (EIF2AK1, EIF4ENIF1), CC-specific transcripts include those involved in carbohydrate metabolism (HYAL1, PFKL, PYGL, MPI), protein metabolic processes (IHH, APOA1, PLOD1) and steroid biosynthetic process (APOA1, CYP11A1, HSD3B1, HSD3B7). Similarly, while transcripts over expressed in OO+CCs were involved in carbohydrate metabolism (ACO1, 2), molecular transport (GAPDH, GFPT1) and nucleic acid metabolism (CBS, NOS2), those over expressed in CCs+OO were involved in cellular growth and proliferation (FOS, GADD45A), cell cycle (HAS2, VEGFA), cellular development (AMD1, AURKA, DPP4) and gene expression (FOSB, TGFB2). This study generated large-scale gene expression data that provide insights into gene function and interactions within and across different pathways that are involved in the maturation of bovine oocytes. Moreover, the presence or absence of oocyte and CC factors during bovine oocyte maturation has a profound effect on transcript abundance in the different cell types, showing the important molecular cross-talk between oocytes and their CCs. This kind of study has not yet been performed in humans.

A more recent study reported the transcriptomic analysis, by using the Affymetrix Bovine Expression Array, of granulosa cells and oocytes from newborn sheep ovaries (primordial, primary, secondary and small antral follicles) isolated by Laser Capture Microdissection (Bonnet et al., 2011). This study will significantly support clinical programs for rescue of fertility (oocyte production potential) in young women affected by ovarian pathologies or undergoing cancer therapy.

2.3.5 Preliminary data and prospective use of equine oocytes

The equine oocyte would make a valid animal model for transcriptomic studies of predictive markers of oocyte quality via analysis of PBs, CCs and GCs. Its peculiarly large PB size and the unique opportunity, due to large follicle size, to perform 1:1 oocyte:somatic cell ratios could allow reliable identification of predictive parameters of oocyte competence by analyzing the PB, CC or GC transcriptome.

Recently, molecular studies preliminary to OMICS applications have been performed in equine embryos. Paris et al., (2011) identified and validated a set of **reference genes** suitable for studying gene expression during equine embryo development. The expression of four carefully-selected reference genes and one developmentally-regulated gene was examined by quantitative PCR in equine in vivo-produced embryos, from the morula to the expanded blastocyst stage. SRP14, RPL4 and PGK1 were identified by geNorm analysis as stably-expressed reference genes suitable for data normalisation. RPL13A expression was less stable and changed significantly during the period of development examined, rendering it unsuitable as a reference gene. As anticipated, CDX2 expression increased significantly during embryo development, supporting its possible role in trophoectoderm specification in the horse. In summary, it was demonstrated that evidence-based selection of potential reference genes aids in validation of stable gene expression in an experimental system, which is particularly useful when dealing with tissues that yield small amounts of mRNA.

Smits et al., (2011) evaluated the difference between **in vivo-** and **in vitro-produced** (IVP) **equine blastocysts** at the genetic level. Suppression subtractive hybridization (SSH) was used to construct a cDNA library enriched for transcripts preferentially expressed in in vivo-derived equine blastocysts compared with IVP blastocysts. Of the 62 different genes identified in this way, six genes involved in embryonic development (BEX2, FABP3, HSP90AA1, MOBKL3, MCM7 and ODC) were selected to confirm this differential expression by RT-quantitative PCR. Five genes were confirmed to be significantly upregulated in in vivo-derived blastocysts (FABP3, HSP90AA1, ODC, MOBKL3 and BEX2), confirming the results of the SSH, however, there was no significant difference in MCM7 expression. Because of their possible importance in embryonic development, the expression of these genes can be used as a marker to evaluate in vitro embryo production systems in the horse, and can be used to compare their roles in embryo development of other species.

Previous studies of functional transcriptomics of individual or associated gene sequences in the **equine oocyte** have been reported by our laboratory (Dell'Aquila et al., 2004 for connexin 43, cyclooxygenase-2 and FSH receptor; Caillaud et al., 2009 for interleukin 1β and its receptors; Dell'Aquila et al., 2008 for the mu opioid receptor; De Santis et al., 2009 for the extracellular calcium-sensing receptor; Lange Consiglio/Cremonesi et al., 2009 for leptin and its ObR receptor) and other groups (Lindbloom et al., 2008 for EGF-like growth factors; Lupole et al., 2010 for ZP genes). To the best of our knowledge, few studies have been performed to date with OMICS technologies in equine reproductive cells or tissues. Fahiminiya et al., (2010) investigated the transcriptome of **granulosa** and **theca cells** from equine follicles at different developmental stages. An equine gene-expression microarray (Agilent technologies Inc., CA, USA) with 44.000 probes was used. Cells were examined from early dominant vs late dominant follicles, and from preovulatory follicles 34 h after injection of crude equine gonadotrophin. It was found that 8349 transcripts were differentially expressed in GC and 2338 in theca cells between preovulatory and late dominant follicles, and that 1602 transcripts were differentially expressed in GC and 8 in theca cells in late dominant vs. early dominant follicles. Thus, it appears that the GC have a highly dynamic nature during the development of dominant follicles. In additional work, Das et al. (2010) analyzed **sperm** and **testis** transcriptomes using the Texas A&M equine whole genome 21351-element oligoarray. Bruemmer et al. (2010) analyzed the **endometrium** transcriptome using the Horse gene expression Agilent microrray for 43000 transcripts. Slough et al., (2011) studied the gene expression (StAR, 3β-HSD, cox, and caspase-3) profile of equine corpus luteum tissue recovered by in vivo biopsy.

OMICS for the Identification of Biomarkers for Oocyte Competence, with Special Reference to the Mare as a
Prospective Model for Human Reproductive Medicine

107

2.4 Oocyte proteome: The direct representation of the oocyte phenotype

Oocyte mRNA is not a direct representation of the factors that drive oocyte phenotype. The
identified mRNA represent potential proteins, but the degree to which the mRNA are being
translated is unknown. Thus, an OMICs goal is the identification and measurement of all
proteins expressed in the oocyte or embryo. There are two main protein-containing oocyte
compartments, the cytoplasm and the zona pellucida. The protein makeup of the oocyte is
important for more than simply evaluating oocyte viability; as noted above, the mammalian
oocyte cytoplasm possesses factors which can reprogram terminally-differentiated germ
cells (sperm) or somatic cells within a few cell cycles. Moreover, it has been suggested that
use of oocyte-derived transcripts may enhance the generation of induced pluripotent stem
cells. The zona pellucida is composed of glycoproteins involved in the sperm-oocyte
interactions which modulate sperm penetration and the fertilization process. Thus,
improving our knowledge of oocyte global protein composition is of great interest.

The main phases of proteomic analysis consist of protein extraction, digestion, separation of
proteins by gel- or non gel-based methods, mass spectrometry evaluation of digested and
separated or electrosprayed peptides, and bioinformatic data analysis (for detailed
proteomic methods, see Seli et al., 2010; Wang et al., 2010 and Arnold & Frohlich, 2011). A
major problem in oocyte proteome analysis is the requirement of large numbers (thousands)
of oocytes. This problem has limited the possibility of performing studies in the human;
studies performed to date have been conducted in the mouse and in large animals.

2.4.1 Proteomic studies in the mouse oocyte

Studies in the mouse have incorporated qualitative proteomic approaches, with the
intention of generating a protein database to be used for the molecular characterization of
the oocyte and developing embryo.

In a pioneering study by Meng et al. (2007), proteomic profiling of mouse mature **COCs** was
performed, using two-dimensional electrophoresis and mass spectrometry. A total of 259
protein spots were identified, corresponding to 156 individual proteins. Functional
classification of the identified proteins, performed manually according to the biological
function of their coding genes, indicated that 12% were involved in cell
signaling/communication, 7% in cell division, 31% in gene/protein expression, 24% in cell
metabolism, 10% in cell structure and motility, 12% in cell/organism defense, and 4% were
unknown.

In a subsequent study by Ma et al., (2008), two-dimensional electrophoresis of mouse
metaphase-II (MII) ooplasmic proteins (the ZP was removed by digestion before protein
extraction) was performed to describe the proteome and phosphoproteome of oocytes
derived from ICR mice. A total of 869 selected protein spots, corresponding to 380 unique
proteins, were identified successfully by mass spectrometry. Of these, 90 protein spots,
representing 53 unique proteins, were stained by Pro-Q Diamond dye, indicating that,
within the MII oocyte cytoplasm, they are in phosphorylated forms. All identified proteins
were bioinformatically annotated and compared to the embryonic stem-cell proteome. A
proteome reference database for the mouse oocyte was established from the protein data
generated in this study (http://reprod.njmu.edu.cn/2d).

A subsequent study (Zhang et al., 2009b) applied one-dimensional sodium dodecyl sulfate polyacrylamide gel electrophoresis and reverse-phase liquid chromatography tandem mass spectrometry to analyze the **mature oocyte proteome** of the mouse in depth. Using this high-performance proteomic approach, the authors successfully identified 625 different proteins from 2700 mature mouse oocytes denuded of their zonae pellucidae. They identified 76 maternal proteins having high levels of mRNA expression both in oocytes and fertilized eggs. Many well-known maternal-effect proteins were included in this subset, including MATER and NPM2. In addition, the observed mouse oocyte proteome was compared with a recently published mouse **embryonic stem cell (ESC) proteome** (Van Hoof et al., 2006, see ref. in Zhang et al., 2009), and 371 overlapping proteins were identified.

In a more recent study by Wang et al. (2010), 7,000 mouse oocytes at different developmental stages, including the **GV stage**, the **MII stage**, and **fertilized oocytes (zygotes)**, were evaluated. The authors successfully identified 2,781 proteins present in GV-stage oocytes, 2,973 proteins in MII oocytes, and 2,082 proteins in zygotes, through semiquantitative mass spectrometry. The results of the bioinformatics analysis indicated that different protein compositions were correlated with oocyte characteristics at different developmental stages. For example, specific transcription factors and chromatin remodeling factors were more abundant in MII oocytes, which may be crucial for the epigenetic reprogramming of sperm or somatic nuclei. These results provided important knowledge for better understanding of the molecular mechanisms associated with early development, and may improve the generation of induced pluripotent stem cells.

A more recent study (Pfeiffer et al., 2011) reported the proteome of MII mouse oocytes to a depth of 3699 proteins, which extends the number of proteins identified to date in mouse oocytes to a comparable size to that of the proteome of undifferentiated mouse ES cells. Twenty-eight oocyte proteins, also detected in ES cells, matched the criteria of the multilevel approach reported in this study to screen for reprogramming factors, namely nuclear localization, chromatin modification, and catalytic activity, thus advancing the definition of "reprogrammome", the set of molecules (proteins, RNAs, lipids, and small molecules) that enable nuclear reprogramming.

2.4.2 Proteomic studies in farm animal oocytes

Studies in farm animals, such as cattle and pigs, have been performed for both qualitative database generation and for quantification of proteome changes during oocyte IVM (bovine: Coenen et al., 2004; Bhojwani et al., 2006; Massicotte et al., 2006; Memili et al., 2007; Berendt et al., 2009; pig: Ellederova et al., 2004; Susor et al., 2007, reviewed by Ma et al., 2008; Zhang et al., 2009; Arnold & Frohlich, 2011).

Berendt et al. (2009) performed two-dimensional gel electrophoresis saturation labeling to detect quantitative differences in the proteomes of immature versus IVM-MII **bovine oocytes**. From 250 ng of sample analyzed per gel, quantitative analysis revealed an average of 2244 spots in pH 4–7 images and 1291 spots in pH 6–9 images. Focusing on the pH 4-7 images, 38 spots with different intensities between oocyte stages were detected. Spots on a gel from 2200 immature oocytes were identified by nano-LC-MS/MS analysis. The ten spots which could be unambiguously identified include the translationally controlled tumor protein, enzymes of the Krebs and pentose phosphate cycles, clusterin, 14-3-3 ε protein and redox enzymes. In addition, the cellular distribution of two differentially-expressed proteins (14-3-3 ε protein, a mediator of Cdc25 phosphatase inhibition, and TCTP, translationally controlled tumor protein) was determined by confocal laser-scanning microscopy. The quantitative and cellular distribution differences of proteins identified in this study may help to identify attractive candidate proteins for oocyte quality evaluation.

To the best of our knowledge, no proteomics studies have been performed to date in the **equine oocyte**. In our group, the functional role of individual proteins involved in the regulation of meiotic maturation has been investigated by means of western blot or immunostaining and confocal microscopy (Dell'Aquila et al., 2008, De Santis et al., 2009; Lange Consiglio et al., 2009). Equine oocytes could be excellent models for oocyte proteomic studies due to the high abundance of maternal proteins accumulated in their large cytoplasm (160 to 180 microns in diameter) during oogenesis. The relatively large cytoplasmic volume is an important feature as it reduces the number of oocytes needed for effective protein extraction, thus increasing the specificity of the proteome analysis. A recent study performed in Xenopus laevis oocytes, chosen due to their abundant ooplasm, identified a number of proteins involved in the regulation of M-phase entry (Kubiak et al., 2011). The equilibrium among activites of these proteins is responsible for the quality of oocytes and the extent of embryo development, via their participation in decision whether to resume meiosis. Identification of cell-cycle control protein activities in mammalian oocytes may have a great impact on the study not only of oocyte quality but also of cancer growth regulation, and thus establishment of targeted therapies.

On the male side, the global proteome of **sperm** and **seminal plasma of fertile stallions** has been investigated (Novak et al., 2010) to determine whether associations exist between the observed proteome and in vivo fertility. Semen was collected throughout the breeding season from 7 stallions at stud in a commercial breeding station. The stallions were bred to a total of 164 mares to determine conception rates. On three occasions during the breeding season, raw semen was obtained from a regular collection and subjected to proteomic analysis using two-dimensional electrophoresis. The semen sample was also assessed for routine semen-quality end points. The first cycle conception rate was negatively related to ejaculate volume ($r = -0.43$, $P = 0.05$) and total IGF1 content (ng) per ejaculate ($r = -0.58$, $P=0.006$), whereas the overall pregnancy rate was positively related to sperm concentration ($r = 0.56$, $P = 0.01$). The abundance of three proteins known to be involved in carbohydrate metabolism in sperm was positively related to fertility. Abundance of cysteine-rich secretory protein 3 (CRISP3) was positively related to first cycle conception rate ($r = 0.495$, $P = 0.027$) and may provide a good marker of fertility. The abundance of four seminal plasma proteins was negatively related to fertility; these were identified as kallikrein-1E2 (KLK2), clusterin, and seminal plasma proteins 1 (SP1) and 2 (SP2). Based on stepwise regression analysis, low levels of clusterin and SP1 in seminal plasma together with abundance of sperm citrate synthase were predictive of fertility ($r=0.77$, $P< 0.0001$). This study identified proteins within sperm and seminal plasma that could serve as biomarkers of semen quality and fertility in stallions, and may present valid models for sperm fertility biomarkers in humans.

2.5 Oocyte lipid fingerprint – Investigating the biological role of structural and reserve lipids

Recently, a new OMIC strategy, namely lipidomics, which utilizes mass spectrometry (MS), chromatography and computer-assisted data analysis, has been proposed. In this approach, lipid molecules are extracted from cells and analyzed by matrix-assisted laser desorption ionization-time of flight (MALDI-TOF) MS (Huang et al., 2005). Like other OMICS, lipidomics is a subject which is both technology-driven and technology-driving, allowing changes in lipid metabolism, including the appearance of new species and the disappearance of others, and compartimentalization of specific lipid species, to be investigated. The underlying fundamentals of different lipidomic experimental approaches and the application of these approaches to the identification of inborn errors of metabolism were reported by Griffiths et al. (2011). Maturing mammalian oocytes, particularly those of farm animals, contain large numbers of cytoplasmic lipid droplets (LDs) whose functional role is still under investigation (Ambruosi et al., 2009). Lipid droplets are discrete organelles present in most cell types and in organisms including bacteria, yeast, plants, insects and animals. Long considered as passive storage deposits, recent proteomic and lipidomic analyses show that LDs are dynamic organelles involved in multiple cellular functions. They serve not only as main reservoirs of neutral lipids such as triglycerides and cholesterol but also contain structural proteins, proteins involved in lipid synthesis and transmembrane proteins (review by Kalantari et al., 2010). A recent study (Ferreira et al., 2010) reported the direct lipid analysis by MALDI-MS of single and intact human, bovine, sheep and fish oocytes. Characteristic lipid profiles, mainly represented by phosphatidylcholines, sphingomyelins and triacylglycerols, were obtained. This study demonstrated that MALDI-MS is capable of providing a reproducible lipid fingerprint from a single oocyte and can be used to investigate developmental modifications or the effects of culture conditions. To our knowledge, no lipidomic studies have been reported to date in the equine oocyte. The equine oocyte, being characterized by polar aggregation of cytoplasmic LDs during maturation, could help to significantly clarify the role of LDs in the maturation and fertilization processes, and in early embryonic development.

2.6 Oocyte glycomic analysis – Post-translational protein-carbohydrate modifications

Glycomics deals with the structure and function of glycans or carbohydrates. Lectin-based diagnostics are the main tool aimed at the detection of diseases associated with alterations of the glycosylation profiles of cells. Lectins are proteins that specifically bind to carbohydrates, of either mono- or oligosaccaridic structure. Certain lectins even recognize cell determinants which are not detected by available antibodies. The increasing use of lectins in biomedical diagnostics is leading rapidly to the development of **lectin/glycan microarrays** which could provide efficient, rapid screening tools to detect normal or altered glycosylation patterns in biological samples. Information on glycomics, concerning methods for use of recombinant and artificial lectins and a recently-launched detection platform using lectin microarrays, as well as their application, were reported by Mislovicova et al., 2009 and Gemeiner et al., 2009. A Glycomics DataBase – a data integration platform for glycans and their structures has been recently created (http://www.glycomics.bcf.ku.edu). To our knowledge no studies have been performed to date using large lectin arrays to evaluate oocyte quality. A recent paper by our research unit (Desantis et al., 2009) reported the use of a 13-lectin panel to evaluate differences of the glycoconjugate pattern between equine oocytes surrounded by compact (viable) or expanded (early atretic) cumulus oophorus. It was found that: 1) equine COCs have a species-specific carbohydrate composition; 2) biosynthesis of glycosylated ZP proteins occurs in both CCs and oocytes; 3) viable (compact) and atretic (expanded) COCs express different lectin-binding patterns in their CCs, ZP and ooplasm. This paper also reviewed numerous studies published on the glycoconjugate pattern of cumulus cells, ZP and ooplasm in several species, including humans. These data confirm that the mare is a good model for evaluation the glycome of oocytes of different quality, developmental stage or functional status, and that the application of lectin arrays could be of great value in evaluating oocyte pathology or the effects of culture conditions.

2.7 Oocyte metabolic profiling – The instantaneous snapshot of oocyte physiology

Following in the wake of OMICS revolutions, new fields of research are emerging. Among them is metabolomics, a field that holds great promise for the study of oocyte and embryo physiology. The metabolome, that is, the compounds produced by the oocyte, provides the natural complement to the genome and proteome. The physicochemical diversity of the metabolome leads to a subdivision of metabolites into compounds soluble in aqueous solutions or those soluble in organic solvents. A complete molecular and quantitative investigation of the latter when isolated from tissue, fluid or cells is a subset of lipidomics (see Section 2.5). A high-priority aim in evaluating oocyte quality is to establish a non-invasive quantification method. Analysis of oocyte metabolism, by evaluating the follicular fluid (FF) or culture media metabolome could be a useful predictor of pregnancy outcome (Sing & Sinclair, 2007; Nel-Themaat & Nagy, 2011). An important aspect limiting this kind of study in humans is the availability of FF aspirated from individual follicles or availability of culture media of individual oocytes. To correlate FF substances with oocyte quality, it is imperative that each follicle is aspirated individually. The procedure of single follicle aspiration is problematic, both for the patient and for the physician, because it requires multiple vaginal punctures (Revelli et al., 2009). Moreover, needle flushing with culture medium after every puncture must be performed, with a standard volume of flushing medium, in order to avoid cross-contamination and to control the dilution of FF substances. In this context, the availability of animal models from which large numbers of individual specimens may be obtained is of great help. A recent review by Revelli et al. (2009) provides an overview on the current knowledge of biochemical predictors of oocyte quality in FF, starting from studies on single biochemical markers and concluding with the most recent studies on metabolomics. Another study (Nagy et al., 2009) evaluated whether near-infrared spectroscopy-generated metabolomic data, obtained from individual oocyte culture media samples, would correlate with nuclear maturity status and subsequent embryo development. Drops (15-20 µl) of in vitro culture media from 3 h culture of individual oocytes recovered from patients undergoing controlled ovarian hyperstimulation were used to obtain a "viability index" of near-infrared spectroscopy metabolomic profiles. Oocytes at different meiotic stages showed significantly different indices, with a higher viability index related to nuclear maturity (MII stage), embryo grade at Day 3 and Day 5 (grade A) and pregnancy rate (human chorionic gonadotropin-positive).

3. Conclusions

OMICS are promising strategies for oocyte quality evaluation in the field of ART, and will lead to a greater understanding of the mechanisms involved in normal oocyte maturation, fertilization and embryo development. It is increasingly clear that large animal models, particularly those species such as horses, whose reproductive management is aimed not only to productivity but also to overcoming infertility, can provide a clinical and biological model for human reproductive phenomena. In addition to their value related to the understanding of human and animal reproduction, "oocyte OMICS" will undoubtedly reveal unexpected and invaluable information that will significantly contribute to the study of stem cell and cancer biology.

4. Acknowledgments

M.E. Dell'Aquila would like to thank all students, technicians, clinicians, breeders and local slaughterhouse staff for their invaluable contributions, given in last 25 years, to research activities on oocyte physiology performed at the Department of Animal Production in Bari. A special thought to Professor Paolo Minoia, our mentor in initiating the exciting adventure of the study of farm animal oocytes. This study was supported by COFIN PRIN MIUR 2008 University of Bari, Italy cod. 2008YTYNKE_001 (Resp. Sci. M.E. Dell'Aquila) and Fondi d'Ateneo 2010 University of Bari, Italy cod. ORBA10YTAK (Resp. Sci. M.E. Dell'Aquila).

5. References

Ambruosi, B.; Lacalandra, G.M.; Iorga, A.I.; De Santis, T.; Mugnier, S.; Matarrese, R.; Goudet, G. & Dell'Aquila M.E. (2009). Cytoplasmic lipid droplets and mitochondrial distribution in equine oocytes: implications on oocyte maturation, fertilization and developmental competence after ICSI. *Theriogenology*, Vol. 71, No. 7, (April 2009), pp. 1093–1104, ISSN 0093-691X.

Antonello, J.; Patassini, C.; Bottacin, A.; Ferlin, A. & Foresta, C. (2011) Ottimizzazione di un protocollo CGH-array su singolo spermatozoo. Proc. I Congress of Federazione Italiana Società Scientifiche della Riproduzione (FISSR, Riccione 26-28 Maggio 2011) pp. 13-14.

Arnold, G.J. & Frohlich, T. (2010). Dynamic proteome signatures in gametes, embryos and their maternal environment. *Reproduction, Fertility and Development*, Vol. 23, No. 1 (December 2010), pp 81-93, ISSN 1031-3613.

Assou, S.; Cerecedo, D.; Tondeur, S.; Pantesco, V.; Hovatta, O.; Klein, B.; Hamamah, S. & De Vos, J. (2009). A gene expression signature by human mature oocytes and embryonic stem cells. *BMC Genomics*, Vol. 8, N. 10 (January 2009), pag. 10, ISSN 1471-2164.

Berendt, F.J.; Frohlich, T.; Bolbrinker, P.; Boelhauve, M.; Gungor, T.; Habermann, F.A.; Wolf, E. & Arnold G.J. (2009) Highly sensitive saturation labeling reveals changes in abundance of cell cycle-associated proteins and redox enzyme variants during oocyte maturation in vitro. *Proteomics* Vol. 9 (February 2009) pp. 550-564, ISSN 1615-9861.

Bonnet, A.; Bevilacqua, C.; Benne, F.; Bodin, L.; Cotinot, C.; Liaubet, L.; Sancristobal, M.; Sarry, J.; Terenina, E.; Martin, P.; Tosser-Klopp, G. & Mandon-Pepin, B. (2011). Trascriptome profiling of sheep granulosa cells and oocytes during early follicular development obtained by Laser Capture Microdissection. *BMC genomics*, Vol 12. (August 2011), pp. 417, ISSN 1471-2164.

Bruemmer, J.E.; Bouma, J.G.; Hess, A.; Hansen, T.R. & Squires, E.L. Gene expression in the equine endometrium during maternal recognition of pregnancy. *Animal Reproduction Science* Vol. 121 (Suppl. 1-2) (2010) S286-S287, ISSN 0378-4320.

Burton, A. & Torres-Padilla, M.E. (2011) Epigenetic reprogramming and development: a unique heterochromatin organization in the preimplantation mouse embryo. *Briefings in Functional Genomics*, Vol. 9, No. 5-6 (December 2010), pp. 444-454, ISSN 2041-2649.

Caillaud, M.; Dell'Aquila, M.E.; De Santis, T.; Nicassio, M.; Lacalandra, G.M.; Goudet, G.; & Gérard, N. (2008). In vitro equine oocyte maturation in pure follicular fluid plus interleukin-1 and fertilization following ICSI. *Animal Reproduction Sciences*, Vol. 106, No. 3-4 (July 2008), pp. 431-439, ISSN 0378-4320.

Callinan, P.A. & Feinberg, A.P. (2006) The emerging science of epigenomics. *Human Molecular Genetics*, Vol. 15, n°1 (April 2006), pp. R95-R101, ISSN 0964-6906.

Carnevale, E.M. (2008). The mare model for follicular maturation and reproductive aging in the woman. *Theriogenology*, Vol. 69, No. 1, (January 2008) pp. 23-30, ISSN 0093-691X.

Choi, Y.H.; Love, L.B.; Varner, D.D. & Hinrichs, K. (2006). Holding immature equine oocytes in the absence of meiotic inhibitors: Effect on germinal vesicle chromatin and blastocyst development after intracytoplasmic sperm injection. *Theriogenology*, Vol. 66, No. 4 (September 2006), pp. 955-63, ISSN 0093-691X.

Choi, Y.H.; Gustafson-Seabury, A.; Velez, J.C.; Hartman, D.L.; Bliss, S.; Riera, F.L.; Roldan, J.E.; Chowdhary, B.; Hinrichs, K. (2010). Viability of equine embryos after puncture of the capsule and biopsy for preimplantation genetic diagnosis. *Reproduction* Vol 140 No. 6 (December 2010), pp. 893-902, ISSN 1470-1626.

Cremer, T. & Zakhartchenko V (2011) Nuclear architecture in developmental biology and cell specialization. *Reproduction, Fertility and Development* Vol. 23, No. 1 (2011), pp. 94-106, ISSN 1031-3613

Dalbiès-Tran, R. & Mermillod, P. (2003) Use of heterologous complementary DNA array screening to analyze bovine oocyte transcriptome and its evolution during in vitro maturation. *Biology of Reproduction*, Vol. 68, No. 1 (January 2003), pp. 252-261, ISSN 0006-3363.

Das P.J.; Vishnoi, M.; Kachroo, P.; Wang, J.; Love, C.C.; Varner, D.D.; Chowdary B.P.; Raudsepp, T. Expression microarray profiling of sperm and testis mRNA of reproductively normal stallions. *Animal Reproduction Science* Vol. 121 (Suppl. 1-2) (2010) S175, ISSN 0378-4320.

Dell'Aquila, M.E.; Albrizio, M.; Maritato, F.; Minoia, P.; Hinrichs, K. (2003). Meiotic competence of equine oocytes and pronucleus formation after intracytoplasmic sperm injection (ICSI) as related to granulosa cell apoptosis. *Biology of Reproduction*, Vol, 68, No. 6 (June 2003) pp. 2065-72, ISSN 0006-3363.

Dell'Aquila, M.E.; Caillaud, M.; Maritato, F.; Martoriati, A.; Gérard, N.; Aiudi, G.; Minoia, P. & Goudet G. (2004). Cumulus expansion, nuclear maturation and connexin 43, cyclooxygenase-2 and FSH receptor mRNA expression in equine cumulus-oocyte complexes cultured in vitro in the presence of FSH and precursors for hyaluronic acid synthesis. *Reproductive Biology and Endocrinology*. Vol. 2 (June 2004), p. 44, ISSN 1477-7827.

Dell'Aquila, M.E.; Albrizio, M.; Guaricci, A.C.; De Santis, T.; Maritato, F.; Tremoleda, J.L.; Colenbrander, B.; Guerra, L.; Casavola, V. & Minoia P. (2008). Expression and localization of the mu-opioid receptor (MOR) in the equine cumulus-oocyte complex and its involvement in the seasonal regulation of oocyte meiotic competence. *Molecular Reproduction and Development*, Vol. 75, No. 8 (August 2008), pp. 1229-46, ISSN 1040-452X.

De Santis, T.; Casavola, V.; Reshkin, S.J.; Guerra, L.; Ambruosi, B.; Fiandanese, N.; Dalbies-Tran, R.; Goudet, G. & Dell'Aquila, M.E. (2009). The extracellular calcium-sensing receptor is expressed in the cumulus-oocyte complex in mammals and modulates oocyte meiotic maturation. *Reproduction*, Vol. 138, No. 3 (September 2009), pp. 439-452, ISSN 1470-1626.

Desantis, S.; Ventriglia, G.; Zizza, S.; De Santis, T.; Di Summa, A.; De Metrio, G. & Dell'Aquila M.E. (2009). Lectin-binding sites in isolated equine cumulus-oocyte complexes: differential expression of glycosidic residues in complexes recovered with compact or expanded cumulus. *Theriogenology*, Vol. 72, No. 3 (August 2009), pp. 300-309, ISSN 0093-691X.

Fahiminiya, S.; Waddington, D.; Donadeu, X. & Gerard, N. (2010) Transcriptomic analysis of dominant follicle development in mares. *Animal Reproduction Science* Vol. 121 (Suppl. 1-2) (2010) S41, ISSN 0378-4320.

Fauser, B.C.; Diedrich, K.; Bouchard, P.; Domínguez, F.; Matzuk, M.; Franks, S.; Hamamah, S.; Simón, C.; Devroey, P.; Ezcurra, D. & Howles, C.M. (2011). Contemporary genetic technologies and female reproduction. *Human Reproduction Update*, doi: 10.1093/humupd/dmr033 (September 2011), ISSN 1355-4786.

Ferreira, C.R.; Saraiva, S.A.; Catharino, R.R.; Garcia, J.S.; Gozzo, F.C.; Sanvido, G.B.; Santos, L.F.A, Lo Turco, E.G.; Pontes, J.H.F.; Basso, A.C.; Bertolla, R.P.; Sartori, R.; Guardieiro, M.M.; Perecin, F.; Meirelles, F.V.; Sangalli, J.R. & Eberlin M.N. (2010). Single embryo and oocyte lipid fingersprinting by mass spectrometry. *Journal of Lipid Research*, Vol. 51, No. 5 (May 2010), pp. 1218-1227, ISSN 0022-2275.

Geraedts, J.; Collins, J.; Gianaroli, L.; Goossens, V.; Handyside, A.; Harper, J.; Montag, M.; Repping, S. & Schmutzler, A. (2010). What next for preimplantation genetic screening? A polar body approach! *Human Reproduction*, Vol. 25, No. 3 (March 2010), pp. 575-577, ISSN 0268-1161.

Geraedts, J.; Montag, M.; Magli, M.C.; Repping, S.; Handyside, A.H.; Staessen, C.; Harper, J.; Schmutzler, A.; Collins, J.; Goossens, V.; Van der Ven, H.; Vesela, K. & Gianaroli, L. (2011). Polar body array CGH for prediction of the status of the corresponding oocyte. Part I: clinical results. *Human Reproduction*, doi: 10.1093/humrep/der294 (September 2011), ISSN 0268-1161.

Gemeiner, P.; Mislovicova, D.; Tkac, J.; Svitel, J.; Patoprsty, V.; Hrabarova, E.; Kogan, G.; Kozar, T. Lectinomics II. (2009). A highway to biomedical/clinical diagnostics. *Biotecnology Advances* Vol. 27, No. 1 (January-February 2009), pp. 1-15, ISSN 0734-9750.

Griffin, D.K.; Wilton, L.J.; Handyside, A.H.; Winston, R.M.K. & Dehanty, J.D.A. (1992). Dual fluorescent in situ hybridisation for simultaneous detection of X and Y chromosome-specific probes for the sexing of human preimplantation embryonic nuclei. *Human Genetics*. Vol. 89, No. 1 (April 1992), pp. 18-22, ISSN 0340-6717.

Griffiths W.J.; Ogundare M.; Williams C.M. & Wang Y. (2011) On the future of "omics":
    lipidomics. *Journal of Inherited Metabolic Diseases*, Vol. 34, No. 3 (June 2011), pp. 583-
    592, ISSN 0141-8955.

Grøndahl, M.L.; Borup R.; Lee Y.B.; Myrhoj, V.; Meinertz, H. & Soresen S. (2009). Differences
    in gene expression of granulosa cells from women undergoing controlled ovarian
    hyperstimulation with either recombinant FSH or highly purified hMG. *Fertility
    and Sterility*, Vol. 91, No. 5 (May 2009), pp. 1820-1830, ISSN 0015-0282.

Grøndahl, M.L.; Yding Andersen, C.; Bogstad, J.; Nielsen, F.C.; Meinertz, H. & Borup, R.
    (2010). Gene expression profiles of single human mature oocytes in relation to age.
    Human Reproduction, Vol. 25, No. 4 (April 2010), pp. 957-968, ISSN 0268-1161.

Hinrichs, K.; Schmidt, A.L.; Friedman, P.P.; Selgrath, J.P. & Martin, M.G. (1993). In vitro
    maturation of horse oocytes: characterization of chromatin configuration using
    fluorescence microscopy. *Biology of Reproduction*, Vol. 48, No. 2 (February 1993), pp.
    363-370, ISSN 0006-3363.

Hinrichs, K. & Williams, K.A. (1997). Relationships among oocyte-cumulus morphology,
    folicular atresia, initial chromatin configuration, and oocyte meiotic competence in
    the horse. *Biology of Reproduction*, Vol. 57, No. 2 (August 1997), pp. 377-384, ISSN
    0006-3363.

Hinrichs, K.; Choi, Y.H.; Love, L.B.; Varner, D.D., Love C.C. & Walckenaer, B.E. (2005)
    Chromatin configuration within the germinal vesicle of horse oocytes: changes post
    mortem and relationship to meiotic and developmental competence. *Biology of
    Reproduction*. Vol. 72, n. 5 (May 2005), pp. 1142-1150, ISSN 0006-3363.

Huang, Y.; Shen, J.; Wang, T.; Yu, Y.K.; Chen, F.F.; Yang, J. (2005). A lipidomic study of the
    effects of N-methyl-N'-nitro-N-nitrosoguanidine on sphingomyelin metabolism.
    Acta Biochim Biophys Sin (Shanghai), Vol. 37, No. 8 (August 2005), pp. 515-524, ISSN
    1672-9145.

Jones, G.M.; Cram, D.S.; Song, B.; Magli, M.C.; Gianaroli, L.; Lacham-Kaplan, O.; Findlay,
    J.K.; Jenkin, G. & Trounson A.O. (2008) Gene expression profiling of human oocytes
    following in vivo or in vitro maturation. *Human Reproduction*, Vol. 23, No. 5 (May
    2008), pp. 1138-1144, ISSN 0268-1161.

Kalantari, F.; Bergeron, J.J. & Nilsson T. (2010). Biogenesis of lipid droplets-how cells get
    fatter. *Molecular Membrane Biology* Vol. 27, No. 8 (November 2010), pp. 462-468,
    ISSN 0968-7688.

Katz-Jaffe, M.G.; McCallie, B.R.; Preis, K.A.; Filipovits, J. & Gardner, D.K. (2009)
    Transcriptome analysis of in vivo vs in vitro matured bovine MII oocytes.
    Theriogenology, Vol. 71, No. 6 (April 2009), pp. 939-946, ISSN 0093-691X.

Klatsky PC, Wessel GM, Carson SA. (2010). Detection and quantification of mRNA in single
    human polar bodies: a minimally invasive test of gene expression during
    oogenesis. *Molecular Human Reproductin*, Vol. 16, No. 12 (December 2010), pp. 938-
    943, ISSN1360-9947.

Kocabas, A.M.; Crosby, J.; Ross, P.J.; Otu, H.H.; Beyhan, Z.; Can, H.; Tam, W.L.; Rosa, G.J.M.;
    Halgren, R.G.; Lim, B.; Fernandez, E. & Cibelli, J.B. (2006). The trascriptome of
    human oocytes. *PNAS*, Vol. 103, No. 38 (September 2006), pp. 14027-14032, ISSN
    1091-6490.

Kubiak, J.Z. (2011). Proteomics of M-phase entry: "Omnen" vs. "More" the battle for oocyte
    quality and beyond. *Folia Histochemica et Cytobiologica*, Vol. 49, No. 1 (2011), pp. 1-7,
    ISSN 0239-8508.

Lange Consiglio, A.; Dell'Aquila, M.E.; Fiandanese, N.; Ambruosi, B.; Cho, Y.S.; Bosi, G.; Arrighi, S.; Lacalandra, G.M. & Cremonesi, F. (2009). Effects of leptin on in vitro maturation, fertilization and embryonic cleavage after ICSI and early developmental expression of leptin (Ob) and leptin receptor (ObR) proteins in the horse. *Reproductive Biology and Endocrinology*, Vol 7. (October 2009), p. 113, ISSN 1477-7827.

Le Bourhis, A.D.; Mullaart, E.; Humblot, P.; Coppieters, W.; Ponsart, C. (2011). Bovine embryo genotyping using a 50k single nucleotide polymorphism chip. *Reproduction, Fertility and Development* Vol.23 (2011), p. 197, ISSN 1031-3613.

Lindbloom, S.M.; Farmerie, T.A.; Clay, C.M.; Seidel, G.E. Jr & Carnevale, E.M. (2008). Potential involvement of EGF-like growth factors and phosphodiesterases in initiation of equine oocyte maturation. *Animal Reproduction Sciences*, Vol. 103, No. 1-2 (January 2008), pp. 187-192, ISSN 0378-4320.

Lupole, R.E.; Bouma, G.J.; Clay, C.M.; Seidel, G.E.Jr & Carnevale E.M. (2010) Effect of mare age on gene expression of equine zona pellucida proteins and sperm binding. *Animal Reproduction Science* Vol. 121 (Suppl. 1-2) (2010) S252-253, ISSN 0378-4320.

Ma, M.; Guo, X.; Wang, F.; Zhao, C.; Liu, Z.; Shi, Z.; Wang, Y.; Zhang, P.; Zhang, K.; Wang, N.; Lin, M.; Zhou, Z.; Liu, J.; Li, Q.; Wang, L.; Huo, R.; Sha, J. & Zhou, Q. (2008). Protein expression profile of the Mouse Metaphase II oocyte. *Journal of Proteome Research* Vol. 7, No. 11 (November 2008), pp. 4821-4830, ISSN 1535-3893.

Magli, M.C.; Montag, M.; Koster, M.; Muzi, L.; Geraedts, J.; Collins, J.; Goossens, V.; Handyside, A.H.; Harper, J.; Repping, S.; Schmutzler, A.; Vesela, K. & Gianaroli, L. (2011). Polar body array CGH for prediction of the status of the corresponding oocyte. Part II: technical aspects. *Human Reproduction*, doi: 10.1093/humrep/der295 (September 2011), ISSN 0268-1161.

Mamo, S.; Carter, F.; Lonergan, P.; Leal, C.V.L.; Al Naib, A.; McGettigan, P.; Mehta, J.P.; Evans, A.C.O. & Fair, T. (2011). Sequential analysis of global gene expression profiles in immature and in vitro matured bovine oocytes: potential molecular markers of oocyte maturation. *BMC Genomics*, Vol 12 (March 2011), pp. 151, ISSN 1471-2164.

Matzuk, M.M. & Lamb, D.J. (2008). The biology of infertility: research advances and clinical challenges. *Nature Medicine* Vol. 14, No. 11(November 2008), ISSN 1078-8956.

Mislovicova, D.; Gemeiner, P.; Kozarova, A. & Kozar, T. (2009). Lectinomics I. Relevance of exogenus plant lectins in biomedical diagnostics. *Biologia*. Vol. 64, No. 1 (2009), pp. 1-19, ISSN 0006-3088.

Meng, Y.; Xiao-hui, L.; Shen, Y.; Fan, L.; Leng, J.; Liu, J. & Sha, J. (2007). The protein profile of mouse mature cumulus-oocyte complex. *Biochimica et Biophysica Acta*, Vol. 1774, No. 11 (November 2007), pp. 1477-1490, ISSN 0304-4165.

Nagy, Z.P.; Jones-Colon, S.; Roos, P.; Botros, L.; Greco, E.; Dasig, J. & Behr, B. (2009). Metabolomic assessment of oocyte viability. *Reproductive BioMedicine Online*. Vol 18, No. 2 (February 2009), pp 219-225, ISSN 1472-6483.

Nel-Themaat, L. & Nagy, Z.P. (2011). A review of the promises and pitfalls of oocyte and embryo metabolomics. *Placenta*, Vol 32, No. Suppl 3 (September 2011), pp. 5257-5263, ISSN 0143-4004.

Novak, S.; Smith, T.A.; Paradis, F.; Burwash, L.; Dyck, M.K.; Foxcroft, G.R. & Dixon, W.T. (2010). Biomarkers of in vivo fertility in sperm and seminal plasma of fertile stallions. *Theriogenology* Vol. 74, No. 6 (October 2010), pp. 956-67, ISSN 0093-691X.

Parfenov, V.; Potchukalina, G.; Dudina, L.; Kostyuchek, K. & Gruzova, M. (1989). Human
    antral follicles: oocyte nucleus and the karyosphere formation (electron microscopic
    and autoradiographic data). *Gamete Research*, Vol. 22, No. 2 (February 1989), pp.
    219-231, ISSN 0148-7280.

Paris, D.B.; Kuijk, E.W.; Roelen, B.A. & Stout, T.A. (2011). Establishing reference genes for
    use in real time quantitative PCR analysis of early equine embryo. *Reproduction
    Fertility and Development*, Vol. 23, No. 2 (January 2011), pp. 353-363, ISSN 1031-3613.

Pfeiffer, M.J.; Siatkowski, M.; Paudel, Y.; Balbach, S.T.; Baeumer, N.; Crosetto, N.; Drexler,
    H.C.A.; Fuellen, G. & Boiani, M. (2011). Proteomic analysis of mouse oocytes
    reveals 28 candidate factors for the "Reprogammome". *Journal of Proteomic Research*,
    Vol. 10 , No. 5 (May 2011), pp. 2140-2153, ISSN 1535-3893.

Regassa, A.; Rings, F.; Hoelker,,M.; Tholen, E.; Looft, C.; Schellander, K. & Tesfaye, D. (2001)
    Transcriptome dynamics and molecular cross-talk between bovine oocyte and its
    companion cumulus cells. *BMC Genomics*, Vol 12 (January 2011), pp. 57, ISSN 1471-
    2164.

Revelli, A.; Delle Piane, L.; Casano, S.; Molinari, E.; Massobrio, M. & Rinaudo, P. (2009).
    Follicular fluid content and oocyte quality: from single biochemical markers to
    metabolics. *Reprodutive Biology and Endocrinology* Vol. 7 (May 2009), p. 40, ISSN
    1477-7827.

Rodrigues-Zas, S.L.; Schellander, K. & Lewin H.A. (2008). Biological interpretation of
    transcriptomic profiles in mammalian oocytes and embryos. *Reproduction*, Vol. 135,
    No. 2 (February 2008), pp. 129-139, ISSN 1470-1626.

Romar, R.; De Santis, T.; Papillier, P.; Perreau, C.; Thelie, A.; Dell'Aquila, M.E.; Mermillod,
    P. & Dalbies-Tran, R. (2011). Expression of maternal transcripts during bovine
    oocyte In vitro maturation is affected by donor age. *Rerpoduction in domestic animals*.
    Vol. 46, No. 1 (February 2011), pp. e23-e30, ISSN 0936-6768.

Seli, E.; Robert, C. & Sirard, M.A. (2010). OMICS in assisted reproduction: possibilities and
    pitfalls. *Molecular Human Reproduction*, Vol. 16, No. 8 (August 2010), pp. 513-530,
    ISSN 1360-9947.

Simarro Fagundes, N.; Michalczechen Lacerda, V.A.; Siqueira Caixeta, E.; Marinheiro
    Machado, G.; de Oliveira Melo, E.; Rumpf, R.; Alves Nunes Dode, M. & Machaim
    Franco, M. (2010) Methylation pattern of a differentially methylated region located
    in the last exon of the *igf2* gene in oocytes from nellore cows *Reproduction Fertility
    and Development* Vol.22, (January 2010) p. 283 ISSN 1031-3613.

Singh, R. & Sinclair, K.D. (2007). Metabolomics: approaches to assessing oocyte and embryo
    quality. *Theriogenology*, 2007, 68 (Suppl I): S56-S62.

Slough, T.L.; Rispoli, L.A.; Carnevale, E.M.; Niswender G.D. & Bruemmer J.E. (2011)
    Temporal gene expression in equine corpora lutea based on serial biopsies in vivo.
    Journal of Animal Sciences, Vol 89, No. 2 (February 2011) pp. 389-396, ISNN 0021-
    8812.

Smallwood, S.A.; Tomizawa, S.; Krueger, F.; Ruf, N.; Carli, N.; Segonds-Pichon, A.; Sato, S.;
    Hata, K.; Andrews, S.R. & Kelsey, G. (2011). Dynamic CpG island methylation
    landscape in oocytes and preimplantation embryos. *Nature genetics*, Vol. 43, n. 8
    (August 2011), pp. 811-815, ISSN 1061-4036.

Smits, K.; Goossens, K.; Van Soom, A.; Govaere, J.; Hoogewijs, M. & Peelman, L.J. (2011). In
    vivo derived horse blastocysts show transcriptional up-regulation of
    developmentally important genes compared with in vitro produced horse

blastocysts. *Reproduction, Fertility and Development*, Vol. 23, No. 2 (2011), pp. 364-375, ISSN 1031-3613.

Thelie, A.; Papillier, P.; Perreau, C.; Uzbekova, S.; Hennequet-Antier, C. & Dalbies-Tran, R. (2009). Regulation of bovine oocyte-specific transcripts during in vitro oocyte maturation and after maternal–embryonic transition analyzed using a transcriptomic approach. *Molecular Reproduction and Development*, Vol. 76, No. 8 (August 2009), pp. 773–782, ISSN 1040-452X.

Treff, N.R.; Su, J.; Kasabwala, N.; Tao, X.; Miller, K.A. & Scott, R.T.J. (2010). Robust embryo identification using first polar body single nucleotide polymorphism microarray-based DNA fingerprinting. *Fertility and Sterility*. Vol. 93, No. 7 (May 2010), pp. 2453-2455, ISSN 0015-0282. (a)

Treff, N.R.; Su, J.; Tao, X.; Levy, B. & Scott, R.T.Jr. (2010). Accurate single cell 24 chromosome aneuploidy screening using whole genome amplification and single nucleotide polymorphism microarrays. *Fertility and Sterility*. Vol. 94, No. 6 (November 2010), pp. 2017-2021, ISSN 0015-0282. (b)

Tosti, E. & Boni, R. (2011). *Oocyte maturation and fertilization: a long history for a short event.* Bentham Books. ISBN 978-1-60805-182-3.

Wain, L.V. & Tobin, M.D. (2011) Copy number variation, In: *Genetic Epidemiology*, M. Dawn Dawn Teare (Ed.), Humana Press, New York, NY - USA Vol. 713 Methods in Molecular Biology pp.167-183, ISSN 1064-3745.

Wang, S.; Kou, Z.; Jing, Z.; Zhang, Y.; Guo, X.; Dong, M.; Wilmut, I. & Gao, S. (2010). Proteome of mouse oocytes at different developmental stages. *Proceedings of the National Academy of Sciences (PNAS) U S A*, Vol. 107, No. 41, (October 2010), pp. 17639-44, ISSN 0027-8424.

White, C.A. & Salamonsen, L.A. (2005). A guide to issues in microarray analysis: application to endometrial biology. *Reproduction*, Vol. 130, No. 1 (July 2005), pp. 1-13, ISSN 1470-1626.

Wells, D. & Patrizio, P. (2008). Gene expression profiling of human oocytes at different maturational stages and after in vitro maturation. *American Journal of Obstetrics and Gynecology* Vol. 198, No. 4 (April 2008), pp. 455.e1-455.e11, ISSN 00029378.

Wood, J.R.; Dumesic, D.A.; Abbot, D.H. & Strauss, J.F. (2007). Molecular abnormalities in oocytes from women with polycystic ovary syndrome revealed by microarray analysis. *The Journal of Clinical Endocrinology and Metabolism*, Vol. 92, No. 2 (February 2007), pp. 705-713, ISSN 0021-972X.

Zhang, P.; Zucchelli, M.; Bruce, S.; Hambiliki, F.; Stavreus-Evers, A.; Levkov, L.; Skottman, H.; Kerkelä, E.; Kere, J. & Hovatta, O. (2009). Transcriptome profiling of human pre-implantation development. *PLoS One*, Vol. 4, No. 11 (November 2009), pp. e7844, ISSN 1932-6203 (a).

Zhang, P.; Ni, X.; Guo, Y.; Guo, X.; Wang, Y.; Zhou, Z.; Huo, R. & Sha, J. (2009). Proteomic-based identification of maternal proteins in mature mouse oocytes. *BMC Genomics*, Vol 10 (August 2009), ISSN 1471-2164 (b).

Meiotic Behaviour of Chromosomes Involved in Structural Chromosomal Abnormalities Determined by Preimplantation Genetic Diagnosis

L. Xanthopoulou and H. Ghevaria
University College London,
UK

1. Introduction

1.1 The origin of structural chromosomal abnormalities

Structural rearrangements can arise premeiotically, during meiosis or during postzygotic mitotic divisions. The main causes leading to chromosomal aberrations are illegitimate recombination due to chromosome misalignment and asymmetric pairing in meiosis and mitosis, defective DNA repair mechanisms and aberrant behaviour at the replication fork (Gardner and Sutherland, 2004).

Chromosome rearrangement breakpoints are not uniformly found throughout the genome (Lupski, 2004) and certain genomic regions, especially subtelomeric and pericentromeric regions (Shaw and Lupski, 2004) are more likely to be involved in a chromosomal rearrangement. This clustering of chromosomal rearrangements around hotspots creates considerable genomic instability (Shaw and Lupski, 2004). Such regions of genomic instability involve low copy repeats (LCRs).

LCRs act as substrates for non-allelic homologous recombination (NAHR), by erroneously facilitating different chromosome regions of the same or of different chromosomes to come together. Chromatin structure can also be involved in the generation of chromosomal aberrations, with areas which are less compacted being more accessible to double strand breaks and DNA damage (Shaw and Lupski, 2004).

There are also non-recurrent rearrangements, whose breakpoints are scattered around the genome and which often occur at unique sequences (Shaw and Lupski, 2004) but even in those cases the breakpoints often involve intronic motifs of smaller repetitive sequences which are usually involved in inducing susceptibility to double strand breaks (Toffolatti et al., 2002). Such rearrangements with unique breakpoints are created by non-homologous DNA end-joining (NHEJ), which is an error prone mechanism (Lieber et al., 2003) of repairing double strand breaks in multicellular organisms (Shaw and Lupski, 2005). NHEJ can result in genomic alterations by generating deletions or duplications through erroneously joining together the ends of double strand breaks from different chromosomes (Pfeiffer et al., 2004).

2. Structural chromosomal abnormalities and meiosis: Focussing on translocations

Structural chromosomal abnormalities involve chromosome breakage followed by the rejoining of chromosome parts into a different configuration. These structural rearrangements can be mainly categorised into translocations (reciprocal or robertsonian), insertions, inversions (pericentric or pericentric), deletions, duplications, ring chromosomes and isochromosomes. In the case of a structural rearrangement, each chromosome pair consists of a normal and a derivative chromosome.

Many structural abnormalities are associated with clinical charactteristics, such as mental retardation, characteristic dysmophic features and often other malformations and developmental delay. For example the cri-du-chat syndrome caused by a terminal deletion on chromosome 5, or the DiGeorge syndrome caused by a microdeletion on the long arm of chromosome 22. Duplication syndromes are also known to be associated with abnormal phenotypes, such as Charcot Marie Tooth disease Type I caused by a duplication in 17p12. Other structural aberrations such as inversions, translocations and ring chromosomes do not cause any abnormalities in balanced carriers when there is no gain or loss of genetic material, however these individuals are faced with reproductive problems.

The most common type of structural chromosomal abnormality in humans is translocation, most specifically reciprocal translocations that are seen in about 1 in 500 live births (Jacobs et al., 1992) and involve the exchange of genetic material between different chromosomes. In carriers of reciprocal translcoations, each chromosome pair consists of a normal and a derivative chromosome, as shown in figure 1i. Carriers are phenotypically normal since there is no loss of genetic material unless the breakpoints disrupt important genes (Gardner and Sutherland, 2004). However they are faced with unfavourable meiotic segregation patterns when the chromosomes involved in the translocation pair up at prophase forming a quadrivalent (also known as pachytene) as shown in figure 1ii. The chromosomal status of the gametes produced by a carrier of a reciprocal translocation depends on the segregation pattern that took place in meiosis. As shown in figure 1iii, the chromosomes in the quadrivalent can segregate in a 2:2 mode, whereby each daughter cell will receive 2 chromosomes, in a 3:1 segregation mode, whereby one daughter cell receives 3 chromosomes and the other daughter call receives just one or there might be complete non disjunction in a 4:0 mode of segregation.

Robertsonian translocations have a prevalence of 1 in 1000 and involve the centric fusion of two acrocentric chromosomes (in 75% of the cases chromosomes 13 and 14 are involved) and the loss of the short arms (Garner and Sutherland, 2004). As a result a derivative chromosome is formed as shown in figure 2i, which comprises of the two long arms of the chromosomes involved in the translocation and at meiosis the chromosomes pair up as shown in figure 2ii. The total chromosome number of carriers of Robertsonian translocations is 45. Although all acrocentric chromosomes have been found to be involved in Robertsonian translocations, rob(13q14q) and rob(14q21q) constitute around 85% of all Robertsonian translocations (Therman et al., 1989).

Balanced carriers of a structural chromosomal rearrangement are usually phenotypically normal since there is no loss of genetic material, unless the breakpoints are within important genes. They are however faced with unfavourable meiotic segregation patterns when the normal and derivative chromosome(s) involved in the abnormality pair up at meiosis I. Carriers therefore are at a high risk of producing unbalanced gametes and hence genetically

Meiotic Behaviour of Chromosomes Involved in Structural Chromosomal Abnormalities Determined by
Preimplantation Genetic Diagnosis

121

abnormal embryos that are associated with recurrent miscarriage, infertility as well as unbalanced offspring (Gardner and Surtherland, 2004).

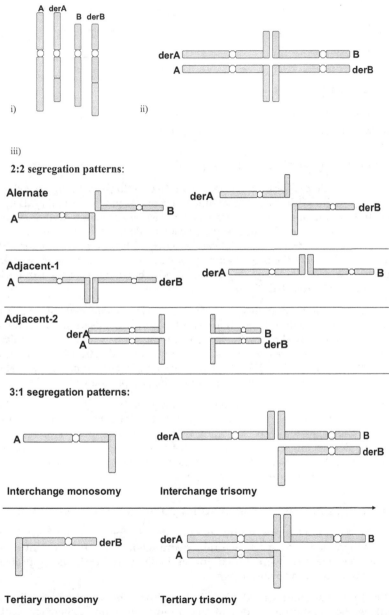

Fig. 1. i) Chromosomes in a cell of a carrier of a reciprocal translocation, ii) Pachytene configuration of the quadrivalent cross during meiosis in a carrier of a reciprocal translocation, iii) possible segregation modes seen at meiosis (4:0 segregants not shown here).

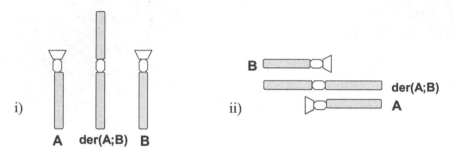

Fig. 2. i) Carrier of a Robertsonian translocation, ii) Pairing up of chromosomes, forming a trivalent during meiosis in a carrier of a Robertsonian translocation.

3. Preimplantation genetic diagnosis (PGD)

Preimplantation genetic diagnosis provides an alternative to prenatal diagnosis for couples at risk of having a child with a specific genetic or chromosomal abnormality. It involves the removal and testing of material from the oocyte or the developing embryo. First and second polar bodies for instance can be analysed in order to determine the genetic status of the oocyte (Verlinsky et al., 1990) or material can be biopsied from embryos generated through *in vitro* fertilization (IVF). In the case of embryo biopsy, one or two single blastomeres can be removed during the cleavage stage or material can be biopsied from the trophectoderm at the blastocyst stage (Dokras et al., 1990).

The biopsied material is then analysed by the polymerase chain reaction (PCR) for single gene disorders (Findlay et al., 1996) or fluorescent in situ hybridisation (FISH) for structural or numerical chromosomal abnormalities (Griffin et al., 1991, Coonen et al., 1998, Conn et al., 1998) whereas recently array comparative genomic hybridization (aCGH) has also been introduced (Alfarawati *et al.*, 2011). In this way unaffected embryos are selected and transferred to the uterus, aiming to establish an unaffected pregnancy.

Couples that are at risk of having a child with a specific single gene or chromosomal abnormality may opt to have PGD in order to avoid an affected pregnancy and to avoid having prenatal diagnosis which is associated with a 1% risk of miscarriage; moreover their decision to have PGD is often linked to the couple's view on pregnancy termination. At the same time other couples that seek PGD have difficulty achieving a pregnancy due to reasons of infertility or subfertility or are victims of repeated pregnancy loss.

4. PGD for structural chromosomal abnormalities: strategies and outcome

PGD provides a unique opportunity to investigate the meiotic behaviour of the chromosomes involved in a structural rearrangement. Until recently structural chromosomal abnormalities were tested for at PGD with FISH, using probes for the chromosomes involved in the translocation.

The main FISH probe strategies used in PGD for recirocal translocations, involve the use of probes flanking the breakpoint in one of the chromosomes, in conjunction with another probe on the other chromosome. In this way on one chromosome one probe distal to the breakpoint (i.e. a subtelomeric probe specific to the segment that is translocated) and one probe proximal to the breakpoint (i.e. a centromeric probe of any other probe that will be

Meiotic Behaviour of Chromosomes Involved in Structural Chromosomal Abnormalities Determined by
Preimplantation Genetic Diagnosis

123

used for the enumeration of that chromosome) are used, together with another probe on the other chromosome, which can either be distal or proximal to the breakpoint on the second chromosome (Scriven et al.1998). So the main two strategies involve using two centromeric and one subtelomeric probe or one centromeric and two subtelomeric probes, where the subtelomeric probes will always be on the segments that are translocated. The aim of these probe strategies is to be able to detect all possible segregation patterns, so they need to be informative for both normal and boh derivative chromosomes.

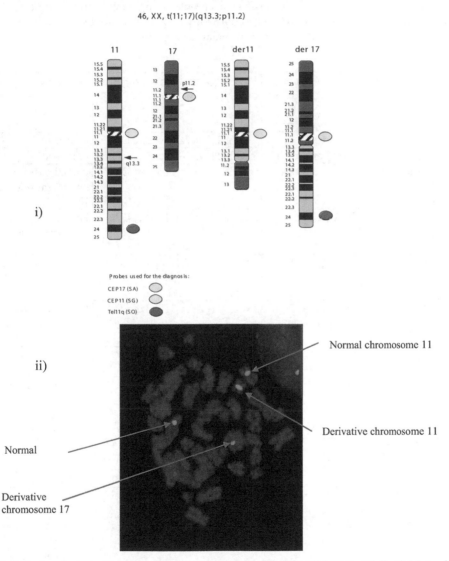

Fig. 3. i) Probe strategy for PGD for translocation 46, XX, t(11;17)(q13.3;p11.2), ii) Metaphase nucleus of the carrier of 46, XX, t(11;17)(q13.3;p11.2) after FISH with the probes for centromere of 11 and 17 and the telomere of 11q.

The probe strategy chosen for PGD for each srtuctural abnormality case is tested prior to clinical application on control lymphocyte slides from the parents of the couple seeking PGD. For example figure 3 below shows an example of a translocation between chromosomes 11 and 17, carried by the female partner: 46, XX, t(11;17)(q13.3;p11.2).

The centromeric probe for chromosome 11 in spectrum green and the centromeric probe for chromosome 17 in spectrum aqua were used together with the subtelomeric probe for 11q in psectrum orange. The probes were tested on parental lymphocytes and in figure 3ii it is clear that the probes chosen pick up the abnormality and allow us to distinuish between the normal and derivative chromosomes on the metaphase spread.

Segregation mode in the gamete		Chromosomes present in the gamete of the carrier parent	FISH signals seen in the embryo	Outcome
2:2 segregation modes	Alternate	11 and 17	2xCEP11, 2xTel11p, 2xCEP17	Balanced
		Der11 and der17	2xCEP11, 2xTel11p, 2xCEP17	Balanced
	Adjacent-1	11 and der17	2xCEP11, 3xTel11p, 2xCEP17	Partial trisomy 11, partial monosomy 17
		Der11 and 17	2xCEP11, 1xTel11p, 2xCEP17	Partial monosomy 11, partial trisomy 17
	Adjacent-2	11 and der11	3xCEP11, 2xTel11p, 1xCEP17	Partial trisomy 11, partial monosomy 17
		17 and der17	1xCEP11, 2xTel11p, 3xCEP17	Partial monosomy 11, partial trisomy 17
3:1 segregation modes	3 ⟶	11, der11 and 17	3xCEP11, 2xTel11p, 2xCEP17	Partial trisomy 11 and 17
	1 ⟶	Der17	1xCEP11, 2xTel11p, 2xCEP17	Partial monosomy 11 partial monosomy 17
	3 ⟶	11, der11, der17	3xCEP11, 3xTel11p, 2xCEP17	Trisomy 11
	1 ⟶	17	1xCEP11, 1xTel11p, 2xCEP17	Monosomy 11
	3 ⟶	Der11, 17, der17	2xCEP11, 2xTel11p, 3xCEP17	Trisomy 17
	1 ⟶	11	2xCEP11, 2xTel11p, 1xCEP17	Monosomy 17
	3 ⟶	11, 17, der17	2xCEP11, 3xTel11p, 3xCEP17	Partial trisomy 11, partial trisomy 17
	1 ⟶	Der11	2xCEP11, 1xTel11p, 1xCEP17	Partial monosomy 11, partial monosomy 17

Table 1. Possible segregation patterns for the carrier of 46, XX, t(11;17)(q13.3;p11.2).

Meiotic Behaviour of Chromosomes Involved in Structural Chromosomal Abnormalities Determined by
Preimplantation Genetic Diagnosis

125

Once it has been established that a particular probe strategy is informative for the translocation chromosomes, the conditions of the FISH protocols are optimised and the couple can commence their stimulation to undergo their IVF cycle. The biopsied samples are then tested at PGD using the optimised protocol and embryos are diagnosed as balanced or unbalanced. Due to the fact that in the majority of cases of the nuclei of the biopsied cells during cleavage stage biopsy are in the interphase stage, it is not possible to distinguish between balanced embryos that will be balanced carriers of the structural aberration and normal embryos (Munne, 2005). By considering the FISH signals for each embryo it is then possible to determine the segregation pattern that took place in the gamete of the carreir parent, as shown in table 1, although crossing over events between the centromere and the breakpoint can complicate the situation further and affect the interpretation of the results (Hulten, 2011). Only biopsied samples with two signals for each probe used will be balanced as two signals indicate diploid status for the loci tested. Any other combination of signals is unbalanced for the translocation chromosomes bue to chromosome malsegregation at meiosis.

Figure 4 shows another example of a reciprocal translocation between chromosomes 10 and 11, 46,XY,t(11;19)(q12.3;q13.1), the probe strategy used at PGD (figure 4ii) and images from single blastomeres that were biopsied on day 3 from cleavage stage embryos (figure 4ii).

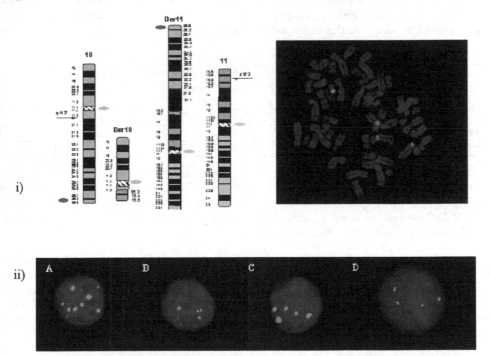

Fig. 4. Probe strategy for translocation 46,XY,t(11;19)(q12.3;q13.1) and FISH on the lymphocytes of the carrier parent and ii) images from the biopsied single cells from cleavage stage embryos at PGD. Only nucleus D is balanced for the translocation, whereas nucleus A is unbalanced and polyploid, nucleus B shows partial monosomy 10q and nucleus C shows partial trisomy 10.

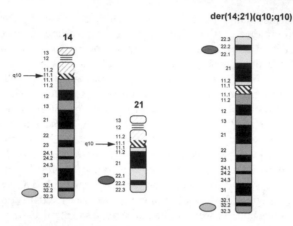

Fig. 5. Probe strategy for PGD for the Robertsonian translocation between chromosomes 14 and 21.

For Robertsonian translocations, the expected segregation patterns can also be worked out as shown in table 2. The resulting gametes are either nullisomic or disomic for either of the chromosomes involved in the translocation, resulting in monosomic or trisomic embryos (Scriven et al., 2001).

Segregation of the translocation chromosomes present in the gamete of the carrier parent	FISH signals seen in the embryo	Outcome
14, 21	2xTel14q, 2xLSI21	Normal
Der14/21	2xTel14q, 2xLSI21	Balanced
14, der14/21	3xTel14q, 2xLSI21	Trisomy 14
21	1xTel14q, 2xLSI21	Monosomy 14
14	2xTel14q,1xLSI21	Monosomy 21
21, der14/21	3xTel14q, 2xLSI21	Trisomy 21

Table 2. Different possible segregation patterns at meiosis for the carrier of the Robertsonian translocation between chromosomes 14 and 21.

Since the first clinical PGD cases for carriers of translocations and structural abnormalities FISH was the method of choice for embryo testing in order to select those embryos that were balanced. A disadvantage of FISH however is that it only gives information for the chromosomes for which probes were used. Additional probes can be included in a subsequent round of hybridization for other chromosomes in order to test for the main chromosomes at risk of being aneuploid. Recently the application of aCGH has been reported in PGD for carriers of reciprocal or Robertsonian translocations as well as inversions (Alfarawati et al., 2011.) which allows the complete enumeration of all chromosome sets in a sample.

Results from PGD cycles for structural abnormalities using array CGH, have revealed aneuploidies for chromosomes other than those involved in the translocation, which could account for the poorer PGD outcome for women of advanced maternal age. Array CGH

Meiotic Behaviour of Chromosomes Involved in Structural Chromosomal Abnormalities Determined by
Preimplantation Genetic Diagnosis

127

can also be used for the detection of other structural abnormalities, provided that the smallest translocated segment is detectable by the resolution of the array platform used. Alternatively when the segments are small FISH can be used. In the case of other structural aberrations, such as inversions, insertions, duplications and deletions, FISH involves the use probes that are included in the segment that is inverted, inserted, duplicated or deleted and one or two other probes on either side (proximally and distally) of that segment in order to be able to detect all different meiotic outcome combinations. Figure 6 below shows examples of strategies used in PGD for an inversion and an intrachromosomal insertion.

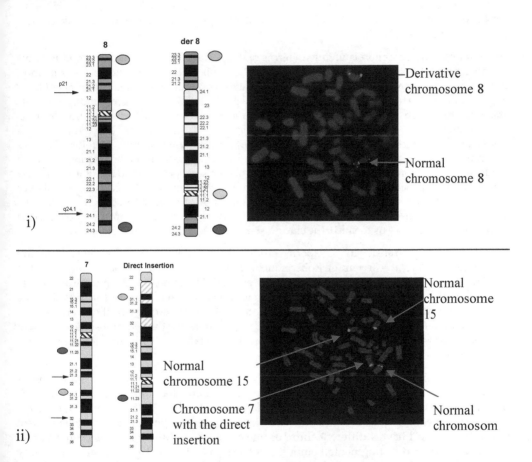

Fig. 6. i) Probe strategy for PGD for an intrachromosomal inversion, 46, XX, inv(8)(p21;q24.1) and FISH on the lymphocytes of the carrier parent, ii) probe strategy for PGD for an intrachromosomal insertion, 46, XY, dir ins (7)(p22q32q31.1) FISH on the lymphocytes of the carrier parent. An extra probe was used in this protocol for chromosome 15 to check for ploidy.

5. Evidence obtained from PGD treatment cycles relevant to the meiotic segregation of chromosomes involved in structural abnormalities

What is unique about studying preimplantation embryos is that all the products from all different modes of segregation from a structural aberration can be seen, which are not viable at later stages of development. Moreover in cases where the structural aberration involves a small segment, unbalanced forms are more likely to be tolerated until later stages of development.

For carriers of reciprocal translocations overall alternate segregation is reported as being the most common segregation pattern (Scriven et al., 2000), followed by adjacent-1 (25%), 3:1 (15%), adjacent-2 (10%) and 4:0 (2%), where most 3:1 segregations came from female carriers (Ogilvie and Scriven, 2002).

However each translocation is unique in terms of the position of the breakpoints and in terms of the length of the translocated segments. As a result different translocations will form different quadrivalents at meiosis I and in each case the configuration of the quadrivalent cross will be different. The quadrivalent configuration partly influences which spindle fibre gets attached to which centromere thus determining the way in which the translocation chromosomes will segregate (Gardner and Sutherland, 2004). The number and the position of chiasmata is also important (Scriven et al., 1998). Each translocation therefore behaves differently at meiosis (Conn et al., 1999) and hence for each translocation the frequency of each mode of segregation will be different and a particular segregation mode might occur at a higher frequency that others (Jalbert et al., 1980). In this way particular translocations will have a prediscposition towards a specific mode of segregation. In order to confirm the meiotic segregation patterns seen on PGD, full follow up analysis is required on the untransferred embryos, which allows us to study the chromosomal ploidy status of those embryos.

For Robertsonian translocations sperm studies have shown that the most common segregation pattern in carriers of Robertsonian translocations is alternate segregation that results in normal or balanced gametes (Ogur et al., 2006), whereas an interchromosomal effect was also seen, as aneuploidy for other chromosomes that were not involved in the translocation were seen. This interchromosomal effect, referring to the translocation chromosomes affecting the recombination and segregation of other chromosomes was also suggested for reciprocal translocations by Estop et al. (2000), but it was not detected in later studies (Oliver-Bonet et al., 2004).

Follow up analysis on untransferred embryos has revealed that cleavage stage embryos show a high level of mosaicism, a situation whereby more than one different cell lines is present in the embryo (Munne et al., 1993, Harper et al., 1995, Delhanty and Handyside, 1995). An extreme form of mosaicism is chaoticism, whereby almost every nucleus present in the embryo will have a different chromosome constitution. Mosaicism has been reported not only in arrested or fragmented embryos but also in embryos of good quality (Delhanty et al., 1997) and has been observed in the embryos of both young and older women (Munne et al., 1995). Different factors are thought to be involved in the formation of mosaic embryos, such as the ovarian stimulation protocol used or the embryo culture conditions, as well as the fact that cell-cycle checkpoints are not fully functional at those early stages of preimplantation embryo development (Delhanty and Handyside, 1995).

A high level of mosaicism has been reported in embryos from reciprocal translocation carriers (Simopoulou et al., 2003), Robertsonian translocation carriers (Conn et al., 1998) as

Meiotic Behaviour of Chromosomes Involved in Structural Chromosomal Abnormalities Determined by
Preimplantation Genetic Diagnosis

129

well as other forms of structural abnormalities such as intrachromosomal insertions (Xanthopoulou et al., 2010). As far as the embryos from translocation carriers are concerned, Iwarsson et al. (2000) reported that the chromosomes involved in the translocation show an even higher degree of mosaicism when compared to control chromosomes. This suggests that the translocation chromosomes not only have a high risk of meiotic malsegregation leading to embryos with unbalanced genomes, but also have a predisposition to segregate unfavourably during the following postzygotic divisions and produce highly mosaic, chaotic embryos (Simopoulou et al., 2003). As a result errors in chromosome segregation during subsequent mitotic divisions can complicate the situation further (Conn et al., 1999). The chaotic nature of the chromosomes in the biopsied samples therefore might therefore produce signals that are not characteristic of a particular segregation pattern. Furthermore Delhanty et al. (1997) observed that there is a patient-related predisposition towards the production of chaotic embryos.

In addition to meiotic malsegregation and postzygotic errors resulting in mosaicism, recently testing embryos at PGD using aCGH has revealed a high level of abnormalities affecting other chromosomes apart from those involved in the structural aberration. More specifically Alfarawati et al. (2011) report that 28.9% of the embryos that were balanced for the aberration chromosomes had an aneuploidy for other chromosomes. Follow up analysis on the untransferred embryos can reveal whether there aneuploidies for other chromosomes are meiotic in origin. Figure 7 below shows results from aCGH PGD cases for reciprocal and Robertsonian translocations. In each case the karyotype of the carrier parent is shown together with the abnormalities detected at PGD. Figure 7i shows a normal, euploid profile, whereas the rest of the images show different abnormalities present, which are highlighted.

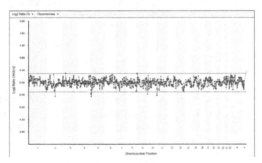

Fig. 7i. Normal euploid aCGH profile of a male embryo

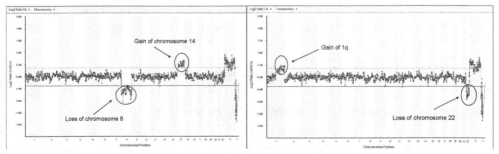

Fig. 7ii. aCGH profiles from cleavage stage embryos from 45, XX, der(13;14)(q10;q10)

The image on the left shows a female embryo with a gain of chromosome 14 as a result of the translocation, but also a loss of chromosome 8. The image on the right shows a female embryo that is balanced for the translocation chromosomes but has a gain of 1q and a loss of chromosome 22.

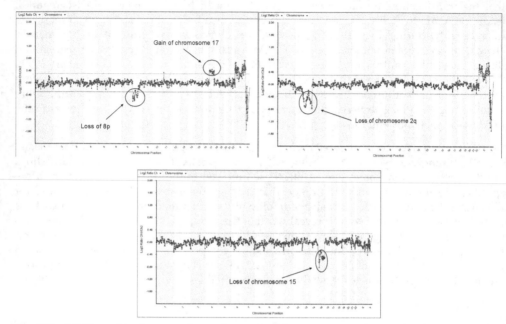

Fig. 7iii. aCGH profiles from trophectoderm after blastocyst stage biopsy for the reciprocal translocation 46, XX, t(8;17) (q21.1;p11.2).

The image on the top left shows a female embryo with a loss of 8p and a gain of chromosome 17 as a result of the translocation, whereas the image on the top right shows a female embryo that was balanced for the translocation chromosomes but had a loss of 2q. The image in the middle shows a male embryo that was balanced for the translocation chromosomes but had loss of chromosome 15.

As a result, apart from unfavourable meiotic segregation, mosaicism and postzygotic errors contribute to the reproductive challenges faces by couples carrying structural abnormalities. Array CGH therefore might be a more appropriate method for PGD for those patients as it allows screening of all chromosomes and therefore aids in choosing those embryos that are viable.

6. Other factors affecting chromosome segregation at meiosis in carriers of reciprocal translocations

6.1 Sex of the carrier parent and segregation mode at meiosis

For couples treated with PGD for reciprocal translocations the overall number of balanced embryos does not seem to be different between male and female carriers (Xanthopoulou et al., 2011). Munne et al. (2000) reported that the meiotic segregation patterns found at female

Meiotic Behaviour of Chromosomes Involved in Structural Chromosomal Abnormalities Determined by
Preimplantation Genetic Diagnosis

131

carriers of Robertsonian translocations were different from those described in male carriers, with females showing a higher level of unbalanced gametes, however this observation was not confirmed by later studies (Munne, 2005). However for one family with an intrachromosomal insertion there seemed to be an effect of the sex of the carrier parent on the mode of segregation (Xanthopoulou et al., 2010).

Chromosomal insertions are rare forms of structural chromosomal abnormalities, that involve breakpoints on the same chromosome and they can be interchromosomal or intrachromosomal depending on whether the material that is broken off from one chromosome is inserted at another site within the same chromosome or is inserted on another chromosome. Moreover insertions can be direct or inverted depending on whether the orientation of the inserted segment is the same or changes with relation to the centromere. An example of an intrachromosomal insertion and the way that chromosomes can segregate at meiosis is shown in figure 8.

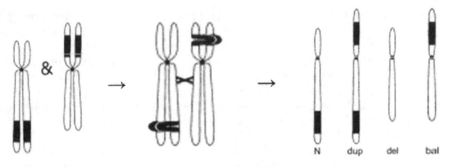

Fig. 8. Chromosome segregation in a carrier of an intrachromosomal insertion at meiosis. Key: N: normal, dupl: duplication of the inserted segment, del: deletion of the inserted segment, bal ins: balanced chromosome that carries the insertion

Balanced carriers of direct intrachromosomal insertions are phenotypically normal but are at risk of unbalanced meiotic segregation due to crossing over. At PGD it is possible to study all possible segregation modes, which might not be viable in later stages of development.

Full follow up analysis on 22 untransferred embryos from a female carrier and 19 embryos from a male carrier of the same intrachromosomal insertion, indicated that the female carrier produced far more balanced embryos (45% versus 16%, Xanthopoulou et al., 2010).

6.2 Maternal age

Advanced maternal age has long been associated with an increase in the rate of aneuploidy, for example trisomy 21, whereas preimplantation embryos from women of advanced maternal age referred for Preimplantation Genetic Screening (PGS) also seem to have an increased level of meiotic errors (Mantzouratou et al., 2007).

Ogilvie and Scriven (2002) reported a higher percentage of 3:1 segregation modes in female carriers of reciprocal translocations, resembrling aneuploidy non-disjunction, but no maternal age associations were made. Advanced maternal age is not considered to increase levels of aneuploidy for the chromosomes involved in a structural abnormality in embryos.

However women of advanced maternal age referred for PGD for reciprocal translocations have reduced chances of achieving a pregnancy (Xanthopoulou et al., 2010) due to the high level of meiotic aneuploidy for chromosomes other than the ones involved in the aberration as well as due to malsegregation of the aberration chromosomes.

7. Conclusion

PGD provides a unique opportunity to study the behaviour of chromosomes involved in structural abnormalities during meiosis. At those early stages of human preimplantation development all the different modes of meiotic segregation which might not be viable at later stages of development can be seen. As described above carriers of structural chromosomal abnormalities face a high risk of malsegregation at meiosis.

8. References

Alfarawati S, Fragouli E, Colls P, Wells D, 2011, First births after preimplantation genetic diagnosis of structural chromosome abnormalities using comparative genomic hybridization and microarray analysis, *Hum Reprod.*;26(6):1560-74.

Cohen O, Cans C, Mermet MA, Demongeot J, Jalbert P, 1994, Viability thresholds for partial trisomies and monosomies. A study of 1,159 viable unbalanced reciprocal translocations, *Hum Genet.*; 93(2):188-94.

Conn CM, Harper JC, Winston RM, Delhanty JD, 1998, Infertile couples with Robertsonian translocations: preimplantation genetic analysis of embryos reveals chaotic cleavage divisions, *Hum Genet.*; 102(1):117-23.

Conn CM, Cozzi J, Harper JC, Winston RM, Delhanty JD, 1999, Preimplantation genetic diagnosis for couples at high risk of Down syndrome pregnancy owing to parental translocation or mosaicism, *J Med Genet.*; 36(1):45-50.

Coonen E, Hopman AH, Geraedts JP, Ramaekers FC, 1998, Application of in-situ hybridization techniques to study human preimplantation embryos: a review, *Hum Reprod Update.*;4(2):135-52.

Delhanty JD, Handyside AH, 1995, The origin of genetic defects in the human and their detection in the preimplantation embryo, *Hum Reprod Update.*; 1(3):201-15.

Delhanty JD, Harper JC, Ao A, Handyside AH, Winston RM, 1997, Multicolour FISH detects frequent chromosomal mosaicism and chaotic division in normal preimplantation embryos from fertile patients, *Hum Genet.*; 99(6):755-60.

Dokras A, Sargent IL, Ross C, Gardner RL, Barlow DH, 1990, Trophectoderm biopsy in human blastocysts, *Hum Reprod.*;5(7):821-5.

Estop AM, Cieply K, Munne S, Surti U, Wakim A, Feingold E, 2000, Is there an interchromosomal effect in reciprocal translocation carriers? Sperm FISH studies, *Hum Genet.*; 106(5):517-24.

Findlay I, Quirke P, Hall J, Rutherford A, 1996, Fluorescent PCR: a new technique for PGD of sex and single-gene defects, *J Assist Reprod Genet.*;13(2):96-103.

Gardner RJM, Sutherland GR, 2004, Chromosome abnormalities and gentic counceling, 3rd Edition, Oxford University Press, New York.

Griffin DK, Handyside AH, Penketh RJ, Winston RM, Delhanty JD, 1991, Fluorescent in-situ hybridization to interphase nuclei of human preimplantation embryos with X and Y chromosome specific probes, *Hum Reprod.*;6(1):101-5.

Meiotic Behaviour of Chromosomes Involved in Structural Chromosomal Abnormalities Determined by
Preimplantation Genetic Diagnosis

133

Harper JC, Coonen E, Handyside AH, Winston RM, Hopman AH, Delhanty JD, 1995, Mosaicism of autosomes and sex chromosomes in morphologically normal, monospermic preimplantation human embryos, Prenat Diagn.; 15(1):41-9.

Hultén MA, 2011, On the origin of crossover interference: A chromosome oscillatory movement (COM) model, Mol Cytogenet.; 8;4:10.

Iwarsson E, Malmgren H, Inzunza J, Ahrlund-Richter L, Sjöblom P, Rosenlund B, Fridström M, Hovatta O, Nordenskjöld M, Blennow E, 2000, Highly abnormal cleavage divisions in preimplantation embryos from translocation carriers, Prenat Diagn.;20(13):1038-47.

Jacobs PA, Browne C, Gregson N, Joyce C, White H, 1992, Estimates of the frequency of chromosome abnormalities detectable in unselected newborns using moderate levels of banding, J Med Genet.;29(2):103-8.

Jalbert P, Sele B, Jalbert H, 1980, Reciprocal translocations: a way to predict the mode of imbalanced segregation by pachytene-diagram drawing, Hum Genet.; 55(2):209-22.

Lieber MR, Ma Y, Pannicke U, Schwarz K, 2003, Mechanism and regulation of human non-homologous DNA end-joining, Nat Rev Mol Cell Bio, 4(9):712-20.

Lupski JR, 2004, Hotspots of homologous recombination in the human genome: not all homologous sequences are equal, Genome Biol.;5(10):242.

Mantzouratou A, Mania A, Fragouli E, Xanthopoulou L, Tashkandi S, Fordham K, Ranieri DM, Doshi A, Nuttall S, Harper JC, Serhal P, Delhanty JD, 2007, Variable aneuploidy mechanisms in embryos from couples with poor reproductive histories undergoing preimplantation genetic screening, Hum Reprod.; 22(7):1844-53.

Munné S, Lee A, Rosenwaks Z, Grifo J, Cohen J, 1993, Diagnosis of major chromosome aneuploidies in human preimplantation embryos, Hum Reprod.; 8(12):2185-91.

Munné S, Alikani M, Tomkin G, Grifo J, Cohen J, 1995, Embryo morphology, developmental rates, and maternal age are correlated with chromosome abnormalities, Fertil Steril.;64(2):382-91.

Munné S, Escudero T, Sandalinas M, Sable D, Cohen J, 2000, Gamete segregation in female carriers of Robertsonian translocations, Cytogenet Cell Genet.; 90(3-4):303-8.

Munne S, 2005, Analysis of chromosome segregation during preimplantation genetic diagnosis in both male and female translocation heterozygotes, Cytogenet Genome Res; 111:305-309.

Mackie Ogilvie C, Scriven PN, 2002, Meiotic outcomes in reciprocal translocation carriers ascertained in 3-day human embryos, Eur J Hum Genet.; 10(12):801-6.

Ogur G, Van Assche E, Vegetti W, Verheyen G, Tournaye H, Bonduelle M, Van Steirteghem A, Liebaers I, 2006, Chromosomal segregation in spermatozoa of 14 Robertsonian translocation carriers, Mol Hum Reprod.;12(3):209-15.

Oliver-Bonet M, Navarro J, Codina-Pascual M, Abad C, Guitart M, Egozcue J, Benet J, 2004, From spermatocytes to sperm: meiotic behaviour of human male reciprocal translocations, Hum Reprod.; 19(11):2515-22.

Pfeiffer P, Goedecke W, Kuhfittig-Kulle S, Obe G, 2004, Pathways of DNA double-strand break repair and their impact on the prevention and formation of chromosomal aberrations, Cytogenet Genome Res., 104(1-4):7-13.

Scriven PN, Handyside AH, Ogilvie C, 1998, Chromosome translocations: segregation modes and strategies for preimplantation genetic diagnosis, Prenat Diagn; 18:1437-1449.

Scriven PN, O'Mahony F, Bickerstaff H, Yeong CT, Braude P, Mackie Ogilvie C, 2000, Clinical pregnancy following blastomere biopsy and PGD for a reciprocal translocation carrier: analysis of meiotic outcomes and embryo quality in two IVF cycles, *Prenat Diagn.;20(7):587-92.*

Scriven PN, Flinter FA, Braude PR, Ogilvie CM, 2001, Robertsonian translocations-- reproductive risks and indications for preimplantation genetic diagnosis, *Hum Reprod.*; 16(11):2267-73.

Shaw CJ, Lupski JR, 2004, Implications of human genome architecture for rearrangement-based disorders: the genomic basis of disease, *Hum Molec Genet*, 13(1):R57-R64.

Shaw CJ, Lupski JR, 2005, Non-recurrent 17q11.2 deletions are generated by homologous and non-homologous mechanisms, *Hum Genet*, 116:1-7.

Simopoulou M, Harper JC, Fragouli E, Mantzouratou A, Speyer BE, Serhal P, Ranieri DM, Doshi A, Henderson J, Rodeck CH, Delhanty JD, 2003, Preimplantation genetic diagnosis of chromosome abnormalities: implications from the outcome for couples with chromosomal rearrangements, *Prenat Diagn; 23(8):652-62.*

Therman E, Susman B, Denniston C, 1989, The nonrandom participation of human acrocentric chromosomes in Robertsonian translocations, *Ann Hum Genet.*; 53(Pt 1):49-65.

Toffolatti L, Cardazzo B, Nobile C, Danieli GA, Gualandi F, Muntoni F, Abbs S, Zanetti P, Angelini C, Ferlini A, Fanin M, Patarnello T, 2002, Investigating the mechanism of chromosomal deletion: characterization of 39 deletion breakpoints in introns 47 and 48 of the human dystrophin gene, *Genomics*; 80(5):523-30.

Verlinsky Y, Pergament E, Strom C, 1990, The preimplantation genetic diagnosis of genetic diseases, *J In Vitro Fert Embryo Transf.;7(1):1-5.*

Xanthopoulou L, Mantzouratou A, Mania A, Cawood S, Doshi A, Ranieri DM, Delhanty JD, 2010, Male and female meiotic behaviour of an intrachromosomal insertion determined by preimplantation genetic diagnosis, *Mol Cytogenet.*; 3(1):2.

Xanthopoulou L, Mantzouratou A, Mania A, Ghevaria H, Ghebo C, Serhal P, Delhanty JD, 2011, When is old too old for preimplantation genetic diagnosis for reciprocal translocations? *Prenat Diagn.* Jun 28. doi: 10.1002/pd.2813. [Epub ahead of print]

8

Insight Into the Molecular Program of Meiosis

Hiba Waldman Ben-Asher and Jeremy Don[*]

*The Mina & Everard Goodman Faculty of Life Sciences, Bar-Ilan University,
Israel*

1. Introduction

"We estimate that >2,300 genes (~4% of the mouse genome) are dedicated to male germ cell-specific transcripts, 99% of which are first expressed during or after meiosis". This quotation from a paper published by Schultz et al. (2003), reflects the tremendous complexity of gamete production, the essence of which is meiosis. Meiosis is a differantiative process in which seemingly contradicting molecular pathways are activated simultaneously. On one hand the regular components and checkpoints of the cell division machinery, which is complex enough by itself, are utilized, but on the other hand a whole array of genes are activated to enable the unique characteristics of the meiotic division, such as partition of homologous chromosomes, and not the sister chromatides, in meiosis I, or executing cell division without a prior DNA duplication in meiosis II. On one hand double strand breaks are deliberately formed to ensure pairing of homologous chromosomes and recombination, but on the other hand a whole array of genes involved in DNA repair and safeguarding genome integrity are alerted. The meiotic complexity is also exemplified by the extensive dependence on a cross-talk between germ cells themselves, and between the differentiating germ cells and their surrounding somatic cells, i.e. Sertoli cells in the testis or granulosa cells in the ovarian follicle. Finally, the complexity of the meiotic process is depicted by the differences between males and females, regarding both the outcome of the meiotic division (four basically similar post meiotic round spermatids in the male versus one functional egg and three polar body cells that degenerate in the female), and its kinetics (a continuous process in the male versus an in-continuous process in the female). It is, therefore, absolutely crucial that the very many different molecular pathways operating during meiosis be tightly concerted and regulated. However, Virginia Hughes, in a paper published in Nature medicine (2008), stated that: "So far, scientists have identified nearly 300 DNA mutations in man with reproductive defects", implying that our understanding of the meiotic molecular network is still very limited, although significant progress has been made since 2008.

Various techniques were applied during the years to study the role of different genes during meiosis. These include: 1) developing spermatogenic cell culture systems and studying the effect of over-expressing / silencing specific genes on entry into and progressing through meiosis *in-vitro* (Farini et al. 2005; Feng et al. 2002; Nayernia et al. 2006); 2) developing genetically modified animal models, mainly mice, (including Knockout models) to study the effect of modification or deletion of a specific gene on the meiotic process *in*-vivo (review in Jamsai & O'Bryan, 2011); 3) utilizing complementation approaches to detect genes with new

---

[*] Corresponding author

meiotic functions, such as Aym1 (Malcov et al. 2004); and 4) utilizing spermatogonial cell transplantation approaches in testicular repopulation studies (Brinster 2002; Brinster & Zimmermann 1994; McLean 2005). These studies contributed greatly to understanding the role of specific genes during meiosis, but it was not until the emergence of the microarray technology and the development of sophisticated bioinformatics tools that large scale studies on meiotic molecular networks and regulation could be executed. Indeed, several microarray studies on meiotic genes were performed (Chalmel et al. 2007; Schlecht et al. 2004; Schultz et al. 2003; Shima et al. 2004; Yu et al. 2003), yielding a huge amount of new information. However, the biological significance of the transcriptomic data obtained in these experiments, in terms of understanding the molecular program of meiosis, is still an ongoing challenge.

Using mouse spermatogenesis as a model system, we recently performed a comprehensive meiotic microarray study (Waldman Ben-Asher et al., 2010). This study was based on the known developmental schedule of the first spermatogenic wave (Bellve et al. 1977; Malkov et al. 1998). According to this developmental schedule, until post-natal age of 7 days (pn d7), the seminiferous tubules within the testis contain only pre-meiotic spermatogonia cells, along side with the somatic Sertoli cells. By pn d10, spermatocytes from the first spermatogenic wave enter prophase I of the meiotic division, and by pn d12 zygotene spermatocytes first appear. At pn d14 and 17, these cells reach the early and late pachytene stage, respectively. At pn d21, post meiotic round spermatids are found and at pn d24 and d27 elongating and elongated haploid spermatids are present in the testis, respectively. Testes from pn d35 mice are expected to contain the entire spermatogenic lineage. Thus, in our microarray study we compared the testicular transcriptomes of pups at: pn d7, pn d10, pn d12, pn d14, and pn d17. In these experiments we were able to clearly define six apparent patterns of gene expression throughout meiosis (Figure 1). Given this as a starting point, we will describe in this chapter the use of several bioinformatic approaches to ascribe biological significance to our results, thus getting new insights into the molecular program of meiosis.

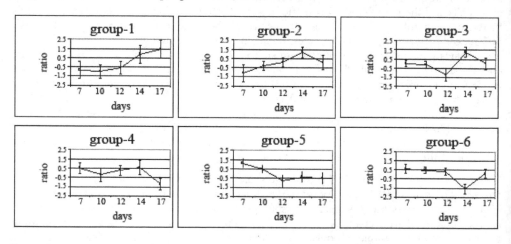

Fig. 1. Six main patterns of gene expression throughout meiosis. Mean expression level of the genes within each group is expressed as the mean ratio between the expression level and the geometric average for each developmental stage. (reproduced with permission from Waldman Ben-Asher et al., 2010).

2. Regulating meiotic gene expression

2.1 Chromosomal localization of genes as an expressional regulatory factor

One of the intriguing findings we have noticed in analyzing our microarray results was that genes from the different expressional groups are not randomly distributed throughout the genome. There are specific chromosomes that preferentially harbor genes from specific expressional groups, whereas other chromosomes are preferentially depleted of genes from specific expressional groups (Waldman Ben-Asher et al., 2010). For this analysis, we first determined, bioinformatically, the chromosomal location of each of the genes within each of the six expression groups that were obtained, and calculated the percentage of genes within each group that appear on a specific chromosome. This, was then, compared to the percentage of annotated genes from the entire mouse genome that are located on each specific chromosome. The statistical significance of the differences between the distribution of the meiotic genes within each group and that of the genes of the entire genome on each chromosome was determined using the confidence interval test, with $p<0.05$ indicating statistically significant differences. To address the randomness of the chromosomal location of genes that are specifically up-regulated or suppressed during the various meiotic stages, 1000 random lists from the entire genome, consisting of the same number of genes as in each of our six expression groups (6,000 lists altogether), were created and the mean distribution of all 1000 randomly sampled lists in each group (expressed as percentage of genes on each chromosome), ±SD, was calculated. We found that the obtained results were basically identical to those obtained with the whole genome distribution, with very small SDs. These results indicated that the distribution of annotated genes from the entire mouse genome along the chromosomes resembles random distribution, rendering the deviations in chromosomal distribution among meiotic genes, statistically and functionally significant. Our results, which are summarized in table 1, indicate that genes from group 1 are enriched on chromosome 11 and on chromosome 17, and are under represented on chromosome X. Genes from group 2 are enriched in chromosomes 3 and 15, and genes from group 3 are enriched in chromosome 11 whereas chromosome 13 is almost completely depleted of genes from this group. The distribution of genes from group 4, do not deviate significantly from the whole genome distribution. Genes from group 5 are under-represented in chromosome 4 and over-represented in chromosome 8, and group 6 genes are over-represented in chromosomes 1 and 6 and under represented in chromosome 15. Furthermore, an in-depth examination of the results in this analysis revealed a mirror-like patterns of expression of groups 2 and 6, with over and under representation on chromosome 15, respectively. This might point at chromosome 15 as containing genes that are especially required for the pachytene stage at day 14. A similar mirror-like patterns of expression exist also in groups 1 and 5 with over and under representation on chromosome 11, respectively (under representation of group 5 genes on chromosome 11 was just above the $p<0.05$ limit), suggesting that this chromosome contains meiotically-regulated genes that are not activated during the first steps of spermatogenesis, but only later, as cells enter and proceed through meiosis (p.n. days 12-17). The functional conclusion from this gene location analysis is that chromosomal location seems to be a factor in regulating gene expression during meiosis.

	Group 1	Group 2	Group 3	Group 4	Group 5	Group 6
Chromo. 1						+ 10% vs. 6%
Chromo. 3		+ 9% vs. 4%				
Chromo. 4					- 1% vs. 7%	
Chromo. 6						+ 9% vs. 5%
Chromo. 8					+ 10% vs. 5%	
Chromo. 11	+ 15% vs. 8%		+ 17% vs. 8%			
Chromo. 13			- 0.1% vs. 4%			
Chromo. 15		+ 9% vs. 3%				- 1% vs. 3%
Chromo. 17	+ 9% vs. 5%					
Chromo. X	- 0.5% vs. 4%					

Table 1. Summary of deviations in distribution of meiotic genes from the six expressional groups, along chromosomes, compared to whole genome/random distribution. All indicated deviations are statistically significant ($p<0.05$). Over-representation is denoted by (+), under-representation is denoted by (-). Percent of meiotic genes versus percent of whole genome genes are indicated. For example, 15% of group 1 genes are located on chromosome 11 versus 8% of whole genome/random distribution.

In an attempt to take the chromosomal location analysis one step forward, we asked, to what extend does genes from the same expressional group that are located on the same chromosome are clustered in the vicinity of each other. The rational for this analysis was that such clustering might enable co-regulation of expression by sharing overall chromatin organization that favors either transcription or silencing. For this analysis we used the DAVID program, a program that identifies functional groups of genes that are enriched in a given dataset compared with their representation in the entire genome (Huang et al., 2009a; Huang et al., 2009b). We, therefore, looked for genes that were clustered to specific cytobands. As shown in table 2, in four of the six groups we found only small clusters (2-5 genes), within specific cytobands, that were statistically significant compared to random distribution along the specific chromosome ($p<0.05$). This suggests that clustering to specific chromosomal regions (cytobands) might, at the most, contribute to regulation of expression at the local level but does not contribute significantly (if at all) to the overall co-regulation of expression within each group.

Group	Cytoband	No. of genes	p-value	Gene symbol	Gene name
1	10 D3	3	0.04	DDIT3	DNA-damage inducible transcript 3
				slc16a7	solute carrier family 16 (monocarboxylic acid transporters), member 7
				USP15	ubiquitin specific peptidase 15
2	17 B1	2	0.01	Ppt2	palmitoyl-protein thioesterase 2
				MSH5	mutS homolog 5 (E. coli)
2	15 A1	2	0.01	GHR	growth hormone receptor
				Lifr	leukemia inhibitory factor receptor
2	3 F2.1	3	0.02	PEX11B	peroxisomal biogenesis factor 11 beta
				TCHH	trichohyalin
				ANP32E	acidic (leucine-rich) nuclear phosphoprotein 32 family, member E
2	15 D3	3	0.04	Gml	GPI anchored molecule like protein
				Hemt1	hematopoietic cell transcript 1
				LRRC6	leucine rich repeat containing 6 (testis)
5	5 C3.1	2	0.01	pgm1	phosphoglucomutase 1
				UGDH	UDP-glucose dehydrogenase
6	4 C3	2	0.01	TYRP1	tyrosinase-related protein 1
				Ptprd	protein tyrosine phosphatase, receptor type, D
6	11 D	5	0.02	PPY	pancreatic polypeptide
				utp18	UTP18, small subunit (SSU) processome component, homolog (yeast)
				NT5C3L	5'-nucleotidase, cytosolic III-like
				copz2	coatomer protein complex, subunit zeta 2
				igfbp4	insulin-like growth factor binding protein 4
6	13 A3.3	3	0.02	LOC636537	Signal sequence receptor 4
				dsp	desmoplakin
				TXNDC5	thioredoxin domain containing 5
6	7 F5	2	0.03	Lsp1	lymphocyte specific 1
				TSPAN3	tetraspanin 32
6	3 G3	3	0.04	Gm11295 or ELOVL6	ELOVL family member 6
				TMEM56	transmembrane protein 56
				NPNT	nephronectin

Table 2. Summary of genes, from the different expressional groups, that are clustered at specific cytobands on the different chromosomes. Cytobands are denoted by the chromosome number followed by the specific cytoband location symbol. The p-value, as calculated by the DAVID program, represents the statistical significance of the clustering, compared to random distribution of the genes along the specific chromosome. Only clusters with p<0.05 were considered in this analysis.

2.2 Common cis-regulatory sequence elements within each expressional group

Unique cis-regulatory elements common to genes within a transcriptional group, if found, may explain co-regulation and similar expression patterns. To address this issue, regarding our six expression groups, we first created a file for each expression group, containing all of the gene promoter sequences in Fasta format. The length of the promoter region was defined as 1200 bp consisting of 1000 bp upstream to the transcription Initiation Site (TIS) and 200 bp downstream the TIS. The promoter region was extracted from the UCSC database, using the table application (NCBI37/mm9 assembly). Next, we used these lists as input in the Genomatix-MatInspector application (Cartharius et al., 2005) to search for matches against transcription factor (TF) recognition motifs. MatInspector is a software tool that utilizes a large library of matrix descriptions for transcription factor binding sites to locate matches in DNA sequences. The output of this application was a list of transcription factor families whose DNA recognition motifs are common to the promoters of the different expression groups. We defined a common TF family binding site as a motif which is

TF	Group1	Group2	Group3	Group4	Group5	Group6
NKXH			*	*	*	*
ETSF	*	*	*	*	*	*
HOXF	*		*	*	*	*
CREB	*	*	*	*	*	*
TBPF			*			
GATA			*	*		
FKHD	*		*		*	*
NR2F	*	*	*	*	*	*
EVI1	*		*	*	*	
MZF1			*			
PAX6		*	*			
CLOX			*	*		
SORY	*					
ZBPF		*			*	*
MYT1				*		
RXRF	*	*		*	*	*
SP1F	*	*		*	*	*
EGRF	*	*		*		*
MYBL	*			*		*
MAZF				*		*
E2FF	*	*		*		*

Table 3. Common TF binding motifs in promoter sequences of genes within each expression group. An asterisk represents the presence of the specific transcription factor target sequence in at least 90% of the genes in that group. In yellow – TF common to all groups. In red - TF common to only one group.

represented in at least 90% of the promoter sequences of each specific group (Table 3). We noticed that three transcription factor families, ETSF (Human and murine ETS1 factors), CREB (cAMP-responsive Element Binding proteins) and NR2F (nuclear receptor subfamily 2 factor), were common to all groups, suggesting they are unlikely to be responsible for the differential expression pattern of any individual group. Of special interest were four TF families which appeared only in one group: TBPF (TATA-binding protein factors) in group 3, MZF1 (Myeloid Zinc Finger 1 factors) also in group 3, SORY (Sox/Sry-sex/testis determining and related HMG box factors) in group 1 and MYT1 (MYT1 C2HC zinc finger protein) in group 4. TATA box binding protein (TBP) is a general transcription factor that plays an important role in transcription initiation of many genes. Various members of the TBP family have been identified, such as the TBP-related factors (TRFs) as well as numerous tissue-specific homologs of TBP-associated factors (TAFs) (Hochheimer & Tjian, 2003). TRF2 (known also as TLP or TRP) has a testis-specific form which is first detectable at pn d14 mouse testis and its level is increased at later stages of testicular development (Sugiura et al., 2003). Our microarray results showed a rather similar pattern of expression for TRF2 (Figure 2 – green line). Interestingly, four other genes of the TBP family (TBP, TAF1b, TAF9(2 probes) ,MED20) were present in our microarray list of meiotic regulated genes, all having the same pattern of increased level of expression from pn d12 (Figure 2). The similar expression pattern of these transcription factors through pn days 12-17 may suggest that they work together through the meiotic phase, and might account, at least in part, for the expression pattern of group 3.

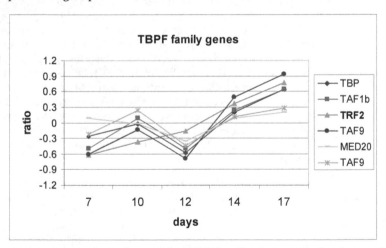

Fig. 2. Expression pattern of members of the TBPF family of transcription factors during meiosis, as obtained in our microarray analysis.

The specific binding motif for the myelin transcription factor 1 (Myt1) family appeared in the promoter sequences of more then 90% of genes of group 4. This family of transcription factors is comprised of three zinc finger genes: Myt1 (known also as Nzf2), Myt1L (known also as Png1), and Myt3 (known also as Nzf3 or St18). These transcription factors belong to the structurally unique CCHHC class, that are expressed predominantly in the developing Central Nervous System, CNS (Romm et al., 2005). Nonetheless, in rat cells, Myt1 was reported to interact with Sin3b, a protein that mediates transcriptional repression by binding to histone deacetylases (HDACs) (Romm et al., 2005). In our microarray results, the second member of the family, Myt1L, showed an increase in its expression between days 12 to 14, suggesting the potential involvement in regulating the characteristic transcriptional repression seen in group 4 between d14 and d17.

The SORY TF family consists of high mobility group (HMG) genes from two subfamilies: HMGA and HMGB. This family includes the SRY gene as well as various SOX genes, all of which function as transcriptional activators. Some of these genes were reported to play a role during spermatogenesis. For example, Sox3 expression was shown to be restricted to type A spermatogonia and to be required for spermatogenesis through a pathway that involves Ngn3 (Raverot et al., 2005). Sox7 and Sox17 were reported to function synergistically in the transcription of the Mouse laminin-α1 gene during differentiation of mouse F9 embryonal carcinoma cells into parietal endoderm cells (Niimi et al., 2004), and Hager et al. (2005) showed that laminin-α chains are vital for spermatogenesis. Notwithstanding, the expression pattern of some members of this family of transcription factors during meiosis, as depicted in our microarray analysis, do not intuitively favor specific involvement in transcriptional regulation of group 1 genes. Nevertheless, further analysis is required before definite conclusions are drawn.

The MZF1 family represents the Myeloid zinc finger protein 1 (also known as Znf42, Mzf2, Zfp98, or Zfp121). It belongs to the Krüppel family of zinc finger proteins, and it was found to play a key role in regulating transcription during differentiation along the myeloid lineage (Yan et al., 2006). These authors also demonstrated that over-expression of MZF1 repressed the ERCC1 promoter activity upon cisplatin exposure, suggesting that MZF1 might be a repressor of ERCC1 transcription. ERCC1 is a critical gene within the nucleotide excision repair pathway and only recently it was shown to play an essential role in DNA damage repair during spermatogenesis related recombination. Deficiency of this gene results in the production of abnormal sperm (Hsia et al., 2003; Paul et al., 2007). Our analysis revealed that only group 3 promoters met the limit of 90% representation of the Mzf1 TF binding site. The Mzf1 expression pattern itself was not revealed by our microarray analysis since it did not pass the stringent selection for genes exhibiting at least two-fold change in expression, compared to the geometric average, at any of the meiotic stages that were tested (Waldman Ben-Asher et al., 2010). However, it is still very well possible that Mzf1 indeed plays a role in repressing expression of meiosis-related genes, such as those of group 3.

Finally, it is, of-course, possible that the differential pattern of expression in each group is a result of a combinatorial co-regulation by several transcription factors. In this context it is noteworthy that none of the expressional groups share the same distribution of common TF motifs in their promoters (Table 3).

3. Functional analysis of gene networks – Apoptosis as a test case

Following the expression kinetics of genes, within specific gene networks, throughout meiosis, enables an insight as to how specific processes are operated and regulated during meiosis. In this study, we used apoptosis as a test case for such an analysis. Apoptosis plays a crucial role during spermatogenesis in general and meiosis in particular. It determines overall testicular cell load, balancing the proportion of the different cell types within the seminiferous tubules, and it plays a role in the removal of aberrantly differentiated meiotic spermatocytes and spermatids during and after meiosis (review in Print & Loveland, 2000).

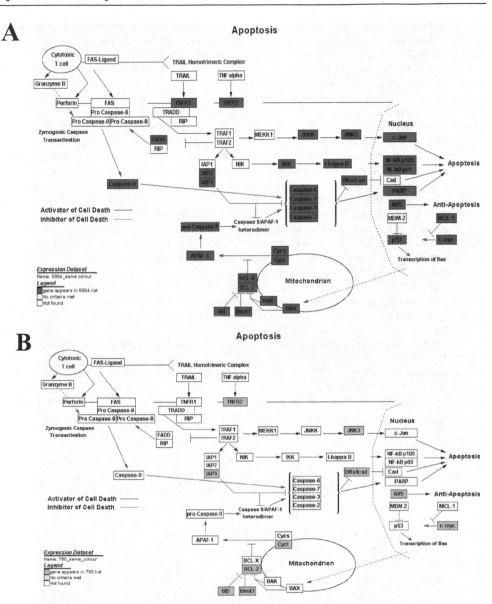

Fig. 3. Apoptotic expression maps highlighting in red genes that appear in our 6864 present sequences (A), and in green genes that appear in our 790 regulated sequences (B). These maps were obtained by applying the "Gene Map Annotator and Pathway Profiler" program to our microarray results.

Moreover, spermatocytes are unique in the sense that they "voluntarily" give up DNA integrity and undergo massive DNA breaks to enable synapsis of homologous chromosomes and crossing-over between them during meiosis. This puts conflicting requirements on the cell. On the one hand, a situation in which each and every chromosome harbors several double strand breaks (DSB) favors activation of the apoptotic pathway. On the other hand, these breaks are physiologically induced and the cells must not be sentenced to death unless breaks are not properly repaired or chromatin is not properly organized. To get an insight as to how these conflicting requirements are balanced, we applied the "Gene Map Annotator and Pathway Profiler" program to our microarray results to characterize the apoptotic pathway during meiosis. Two expression maps were used: one representing the 6864 present sequences (genes whose expression was detected in our microarray analysis but did not pass the two-fold change selection), and the other representing the 790 regulated sequences (Figure 3). 33 genes from our "present" sequences, and 10 genes from the regulated list, lighted-up using this program (Figure 3A-B). These 10 genes included TNFR2, Bid, BimEl, c-Myc and CytCt (a testis specific isoform of cytochrome C), which have a generally accepted pro-apoptotic function, and IAP3, Bcl-2, Dffa and ATF5 generally known as anti-apoptotic genes. The tenth gene, JNK3, is part of the more general MAP kinase signal transduction pathway that can either promote apoptosis or survival through activation of c-Jun (Ham et al., 2000; Kennedy & Davis, 2003). Following the specific expression pattern of these 10 regulated genes (Figure 4), it is apparent that towards the zygotene stage (pn d12), the caspase inhibitor IAP3 is up-regulated, whereas CytCt level is low, a pattern that restricts apoptosis. It is also apparent that at early pachytene (pn d14) the anti-apoptotic gene, Bcl-2, is up-regulated together with the anti apoptotic factor Dffa (Inhibitor of Caspase Activated DNase - ICAD), and BimEl, a mild negative regulator of Bcl-2. In contrast, as a mirror image, the pro-apoptotic genes Bid, which negatively regulates Bcl-2, and TNFR2, together with the anti-apoptotic transcription factor ATF5   (known also as ATFx), JNK3, and the caspase inhibitor IAP3 are down-regulated. This pattern is reversed by day 17. Thus, Bcl-2, Dffa and BimEl are down-regulated whereas Bid, TNFR2, IAP3, JNK3 and ATF5 are up-regulated. Note also that CytCt is up-regulated between pn d12 and pn d17.

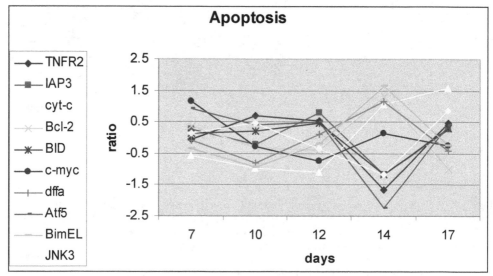

Fig. 4. Specific expression patterns of the ten genes that were highlighted in the apoptotic expression map of the 790 regulated sequences.

By drawing two maps, one for the 6864 present sequences and one for the 790 regulated sequences, we could determine two groups of apoptotic genes, the operational background genes and the actively regulated genes. The operational background genes are those whose transcript level does not change much as meiosis proceeds, but if needed, are available to execute apoptosis. The actively regulated genes are those pro and anti apoptotic genes whose transcript levels fluctuate significantly during the various meiotic stages and create a delicate balance between apoptosis and survival. DSB first appear just before zygotene, between pn d10 and pn d12, to enable synapsis of homologous chromosomes and crossing-over (review in Hochwagen & Amon, 2006). At this stage, elevated levels of the caspase inhibitor, IAP3, and low levels of CytCt seem to restrict apoptosis. As cells progress to early pachytene, high levels of Bcl2 and Dffa, together with low levels of the negative regulator of Bcl2, Bid, seem to protect cells from apoptosis and facilitate crossing-over, subsequent repair and chromatin organization. At this stage CytCt increases, IAP3 decreases, and elevated c-myc might put cells on stand-by to execute apoptosis if something goes wrong. By pn d17, representing late pachytene, DSB are repaired, and any un-repaired cell must undergo apoptosis. This is reflected by the mirror image where high CytCt , elevated Bid, TNFR2 and ATF5 together with down regulation of Bcl2 and Dffa are apparent. These results suggest that during meiosis, a delicate interplay between anti and pro-apoptotic genes and their relative abundance in a given cell determine its fate to life or death.

4. Comparing transcriptomes - A lesson to be learned

One way to ascribe biological significance to microarray results is to compare data obtained in parallel experiments on different differentiative systems sharing common molecular processes. Given that B-cell differentiation and meiosis both share DNA rearrangement processes (V(D)J recombination and meiotic recombination, respectively) we reasoned that novel insights could be obtained by comparing our meiosis microarray results to a B-cell differentiation database. Hoffmann et al. (2003) have classified the differentially expressed genes during murine B cell development into 20 clusters according to their expression pattern along the 5 differentiative stages: Pre-BI, Large Pre-BII, Small Pre-BII, Immature B and Mature B cells, and used this cluster classification to compare gene expression between parallel developmental stages of B cells and T cells. We focused our attention on genes that were highly expressed in either Pre BI cells (clusters 1, 2, 3 and 5, in Hoffmann et al, 2003), which undergo V(D)J recombination of the heavy chain (especially V to DJ rearrangement), or in small Pre BII and immature B cells (clusters 9, 10, 11, 12, 16 and 17, in Hoffmann et al, 2003) undergoing a second wave of rearrangement of the light chain (V_L to J_L). These genes were compared to meiotic genes up-regulated towards early pachytene (d14) when meiotic recombination occurs (groups 1, 2 and 3, in this study). For the comparison, the accession numbers of the 390 sequences contained within the three relevant meiotic clusters, as well as of the 955 sequences consisting of the relevant B-cell differentiation clusters (obtained from supplementary data provided by the authors in Hoffmann et al, 2003), were all translated to new Affymetrix accession numbers to form a common identification base. Following this analysis, 11 genes emerged from the cross between the meiotic genes and the Pre BI specific genes, and additional 10 genes from the comparison between the meiotic genes and the genes up-regulated in small Pre BII and immature B cells (Figure 5). A more in-depth observation at some of these genes raises interesting insights as to some of the molecular pathways operating in these processes.

4.1 Rad54l and HOP2

Up-regulation of Rad54 and Hop2 (genes characteristic of the homologous recombination DNA repair pathway) during meiosis was not unexpected since the heterodimer Hop2-Mnd1, as well as Rad54, are known to physically interact with the recombinases Rad51 or Dmc1 during meiotic recombination and to stimulate their activity by facilitating the DNA-strand-invasion step, a key step in the homologous recombination process (Petukhova et al, 2005; Sung & Klein, 2006). On the other hand, V(D)J recombination during B-lymphocyte differentiation is thought to occur through the Non-Homologous End Joining (NHEJ) pathway, and hence up-regulation of these genes was less expected. Moreover, Essers et al., (1997) showed that RAD54-/- mice are viable and exhibit apparently normal V(D)J and immunoglobulin class-switch recombination. Nevertheless, up-regulation of these genes specifically during V(D)J recombination suggests that they might indeed play some role in NHEJ processes, and that in the absence of Rad 54 there might be compensating genes that function. If this is the case, Rad54 joins other homologous recombination DNA repair genes, such as the MRN complex (Mre11, Rad50 and Nbs1) and BRCA1, that were found to play a role in the NHEJ pathway as well (Durant & Nickoloff, 2005; Sancar et al, 2004).

4.2 Mog1 and Ranbp5

These two genes are involved with Ran-GTP-dependent nuclear / cytoplasmic transport of proteins. Mog1 is a nuclear protein that stimulates the release of GTP from Ran, forming a Mog1-Ran complex which stabilizes Ran in a nucleotide-free form thereby modulating nuclear levels of RanGTP (Steggerda & Paschal, 2000; Baker et al, 2001). Ranbp5 is an importin β related protein (also known as importin β3) that acts in a nucleocytoplasmic transport pathway that is distinct from the importin-alpha-dependent import of proteins (Deane et al, 1997). Both genes were previously reported to be expressed during spermatogenesis (Li et al, 2005; Loveland et al, 2006) but the fact that both are up-regulated during DNA rearrangement processes might hint that their target proteins for nucleocytoplasmic transport are involved with DNA rearrangement.

4.3 p107

One process that a cell undergoing DNA rearrangement must avoid is cell division. It is, therefore, expected that during physiological rearrangement processes cells would repress cell cycle promoting genes. p107, a member of the Rb pocket protein family of cell cycle regulators, forms repressive complexes with either E2F4 or E2F5 (Iaquinta & Jacqueline, 2007). Such complexes have been detected by ChIP analyses in many E2F-responsive promoters of G0 cells, ensuring they do not divide. Moreover, recruitment of HDACs (histone deacetylases) to these complexes further ensures that these important cell cycle genes stay silent (Cobrinik, 2005). Up-regulation of p107 in differentiating meiotic and B-cells, might, therefore, play a role in silencing cell division genes until DNA rearrangement processes has been successfully completed.

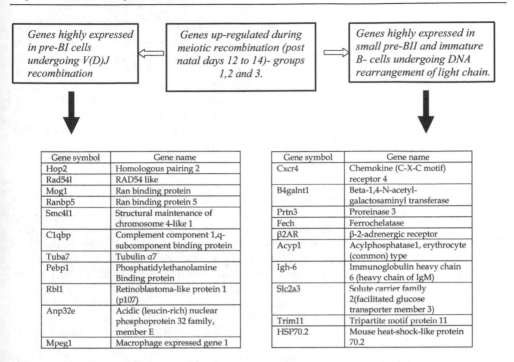

Fig. 5. Comparison between genes up-regulated towards the early pachytene stage (pn d14) where spermatocytes undergo meiotic recombination (groups 1, 2 and 3 in this study), and genes up-regulated in pre-BI or in small pre-BII and immature B-cells, undergoing V to DJ rearrangement of the heavy chain and V_L to J_L rearrangement of the light chain, respectively, during B-cell differentiation. The "B-cell differentiation" data was obtained from microarray data sets and clustering as reported by Hoffmann et al. (2003).

4.4 SMC4

In eukaryotes, the Structural Maintenance of Chromosome (SMC) proteins constitute a family of six highly conserved members of chromosomal ATPases, involved in chromosomal structural dynamics (review in Hirano, 2006). These SMC proteins form different complexes based on three SMC heterodimers. SMC1 and SMC3 form a heterodimer that, together with two other non–SMC subunits, form the cohesion complex which keeps sister chromatids together from S-phase until anaphase, when they are separated into two daughter cells. SMC2 and SMC4, form a heterodimer that together with three other non–SMC subunits compose the condensin complex, which plays an important role in mitotic/meiotic chromosomes condensation, as well as in non-mitotic chromatin condensation processes. A third pair of SMC subunits, SMC5 and SMC6, is thought to be essential for genomic integrity and DNA damage response (Hirano, 2006; De Piccoli et al., 2009). These latter SMC sub-units were reported to be highly expressed in the testes of mammals, together with a recently identified meiosis-specific SMC1 related protein (SMC1β) that was suggested to be crucial for completion of meiosis in mammals (Revenkova et al, 2004; Hirano, 2006). It is therefore interesting that among all SMC proteins, it is the SMC4 related protein that was identified in our comparison. This might suggest that SMC4, as part of the condensin complex is important for the DNA rearrangement processes. Indeed, some DNA repair roles have recently been attributed to cohesins and condensins, in addition to their traditional function, with condensins being involved specifically with single-strand break repair (Coldecott 2008; De Piccoli, 2009). This might imply that although DNA rearrangement consists basically of double strand breaks, single-strand break repair processes might also take place during rearrangement. Alternatively, condensins might play a role in DNA repair processes other then that of single-strand breaks. It is also possible, of-course, that SMC4 plays an as yet unknown role that has not been characterized to date.

4.5 Cxcr4

This gene encodes the chemokine receptor 4, a G-protein-coupled receptor for the CXCL12 chemokine (known also as SDF-1). Upon activation, this receptor mediates several biological activities, among which are the migration of primordial germ cells to the gonads (Molyneaux et al, 2003; Stebler et al, 2004), retention of primordial follicles in an un-activated state in the neonatal mouse ovary (Holt et al, 2006), and the retention of differentiating B cells in the bone marrow until maturation (Palmesino et al, 2006). Upon stimulation, CXCR4 has also been reported to induce the MAP-kinase cascade and the PI3/PKB pathway, which may elicit an anti-apoptotic response (Palmesino et al, 2006). The activated expression of Cxcr4 in both differentiating B lymphocytes in the bone marrow and during meiosis in the testis might suggest the intriguing possibility that it plays a somewhat similar role in the testis, i.e. retention of spermatocytes within the seminiferous epithelium until maturation (completion of meiosis and spermiogenesis) has been completed. Alternatively, it is also possible that it acts as a survival factor during meiosis.

4.6 β-2-adrenergic receptor (β2AR)

A disturbing unresolved discrepancy exists between the important role ascribed to the follicle-stimulating hormone (FSH) during spermatogenesis and the apparent lack of phenotype seen in FSH KO mice (review in Huhtaniemi, 2006). FSH binds to and activates the FSH receptor (FSHR) on Sertoli cells, which in turn utilize the cAMP / PKA signaling pathway to activate the CREB transcription factor. CREB activation is crucial for the ability of Sertoli cells to nurture primary spermatocytes and to support their survival. Expression of a dominant negative form of CREB in Sertoli cells resulted in apoptosis of spermatocytes (review in Don & Stelzer, 2002). Our results regarding the expression of β2AR, might provide at least a partial explanation to this discrepancy. On the one hand the β2AR was shown to activate the cAMP- PKA- CREB pathway in B-cells (Kin & Sanders, 2006). On the other hand, it was reported to be expressed in Sertoli cells of immature rats (Jacobus et al, 2005), although there are no data available as to its expression in Sertoli cells of mature animals. Our results demonstrating up-regulation of β2AR during meiosis suggest it might activate the cAMP- PKA- CREB pathway in Sertoli cells and hence compensate, at least partially, for the absence of FSH in the KO models. This hypothesis must, however, be experimentally tested.

In conclusion, the comparison between genes activated during B-cell differentiation and meiotic differentiation has focused our attention on several common genes, some of which shed light on novel molecular aspects of spermatogenesis in general, and of meiosis in particular.

5. Conclusions

The microarray technology has revolutionized the area of gene expression research by providing enormous amounts of transcriptome / proteome / phosphoproteome data, and enabling comparison between data sets representing the same tissue in different organisms, different tissues within the same organism or different treatments or conditions within a specific tissue or cell-type. The challenge in analyzing such experiments is to put these data in order and to extract the biological significance of it. In this study we used various bioinformatics tools in an attempt to ascribe biological significance to our microarray results, comparing the transcriptome of the mouse testis at five post-natal developmental ages representing different meiotic stages of the first spermatogenic wave. We found that chromosomal location of genes (but not clustering within a specific chromosome) could be a factor in determining specific patterns of gene expression during meiosis. Furthermore, we determined the distribution of common TF binding motifs in promoter sequences of genes within each of the six expressional groups that were determined (representing six major patterns of expression), pointing at specific transcription factors (or combination of transcription factors) that might contribute to the co-regulation of gene expression within each group. Expression kinetic analysis of gene networks is an important way of ascribing biological significance to microarray results. Using apoptosis as a test case we demonstrated herein how by a timely interplay between pro and anti apoptotic genes the delicate balance between the need to enable DNA breaks for pairing and recombination and the need to discard cells that their DNA has not been properly repaired, is kept. Finally, by comparing genes that are up-regulated during meiotic recombination, to genes up-regulated during DNA rearrangement in differentiating B-cells, we were able to get some new ideas regarding genes and molecular pathways operating during meiosis. Nevertheless, we have described only the tip of the iceberg of what could be concluded from our data, as well as from data obtained in corresponding studies executed by other groups, and most importantly, by the combined analyses of all these data sets. Further analysis and interpretations must await further studies.

6. References

Baker, R.P., Harreman, M.T., Eccleston, J.F., Corbett, A.H., & Stewart, M. (2001). Interaction between Ran and Mog1 is required for efficient nuclear protein import. *Journal of Biological Chemistry* 276, 41255-41262.

Bellve, A.R., Cavicchia, J.C., Millette, C.F., O'Brien, D.A., Bhatnagar, Y.M., & Dym, M. (1977). Spermatogenic cells of the prepuberal mouse. Isolation and morphological characterization. *Journal of Cell Biology* 74, 68-85.

Brinster, R.L. (2002). Germline stem cell transplantation and transgenesis. *Science* 296, 2174-2176.

Brinster, R.L., & Zimmermann, J.W. (1994). Spermatogenesis following male germ-cell transplantation. *Proceedings of the National Academy of Sciences USA* 91, 11298-11302.

Cartharius, K., Frech, K., Grote, K., Klocke, B., Haltmeier, M., Klingenhoff, A., Frisch, M., Bayerlein, M., & Werner, T. (2005). MatInspector and beyond: promoter analysis based on transcription factor binding sites. *Bioinformatics* 21, 2933-2942.

Chalmel, F., Rolland, A.D., Niederhauser-Wiederkehr, C., Chung, S.S., Demougin, P., Gattiker, A., Moore, J., Patard, J.J., Wolgemuth, D.J., Jégou, B., & Primig, M. (2007). The conserved transcriptome in human and rodent male gametogenesis. *Proceedings of the National Academy of Sciences USA* 104, 8346-8351.

Cobrinik, D. (2005). Pocket proteins and cell cycle control. *Oncogene* 24, 2796-2809.

Coldecott KW (2008) Single-strand break repair and genetic disease. *Nature Reviews Genetics* 9, 619-631.

Deane, R., Schafer, W., Zimmermann, H.P., Mueller, L., Gorlich, D., Prehn, S., Ponstingl, H., & Bischoff, F.R. (1997). Ran-binding protein 5 (RanBP5) is related to the nuclear transport factor importin-beta but interacts differently with RanBP1. *Molecular and Cellular Biology* 17, 5087-5096.

De Piccoli, G., Torres-Rosell, J., & Aragon, L. (2009). The unnamed complex: what do we know about Smc5-Smc6? *Chromosome Research* 17, 251-263.

Don, J., & Stelzer, G. (2002). The expanding family of CREB/CREM transcription factors that are involved with spermatogenesis. *Molecular and Cellular Endocrinology* 187, 115-124.

Durant, S.T., & Nickoloff, J.A. (2005). Good timing in the cell cycle for precise DNA repair by BRCA1. *Cell Cycle* 4, 1216-1222.

Essers, J., Hendriks, R.W., Swagemakers, S.M.A., Troelstra, C., deWit, J., Bootsma, D., Hoeijmakers, J.H.J., & Kanaar, R. (1997). Disruption of mouse RAD54 reduces ionizing radiation resistance. *Cell* 89, 195-204.

Farini, D., Scaldaferri, M.L., Iona, S., La Sala, G., & De Felici, M. (2005). Growth factors sustain primordial germ cell survival, proliferation and entering into meiosis in the absence of somatic cells. *Developmental Biology* 285, 49-56.

Feng, L.X., Chen, Y., Dettin, L., Pera, R.A., Herr, J.C., Goldberg, E., & Dym, M. (2002). Generation and *in vitro* differentiation of a spermatogonial cell line. *Science* 297, 392-395.

Hager, M., Gawlik, K., Nystrom, A., Sasaki, T., & Durbeej, M. (2005). Laminin alpha 1 chain corrects male infertility caused by absence of laminin alpha 2 chain. *American Journal of Pathology* 167, 823-833.

Ham, J., Eilers, A., Whitfield, J., Neame, S.J., & Shah, B. (2000). c-Jun and the transcriptional control of neuronal apoptosis. Biochem. Pharm. 60, 1015-1021.

Hirano T. (2006) At the heart of the chromosome: SMC proteins in action. *Nature Reviews Molecular Cell Biology* 7, 311-322.

Hochheimer, A., & Tjian, R. (2003). Diversified transcription initiation complexes expand promoter selectivity and tissue-specific gene expression. *Genes & Development* 17, 1309-1320.

Hochwagen, A., & Amon, A. (2006). Checking your breaks: Surveillance mechanisms of meiotic recombination. *Current Biology* 16, R217-R228.

Hoffmann, R., Bruno, L., Seidl, T., Rolink, A., & Melchers, F. (2003). Rules for gene usage inferred from a comparison of large-scale gene expression profiles of T and B lymphocyte development. *Journal of Immunology* 170, 1339-1353.

Holt, J.E., Jackson, A., Roman, S.D., Aitken, R.J., Koopman, P., & McLaughlin, E.A. (2006). CXCR4/SDF1 interaction inhibits the primordial to primary follicle transition in the neonatal mouse ovary. *Developmental Biology* 293, 449-460.

Hsia, K.T., Millar, M.R., King, S., Selfridge, J., Redhead ,N.J., Melton, D.W., & Saunders, P.T.K.(2003). DNA repair gene Ercc1 is essential for normal spermatogenesis and oogenesis and for functional integrity of germ cell DNA in the mouse. *Development* 130, 369-378.

Huang, D.W., Sherman, B.T., & Lempicki, R.A. (2009a). Systematic and integrative analysis of large gene lists using DAVID Bioinformatics Resources. *Nature Protocols* 4, 44-57.

Huang, D.W., Sherman, B.T., & Lempicki, R.A. (2009b). Bioinformatics enrichment tools: paths toward the comprehensive functional analysis of large gene lists. *Nucleic Acids Research.* 37, 1-13.

Hughes, V. (2008). Geneticists crack the code of infertility. *Nature Medicine* 14, 1174.

Huhtaniemi, I. (2006). Mutations along the pituitary - gonadal axis affecting sexual maturation: Novel information from transgenic and knockout mice. *Molecular and Cellular Endocrinology* 254, 84-90.

Iaquinta, P.J., & Jacqueline, A.I. (2007). Life and deathdecisions by the E2F transcription factors. *Current Biology* 19, 649-657.

Jacobus, A.P., Rodrigues, D.O., Borba, P.F., Loss, E.S., & Wassermann, G.F. (2005). Isoproterenol opens K+ (ATP) channels via a beta(2)-adrenoceptor-linked mechanism in sertoli cells from immature rats. *Hormone and Metabolic Research* 37, 198-204.

Jamsai, D., & O'Bryan, M.K. (2011). Mouse models in male fertility research. *Asian Journal of Andrology* 13, 139-151.

Kennedy, N.J., & Davis, R.J. (2003). Role of JNK in tumor development. *Cell Cycle* 2, 199-201.

Kin, N.W., & Sanders, V.M. (2006). It takes nerve to tell T and B cells what to do. *Journal of Leukocyte Biology* 79, 1093-1104.

Li, B., Nair, M., Mackay, D.R., Bilanchone, V., Hu, M., Fallahi, M., Song, H., Dai, Q., Cohen, P.E., & Dai, X. (2005). Ovol1 regulates meiotic pachytene progression during spermatogenesis by repressing Id2 expression. *Development* 132, 1463-1473.

Loveland, K.L., Hogarth, C., Szczepny, A., Prabhu, S.M., & Jans, D.A. (2006). Expression of nuclear transport importins beta 1 and beta 3 is regulated during rodent spermatogenesis. *Biology of Reproduction* 74, 67-74.

Malcov, M., Cesarkas, K., Stelzer, G., Shalom, S., Dicken, Y., Naor, Y., Goldstein, R.S., Sagee, S., Kassir, Y., & Don, J. (2004). Aym1, a mouse meiotic gene identified by virtue of its ability to activate early meiotic genes in the yeast *Saccharomyces cerevisiae*. *Developmental Biology* 276, 111-123.

Malkov, M., Fisher, Y., & Don, J. (1998). Developmental schedule of the postnatal rat testis determined by flow cytometry. *Biology of Reproduction* 59, 84-92.

McLean, D.J. (2005). Spermatogonial stem cell transplantation and testicular function. *Cell and Tissue Research* 322, 21-31.

Molyneaux, K.A., Zinszner, H., Kunwar, P.S., Schaible, K., Stebler, J., Sunshine, M.J., O'Brien, W., Raz, E., Littman, D., Wylie, C., & Lehmann, R. (2003). The chemokine SDF1/CXCL12 and its receptor CXCR4 regulate mouse germ cell migration and survival. *Development* 130, 4279-4286.

Nayernia, K., Nolte, J., Michelmann, H.W., Lee, J.H., Rathsack, K., Drusenheimer, N., Dev, A., Wulf, G., Ehrmann, I.E., Elliott, D.J., Okpanyi, V., Zechner, U., Haaf, T., Meinhardt, A., & Engel, W. (2006). In *vitro*-differentiated embryonic stem cells give rise to male gametes that can generate offspring mice. *Developmental Cell* 11, 125-132.

Niimi, T., Hayashi, Y., Futaki, S., & Sekiguchi, K. (2004). SOX7 and SOX17 regulate the parietal endoderm-specific enhancer activity of mouse laminin alpha 1 gene. *Journal of Biological Chemistry* 279, 38055-38061.

Palmesino, E., Moepps, B., Gierschik, P., & Thelen, M. (2006). Differences in CXCR4-mediated signaling in B cells. *Immunobiology* 211, 377-389.

Paul, C., Povey, J.E., Lawrence, N.J., Selfridge, J., Melton, D.W., & Saunders, P.T. (2007). Deletion of genes implicated in protecting the integrity of male germ cells has differential effects on the incidence of DNA breaks and germ cell loss. *PLoS. ONE.* 2, e989.

Petukhova, G.V., Pezza, R.J., Vanevski, F., Ploquin, M., Masson, J.Y., & Camerini-Otero, R.D. (2005). The Hop2 and Mnd1 proteins act in concert with Rad51 and Dmc1 in meiotic recombination. *Nature Structural & Molecular Biology* 12, 449-453.

Print, C.G., & Loveland, K.L. (2000). Germ cell suicide: new insights into apoptosis during spermatogenesis. *Bioessays* 22, 423-430.

Raverot, G., Weiss, J., Park, S.Y., Hurley, L., & Jameson, J.L. (2005). Sox3 expression in undifferentiated spermatogonia is required for the progression of spermatogenesis. *Developmental Biology* 283, 215-225.

Revenkova, E., Eijpe, M., Heyting, C., Hodges, C.A., Hunt, P.A., Liebe, B., Scherthan, H., & Jessberger, R. (2004). Cohesin SMC1 beta is required for meiotic chromosome dynamics, sister chromatid cohesion and DNA recombination. *Nature Cell Biology* 6, 555-562.

Romm, E., Nielsen, J.A., Kim, J.G., & Hudson, L.D. (2005). Myt1 family recruits histone deacetylase to regulate neural transcription. *Journal of Neurochemistry* 93, 1444-1453.

Sancar, A., Lindsey-Boltz, L.A., Unsal-Kacmaz, K., & Linn, S. (2004). Molecular mechanisms of mammalian DNA repair and the DNA damage checkpoints. *Annual Review of Biochemistry* 73, 39-85.

Schlecht, U., Demougin, P., Koch, R., Hermida, L., Wiederkehr, C., Descombes, P., Pineau, C., Jegou, B., & Primig, M. (2004). Expression profiling of mammalian male meiosis and gametogenesis identifies novel candidate genes for roles in the regulation of fertility. *Molecular Biology of the Cell* 15, 1031-1043.

Schultz, N., Hamra, F.K., & Garbers, D.L. (2003). A multitude of genes expressed solely in meiotic or postmeiotic spermatogenic cells offers a myriad of contraceptive targets. *Proceedings of the National Academy of Sciences U S A* 100, 12201-12206.

Shima, J.E., McLean, D.J., McCarrey, J.R., & Griswold, M.D. (2004), The murine testicular transcriptome: characterizing gene expression in the testis during the progression of spermatogenesis. *Biology of Reproduction* 71, 319-330.

Stebler, J., Spieler, D., Slanchev, K., Molyneaux, K.A., Richter, U., Cojocaru, V., Tarabykin, V., Wylie, C., Kessel, M., & Raz, E. (2004). Primordial germ cell migration in the chick and mouse embryo: the role of the chemokine SDF-1/CXCL12. *Developmental Biology* 272, 351-361.

Steggerda, S.M., & Paschal, B.M. (2000). The mammalian Mog1 protein is a guanine nucleotide release factor for Ran. *Journal of Biological Chemistry* 275, 23175-23180.

Sugiura, S., Kashiwabara, S., Iwase, S., & Baba, T. (2003). Expression of a testis-specific form of TBP-related factor 2 (TRF2) mRNA during mouse spermatogenesis. *The Journal of Reproduction and Development* 49, 107-111.

Sung, P., & Klein, H. (2006). Mechanism of homologous recombination: mediators and helicases take on regulatory functions. *Nature Reviews Molecular Cell Biology* 7, 739-750.

Waldman Ben-Asher, H., Shahar, I., Yitzhak, A., Mehr, R., & Don, J. (2010). Expression and Chromosomal Organization of Mouse Meiotic Genes. *Molecular Reproduction and Development* 77, 241-248.

Yan, Q.W., Reed, E., Zhong, X.S., Thornton, K., Guo, Y., & Yu, J.J. (2006). MZF1 possesses a repressively regulatory function in ERCC1 expression. *Biochemical Pharmacology* 71, 761-771.

Yu Z., Guo, R., Ge, Y., Ma, J., Guan, J., Li, S., Sun, X., Xue, S., & Han, D. (2003). Gene expression profiles in different stages of mouse spermatogenic cells during spermatogenesis. *Biology of Reproduction* 69, 37-47.

Part 2

Molecular and Cytogenetic Studies of Plant Meiosis

9

Meiotic Behavior in Intra- and Interspecific Sexual and Somatic Polyploid Hybrids of Some Tropical Species

Maria Suely Pagliarini,
Maria Lúcia Carneiro Vieira and Cacilda Borges do Valle
University of Maringá, Department of Cell Biology and Genetics,
University of Sao Paulo, College of Agriculture 'Luiz de Queiroz',
Department of Genetics, Embrapa Beef Cattle,
Brazil

1. Introduction

Hybridization, as stated by the plant evolutionist G. Ledyard Stebbins, can be viewed as a reunion between differentiated genetic materials. Plant intra and interspecific hybridization is a common means of extending the range of variation beyond that displayed by the parental species. Hybridization is a strong evolutionary force which can potentially reshape the genetic composition of populations and create novel genotypes that facilitate adaptation to new environments (Stebbins, 1950).

Interspecific hybridization provides information on phylogenetic relationships between any two species giving clues with regard to evolutionary patterns. Often generation of such information is based on cross compatibility, chromosome association, and pollen fertility. Such information also helps in developing breeding strategies for introgression of genes from related species into economically useful species. As it creates genetic variation, it has great potential for plant improvement (Goodman et al., 1987; Choudhary et al., 2000; Sain et al., 2002).

For certain crops, plant breeders in the 20th century have increasingly used interspecific hybridization for gene transfer from a non-cultivated plant species to a crop variety in a related species. Goodman et al. (1987) presented a list of species in which gene transfer have been successful. Wild relatives may be sources of useful traits for the improvement of crops. From a plant breeding point of view it is desirable to document the possibility of transferring traits to a crop plant from its wild relatives through conventional sexual hybridization. Sexual exchanges between species as sources of genetic variability to improve crops have been made possible during the last century by the discovery of efficient ways to circumvent the natural barriers to genetic exchange (Goodman et al., 1987). However, inherent problems of specific introgression such as hybrid instability, infertility, non-Mendelian segregations, and low levels of intergenomic crossing-over can constitute important limitations (Stebbins, 1950). Moreover, features associated with polyploidy or ploidy dissimilarity between species may result in additional constraints for interspecific gene flow (Rieseberg et al., 2000).

After a hybrid plant has been successfully recovered, differences in the number or compatibility of parental chromosomes may cause sterility. Cytogenetic manipulations have been instrumental in obtaining stable gene transfers. Sterility may result from incomplete or unstable pairing of chromosomes during cell division. For a desired gene from the donor to be incorporated into a chromosome of the crop variety, recombination must take place. If the two species are closely related, natural pairing and recombination may occur (Goodman et al., 1987). High pairing affinity contributes so that once the barriers separating the species are overcome the gene pools of the two genera are interchangeable (Zwierzykowski et al., 1999).

Until recently, the results of interspecific hybridization could only be studied in a fairly indirect manner. One method was to analyze the phenotype of hybrids, such as the symmetry of morphological characters or the viability of pollen or seed. Alternatively, meiosis in hybrids could be studied by light microscopy and the degree of differentiation between hybridizing taxa estimated by analyses of chromosome pairing behavior and meiotic abnormalities (Rieseberg et al., 2000). Although both of these approaches have been extremely valuable, they can only provide glimpses into the complex interactions of alien genes and genomes following genetic recombination.

Cytological analyses are usually performed to evaluate the meiotic process in experimental hybrids. Species with close genetic affinity produce hybrids with regular chromosome pairing, while the hybrids of those more distantly related species have meiotic irregularities and are sterile (Marfil et al., 2006). In diploid interspecific hybrids, the meiotic analysis of chromosome association in the F_1 generation shows the genetic homology between the respective pairs of chromosomes. However, in interspecific tetraploid or hexaploid hybrids, chromosome pairing is affected by the number and similarity among genomes.

In this chapter, we will describe some examples of tropical species for which microsporogenesis was analyzed under light microscopy. Particularly, we will discuss the meiotic behavior in interspecific sexual tetraploid hybrids of *Brachiaria ruziziensis* x *B. brizantha* and in intraspecific sexual hexaploid hybrids of *B. humidicola*. The meiotic behavior of some artificial polyploids, namely somatic hybrids obtained through protoplast fusion in *Citrus* and *Passiflora* species will be discussed.

2. Some considerations about the *Brachiaria* genus

Brachiaria (Syn. *Urochloa* P. Beauv.), a genus of African origin, consists of about 100 species distributed across tropical and subtropical regions of the world, most of which are apomictic. It is the single most important grass genus for tropical pastures, widely adopted due to widespread adaptation to poor and acid soils commonly found in the tropics (Miles et al., 1996). Some *Brachiaria* species were introduced into South America during the second half of the 20th century and are currently used over millions of hectares as pastures for both beef and dairy cattle (Boddey et al. 2004). In Brazil, there are ten registered cultivars listed on the National Service for Cultivar Protection. Among them, only one (cv. Mulato II) is a hybrid. The others are derived from selection of the natural variability found in the genus. *Brachiaria brizantha* and *B. decumbens* are considered pivotal species in the genus since the first contains accessions resistant to the major insect-pest, spittlebugs, and the second, although lacking resistance to the insect, is well adapted to infertile, acid soils generally found in tropical savannas (Keller-Grein et al., 1996). The gene pools in the genus *Brachiaria* are not yet well defined. A throrough taxonomic study of a large germplasm collection classified *Brachiaria brizantha*, *B. decumbens* and *B. ruziziensis* in the same taxonomic group together with four other species (Renvoize et al., 1996). These species are cross-compatible. *Brachiaria* breeding is difficult to accomplish because of differences in chromosome numbers among accessions and lack of sexuality in the most important species. The majority of accessions are polyploid, mainly tetraploid (Mendes-Bonato et al., 2002; 2006a; Utsunomiya et al., 2005; Risso-Pascotto et al., 2005; 2006; Pagliarini et al., 2008) and polyploidy in the genus is correlated with apomixis (Valle & Savidan, 1996).

The *Brachiaria* germplasm collection existing at Embrapa Beef Cattle Research Center (Campo Grande, Mato Grosso do Sul, Brazil) was collected in eastern and southeastern wild tropical African savannas in the mid-1980s by the International Center for Tropical Agriculture (CIAT, Colombia) and, after, introduced to Brazil by the Embrapa Genetic Resources & Biotechnology (Cenargen/Embrapa, Brasilia, DF, Brazil). After quarantine, the *Brachiaria* collection was transferred to Embrapa Beef Cattle to serve as the basis for the breeding program.

3. Hybridization in the genus *Brachiaria*

In the existing *Brachiaria* germplasm collected in Africa, considerable genetic variation is available and much more is presently being released using sexuality to access heterozygosity otherwise fixed by apomixis (Valle & Pagliarini, 2009). Until three decades ago, the genetic improvement of *Brachiaria* species depended entirely on selection among naturally existing germplasm. Apomixis among polyploid accessions made genetic recombination impossible to exploit (Valle & Savidan, 1996). Fully sexual genotypes have since then been found, either within the species themselves or among close relatives, but generally among diploids.

Artificial hybridization in *Brachiaria* has been tried at least since the early 1970s, when Ferguson and Crowder (1974) attempted to produce hybrids by pollinating diploid (2n = 18) sexual *B. ruziziensis* with pollen from a tetraploid (2n = 36) apomictic *B. decumbens*. This early attempt was unsuccessful, and the authors suggested duplication of the chromosome number to overcome the ploidy barrier. This indeed was the right direction in *Brachiaria* breeding when an obligate sexual tetraploid was produced by the colchicine treatment of a sexual diploid *B. ruziziensis* (Swenne et al., 1981). The Belgian material became the basis of breeding in *Brachiaria*. In Brazil, the first attempt at the interspecific hybridization using the Belgian material was undertaken in 1988 at Embrapa Beef Cattle, using tetraploid apomictic accessions of *B. brizantha* and *B. decumbens* as pollen donors (Valle & Savidan, 1996).

3.1 Interspecific hybrids

3.1.1 *Brachiaria ruziziensis* (sexual, artificial tetraploid) x *B. decumbens* (apomictic, natural tetraploid)

Ndikunama (1985) reported the first successful interspecific hybridization done in Belgium, using the artificial tetraploid sexual genotype of *B. ruziziensis* as the female parent and apomictic accessions of *B. decumbens* and *B. brizantha* as pollen donors. New attempts were done by Lutts et al. (1991, 1994). They concluded that the hybrids produced significantly more seed when B. *ruziziensis* was pollinated by B. *decumbens* than when crossed to B. *brizantha* and the seedlings were also less vigorous. At Embrapa Beef Cattle (Brazil), several hybrids were also obtained from B. *ruziziensis* x B. *decumbens* crosses, but lack of spittlebug resistance among other undesirable phenotypic traits made them unfit as candidates for a new cultivar. Thus, efforts were not spent in the analysis of their microsporogenesis.

3.1.2 *Brachiaria ruziziensis* (sexual, artificial tetraploid) x B. *brizantha* (apomictic, natural tetraploid)

Several hybrids between B. *ruziziensis* x B. *brizantha* were selected to continue in the breeding program based on their phenotypical performance, including seed production to attend the Brazilian as well as the export market. Meiotic behavior was analyzed in 11 hybrids obtained from these crosses. The mean frequency of abnormalities in the hybrids was variable (Table 1). Only in one hybrid, the mean of abnormalities was low (18.2%) (Felismino et al., 2010). In the other hybrids, the mean of meiotic abnormalities ranged from 44.9% to 69.1% (Risso-Pascotto et al., 2005; Fuzinatto et al. 2007, Adamowski et al. 2008, Felismino et al. 2010; Felismino, 2011). The main meiotic abnormalities were those related to irregular chromosome segregation due polyploidy. Precocious migration of univalents to the poles at metaphase I (Fig. 1 c), lagging chromosomes at anaphase I, and micronuclei at telophase I (Fig. 1 d) were recorded in the first meiotic division of most hybrids. In the second division, abnormalities were the same; micronuclei at prophase II (Fig. 1 e), precocious migration of chromatids to the poles at metaphase II (Fig. 1 f), lagging chromatids at anaphase II (Fig. 1 g), that generated micronuclei at telophase II (Fig. 1 h), that, in some cases, remained in the tetrad (Fig. 1i) were recorded.

Hybrids	Mode of reproduction	Mean of meiotic abnormalities	References
Hb S05	Sexual	67.4%	Risso-Pascotto et al., 2005
Hb A14	Apomictic	65.1%	Risso-Pascotto et al., 2005
HBGC076	Sexual	53.3%	Fuzinatto et al., 2007
HBGC009	Sexual	50.9%	Fuzinatto et al., 2007
HBGC014	Sexual	46.5%	Fuzinatto et al., 2007
H34	Sexual	69.1%	Adamowski et al., 2008
H27	Sexual	56.1%	Adamowski et al., 2008
H17	Apomictic	44.9%	Adamowski et al., 2008
HBGC313	Sexual	55.6%	Felismino et al., 2010
HBGC315	Sexual	48.3%	Felismino et al., 2010
HBGC324	Sexual	18.2%	Felismino et al., 2010
HBGC325	Apomictic	46.1%	Felismino et al., 2010
Hb 331	Apomictic	64.6%	Felismino, 2011
Hb 336	Sexual	49.8%	Felismino, 2011

Table 1. Percentage of meiotic abnormalities recorded in interspecific *Brachiaria* hybrids

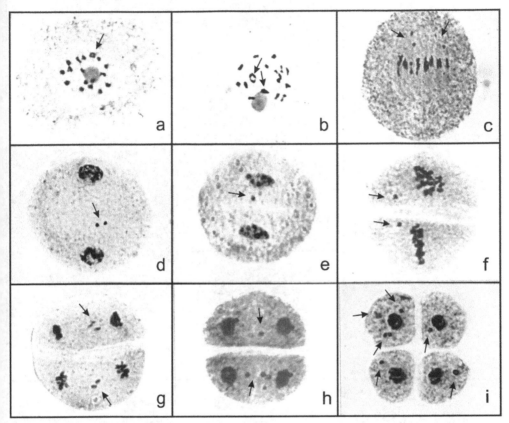

Fig. 1. Chromosome pairing and irregular chromosome segregation in interspecific *Brachiaria* hybrids. a, b) Meiocytes in diakinesis showing one (a) and two (b) quadrivalents (arrows). c) Metaphase I with precocious chromosome migration to one pole (arrows). d) Telophase I with two micronuclei (arrow). e) Early prophase II with a micronucleus (arrow). f) Metaphase II with precocious chromosome migration to one pole in both cells (arrows). g) Late anaphase II with laggard chromosomes in both cells (arrows). h) Telophase II with micronuclei in both cells (arrows). Tetrad of microspores with several micronuclei in the four microspores (i).

The meiotic behavior of micronuclei was variable among meiocytes and hybrids. In some cells, the micronuclei originated in the first meiotic division were separated by an abnormal cytokinesis, generating microcytes of different sizes and chromosome contents (Fig. 2 a – d) that entered the second division. At the end, the meiotic products depended on the number and the result of meiotic behavior of micronuclei in the first and in the second meiotic division. The end products were a tetrad with micronuclei in one, two, three or the four microspores (Fig. 2 e to g) up to a tetrad with several micronuclei in the microspores.

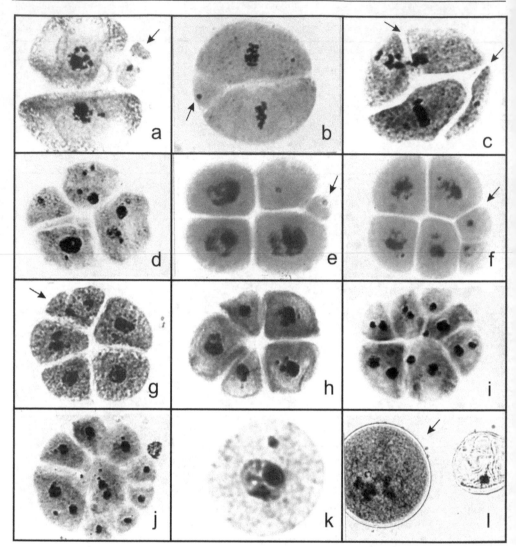

Fig. 2. Abnormal cytokinesis and the meiotic product formation in interspecific *Brachiaria* hybrids. a – c) Abnormal cytokinesis at the end of the first division leading to microcyte formation in prophase II (a) and metaphases II (b, c) (arrows) .d - g) Tetrads with micronuclei (d) and microcyte (e - g) (arrows). h) Pentad. i) Polyads with micronuclei and microcytes. k) Microspore with micronuclei (arrow). l) Viable (arrow) and sterile pollen grains.

Micronuclei were separated as microcytes with different amount of chromatin all the way to polyads with microspores and microcytes of different sizes (Fig. 2 h to j). The result was microspores of different sizes and with (Fig. 2 k) or without micronuclei. As these meiotic products are unbalanced, pollen sterility was high (Fig. 2 l).

3.1.3 *Brachiaria brizantha* (sexual, diploid) x *B. decumbens* (sexual, diploid) – tetraploidized hybrids

The first successful attempt in obtaining interspecific hybrids between *B. brizantha and B. decumbens* was recently accomplished and microsporogenesis was analyzed by Souza-Kaneshima et al. (2010). These constitute the two most widely used tropical forage species for cultivated pastures and support both beef and dairy cattle production in the tropics. Two apomictics cultivars – *B. brizantha* cv. Marandu and *B. decumbens* cv. Basilisk – cover more than 100 million hectares of cultivated pastures throughout Latin America and Southeast Asia. Artificial hybridization between two diploid (2n = 2x = 18) sexual accessions (*B. brizantha*, B105 x *B. decumbens*, D004) was performed in the greenhouse. Only three hybrids were recovered. One of them was treated with colchicine in tissue culture and two amphidiploid plants were obtained (Simioni & Valle, 2009).

The meiotic behavior was quite normal in the female (B105) and male (D004) genitors. In the diploid hybrid, genome separation was detected in a high number of cells since pachytene but chromosome association at diakinesis could not unfortunately be evaluated. At pachytene, both parental genomes were distantly positioned in the cytoplasm. The inability to share the same position inside the cell was also detected in metaphase I, when two metaphase plates were organized. In the following phases of meiosis, genome separation was not found among meiocytes, but polyads were recorded as meiotic products. These abnormalities increased the frequency of abnormal cells to 17.8% in the diploid F1 generation.

The same abnormalities recorded in the diploid hybrid were detected also in the amphidiploid hybrids CH4-8 and CH4-100, but in higher frequencies. In amphidiploids there was a predominance of bivalents at diakinesis, but one to three quadrivalents were recorded among meiocytes. Genome separation was also detected. Metaphase I with two metaphase plates and doubled anaphases I, giving rise to tetranucleated telophase I were recorded. In the second division, four cells, instead the two normal ones were observed. Polyads were also recorded among the meiotic products.

In the amphidiploid plants, however, another abnormality related to spindle organization was detected in high frequency. In the affected cells, meiosis was normal until diakinesis. From this phase on, chromosomes were chaotically dispersed in the cytoplasm because of absence of a spindle organizing center in the poles. In these cells, anaphase I did not occur and chromosomes, alone or in groups, generated several telophase nuclei of different sizes. In these cells, the first cytokinesis was abnormal, dividing the meiocyte in more than two cells. The plane of cytokinesis was apparently determined by the position of the telophase nuclei. In the second division, meiocytes with multiple spindles were abundantly recorded, generating multinucleated cells in telophase II. In both amphidiploid hybrids, meiotic products were highly abnormal, with several micronuclei and microcytes in the tetrads generating pollen grains of different sizes. The percentage of abnormal cells in the tetraploid progenies was similar, 49.2% in CH4-8 and 50.8% CH4-100.

4. Intraspecific *Brachiaria* hybrids

4.1 *Brachiaria humidicola* (sexual, natural hexaploid) x *B. humidicola* (apomictic, natural hexaploid)

Genetic variation and the means to manipulate it are the primary requirements in a breeding program. Interspecific *Brachiaria* hybrids between *B. ruziziensis* x *B. brizantha* have shown a considerable range of meiotic abnormalities which could impair pollen viability and seed production. In interspecific tetraploid hybrids within the genus, the problem can be compounded by the issues of genome affinity (Risso-Pascotto et al., 2005; Mendes-Bonato et al., 2006 b; Fuzinatto et al., 2007; Adamowski et al., 2008; Felismino et al., 2010; Souza-Kaneshima et al., 2010). Theoretically, intraspecific hybridization should produce fewer meiotic abnormalities than interspecific hybridization, thus efforts were directed at identifying sexually compatible genitors to be crossed to the natural apomicts within the same species. The discovery of a sexually reproducing, 36-chromosome accession of *B. humidicola* (H031, BRA005100) in Burundi (Africa) by Valle & Glienke (1991) opened new opportunities for the exploitation of genetic variability within this species. *Brachiaria humidicola* is adapted to poorly drained soils (Argel & Keller-Grein, 1996) and shows desirable agronomic characteristics. This accession was crossed with an apomictic cultivar (BRS Tupi) with the same ploidy level, originating from the same African region. BRS Tupi has excellent productivity, good nutritive value, and performance under grazing, and it is being released as a new *Brachiaria* cultivar for the tropics. From among 361 progeny obtained, 50 were selected based on vigor and overall suitability for further breeding work. In the female parent (H031), meiosis was somewhat irregular, with 16.3% of abnormal tetrads, whereas the male (cv. BRS Tupi) meiosis was very regular, with only 3.1% of abnormal tetrads. Among hybrids (sexual and apomictic), the percentage of abnormal tetrads ranged from 15.8% to 98.3%. Among the hybrids, high frequencies of meiotic abnormalities were unexpected both because they were intraspecific hybrids and because both parents' meioses were relatively stable.

The frequency of abnormalities at metaphase I in the hybrids was, in general, lower than that at metaphase II. The meiotic phases more affected by irregular chromosome segregation were anaphase II and telophase II. Metaphase II was affected by chromosomes outside the plate as well as anaphase II showing a great frequency of several lagging chromosomes. This suggests that the parental genomes did not display the same meiotic rhythm in the second division, generating several micronuclei in telophase II which remained in the tetrads. Problems related to differences in the meiotic rhythm have also been frequently recorded among interspecific tetraploid *Brachiaria* hybrids (Mendes-Bonato et al., 2006 b; Adamowski et al., 2008; Souza-Kaneshima et al., 2010). In general, the absence of genome affinity can result in chromosome elimination of one of the parents by asynchrony during meiotic phases.

5. Chromosome association and gene introgression

The knowledge about the similarity of the genomes in different species provides a means by which evolutionary relationships can be assessed. In addition, it offers an important starting point in alien introgression programs. According to King et al. (1999), the similarity of the genomes of the crop to those of related species is important as this permits an estimation of the frequency of recombination that will occur in interspecific hybrids. Thus, this information allows predictions to be made of the likelihood of transferring specific target genes from one species to another.

Multivalent chromosome association at diakinesis revealed genome affinity between both parental species in the hybrids, suggesting some possibility for gene introgression. Analyses of meiocytes at diakinesis in the 11 interspecific tetraploid *Brachiaria* hybrids showed that some chromosome association can occur among the parental genomes, but in low frequency.

No more than four quadrivalents were found in these hybrids (Fig. 1 a, b), but one tetravalent was the most common multiple chromosome association (Risso-Pascotto et al., 2005; Fuzinatto et al., 2007; Adamowski et al., 2008; Felismino et al., 2010; Felismino, 2011). The occurrence of quadrivalents among *B. ruziziensis* x *B. brizantha* shows that both species are closely related and their gene pools are interchangeable. Thus, we can assume that gene introgression can occur in the hybrids, but in low frequency. Recombination was expected in these hybrids because these species belong to the same taxonomic/genomic group – group 5 (Renvoize et al., 1996; Valle & Savidan, 1996). However, the occurrence of chromosome pairing per se does not ensure that the desirable genes are being introgressed into the hybrid because of the probability of the chromosomes involved containing such genes. According to Goodman et al. (1987), even when successful, interspecific hybrids generally present other problems: (1) even after several cycles of selection, the recombination will frequently not separate tightly linked genes, and undesirable traits may be carried along with the desirable one; (2) the inheritance and expression of the desirable introgressed gene may be unpredictable in the new genetic background.

6. Genome affinity

Genome separation and genome elimination are common among interspecific hybrids. Shwarzacher et al. (1992) reported several examples among plants where the phenomenon occurred. According to Luckens et al. (2006), the formation of natural allopolyploids requires the adaptation of two nuclear genomes within a single cytoplasm, which may involve programmed genetic changes during the first generation following genome fusion.

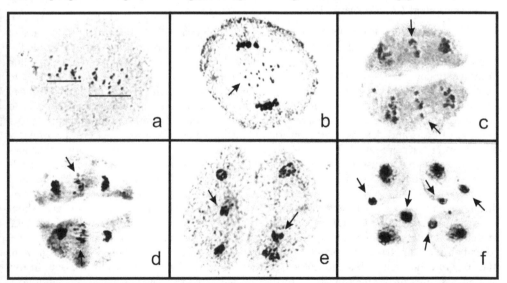

Fig. 3. Aspects of genome affinity and chromosome elimination observed in some interspecific hybrids. a) The parental genomes separated into two metaphase plates in the meiocyte. b) Meiocyte in anaphase I with a laggard genome (arrow). c, d) Meiocytes in anaphase II with a laggard genome (arrow). Meiocyte in telophase II with large micronuclei resulting from the laggard genome (arrow). f) Tetrad of microspores with large micronuclei (arrows).

Genome similarities in the genus *Brachiaria* were not studied from the cytogenetic point of view. Karyological studies in the genus were performed for a few species and accessions (Bernini & Marin-Morales, 2001), and genomes were not yet described. In the last decade, however, evaluation of the meiotic behavior in some hybrids between *B. ruziziensis* x *B. brizantha* (Risso-Pascotto et al., 2004; Mendes-Bonato et al., 2006b; Adamowski et al., 2008; Fuzinatto et al., 2007; Felismino et al., 2010; Felismino, 2011) has revealed that genomes of parental species not always show perfect affinity. Lacking of genome affinity during meiosis was revealed by asynchrony in the meiosis time (Fig. 3 b to f) (Risso-Pascotto et al., 2004; Adamowski et al., 2008) or separation of the two parental genomes into two metaphase plates (Fig. 3 a) (Mendes-Bonato et al., 2006b). The lack of genome affinity between *B. ruziziensis* and *B. brizantha* seems to be genotype-specific.

7. Artificial polyploids obtained through protoplast fusion, a case study involving *Passiflora* species

The genus *Passiflora* L. (Passifloraceae), which includes the commercial species of passionflowers, is a large and widespread genus consisting primarily of tropical species. Currently the taxonomy of the genus designates four subgenera *Astrophea* (with 57 species), *Deidamioides* (13) *Decaloba* (214) and *Passiflora* (236), in which are included the typical passion fruits (Ulmer & MacDougal, 2004). Most are vines, though some representatives are shrubs or trees. South America is the center of diversity for most of the *Passiflora* species; around 40 are indigenous to Asia and South Pacific Islands. Passion vines are evergreen climbers, grown for their edible fruits, and several species are cultivated for their unusual and beautiful flowers (Vanderplank, 1996).

Several species are important for their nutritional and pharmacological properties, but some of them are used as ornamental plants. Among all, about 50 species bear edible fruits but only the two forms of *Passiflora edulis,* i.e., the purple and the yellow ones are considered to be of value in international commerce. Although small-scale, the juice is manufactured to be exported to EU countries. The wide genetic variability, known to exist within and between *Passiflora* species can be exploited in breeding programs in order to obtain improved populations or to search for genes of interest. Commercial passion fruit varieties available are susceptible to a number of pests and diseases with considerable negative effects on production (for a review see Vieira & Carneiro, 2004; Zerbini et al., 2008). With the purpose of producing rootstocks resistant to soil-borne diseases, interspecific protoplast fusion was performed, and the hybrids between the yellow passion fruit, *P. edulis* f. *flavicarpa* and two wild species *P. amethystina* and *P. cincinnata,* all with 2n=2x=18 were obtained (Dornelas et al. 1995; Vieira & Dornelas, 1996). Briefly, all the hybrid cells were produced by PEG-mediated protoplast fusion and shoot regeneration occurred via indirect organogenesis. Selection was performed based on total protein and isoenzyme electrophoretic patterns at the callus stage (Vieira & Dornelas, 1996). After being acclimatized, the plantlets were transferred to the field where the vines grew to maturity and flourished, but did not set normal fruits, even when artificially crossed. In this section, we present the meiotic study of these somatic hybrids carried out in our laboratory (Barbosa & Vieira, 1997; Barbosa et al. 2007).

Firstly, we will discuss the meiotic behavior of the two parental diploid species, *P. edulis* f. *flavicarpa* (E) and *P. amethystina* (Am) and their four somatic hybrids denoted (E +Am) # 12, # 13, # 28 and # 35, each derived from different calli. Flower buds were collected from the vines, fixed, and slides were prepared using standard protocols. For segregation analysis, cells were studied in the diakinesis, and pollen viability (V) was determined according to Alexander (1980), which employs malachite green and fuchsinic acid.

As expected, the meiosis of the parental species showed nine bivalents (9II). In *P. edulis* f. *flavicarpa*, one or two bivalents were associated with the nucleolus in 33% or 67% of the cells, respectively, indicating the presence of two nucleolar organizing regions (NOR's). In *P. amethystina*, two bivalents were associated with the nucleolus in all cells analyzed. Chromosome pairing in the parental species was regular, although some laggard chromosomes and anaphase bridges were observed (less than 1%). The analysis of somatic hybrids showed 4x = 36 chromosomes, with four NOR's; no aneuploids were present. Meiosis was irregular with high frequencies of bridges and laggards in both divisions (Table 2). Interestingly, clear differences in the meiotic behavior among the hybrids were observed. The hybrids (E + Am) # 12 and # 28 had more regular meiosis than (E + Am) # 13 and # 35. Meiotic figures in hybrid cells display bivalents and a few univalent or multivalent. The configurations observed at diakinesis are presented on Table 2. At least 14 II were observed in 96.7% of the cells of the hybrids (E + Am) # 12 and # 35 meanwhile, the same minimum of bivalents was detected in 100% of the cells of the hybrids (E +Am) # 13 and # 28. As expected, the two sets of homologous chromosomes pair preferentially in the hybrids, reflecting the level of homology between the parental chromosomes. Moreover, the frequencies of cells with quadrivalents were 73.3% in (E + Am) # 13, 83.3% in # 12 and, 93.3% in (E + Am) # 28 and # 35.

Hybrid # 28 shows the lowest number of bridges in anaphase cells (I and II) and laggard chromosomes. The hybrid (E + Am) # 35 differs from the other three, by the largest number of cells with univalents (33.3%) which explains the high level of laggard chromosomes and chromatids in the subsequent division stages (Table 2). The present data support that recombination of the chromosomes of the two parental species occurred in the hybrids, mainly in the hybrid # 28 and # 35, since most of their cells showed the occurrence of quadrivalents. In view of the presence of quadrivalents in all hybrid genotypes analyzed, *P. edulis* f. *flavicarpa* and *P. amethystina* may possess some chromosome homoeology. Later on, we observed that *P. edulis* f .*flavicarpa* and *P. amethystina* show very similar karyotypes (Cuco et al., 2005).

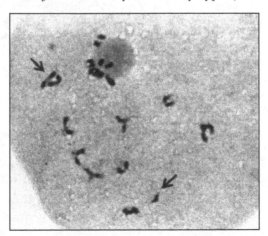

Fig. 4. Meiotic behaviour at diakinesis of the somatic hybrid (E + Am). Note the presence of bivalents, a quadrivalent (arrow) and a univalent (arrow).

Plant material	12II +4I +2IV	13II +2I +2IV	14II +4I +1IV	14II +2IV	15II +2I +1IV	16II +1IV	17II +2I	18II	Laggards² MI	AI	Bridges	Laggards² MII	AII	Chromatid Bridges
(E)	−	−	−						0.99	0.00	0.65	0.38	0.0	0.25
(Am)	−	−	−						0.44	0.00	0.79	0.68	0.0	0.23
(E+Am) #12	3.3	0.0	3.3	26.7	10.0	40.0	10.0	6.7	1.15	0.00	2.12	2.47	0.00	7.11
(E+Am) #13	0.0	0.0	0.0	16.7	13.3	43.3	6.7	20.0	5.90	0.00	3.83	4.10	1.92	4.39
(E+Am) #28	0.0	0.0	0.0	30.0	20.0	43.3	0.0	6.7	1.15	0.00	2.09	1.15	0.00	1.37
(E+Am) #35	0.0	3.3	0.0	40.0	23.3	26.7	6.7	0.0	8.11	7.57	12.37	13.16	3.95	16.05

[1]Number of cells analyzed = 30; [2]MI and MII, first and second metaphase, respectively, AI and AII, first and second anaphase, respectively.

Table 2. Meiotic behavior (in percentages) in *P. edulis* f. *flavicarpa* (E), *P. amethystina* (Am) and their four somatic hybrids (E+Am). Modified from Barbosa & Vieira (1997).

On average, all the materials analyzed presented high levels of pollen viability, ranging from 72.9% up to 88.2%. The hybrid showing the lowest mean viability is the one that presents the higher percentage of meiotic irregularities. In contrast, the high levels of mean viability found for the parental species (>95%) and the remaining hybrid plants were related to the low percentage of meiotic irregularities observed (Table 2). The correlations between pollen viabilities and meiotic irregularities were high and negative.

Similarly, the meiotic behavior of four somatic hybrids were examined and compared with their corresponding diploid fusion parents, *P. edulis* f. *flavicarpa* (E) and *P. cincinnata* (C).The meiotic behavior revealed relatively high stability, with most of the hybrid cells showing 18 bivalents. Some instability, such as a quadrivalent configuration was also recorded which has been interpreted as an interchange that occurred in the progenitors more than as a result of *in vitro* culture or chromosome reorganization in the new genome. Even in low frequencies, the occurrence of univalents resulted in misdivision, laggard and micronucleus formation. High values of pollen viability (>70%) were found in the diploid parents as well as in the hybrid plants. As expected, the course of meiosis was regular in the parental cells. Some chromosome bridges and laggards were observed, possibly as a consequence of the formation of univalents, though they were not actually recorded. These abnormalities occurred at a very low frequency, i.e. <2% of the cells.

As expected, the meiosis of the parental species showed nine bivalents (9II), two of them were attached to the nucleolus in 67% of the cells at diakinesis. In *P. cincinnata*, this percentage was higher (73%).The four hybrid plants were 4x=36; the hybrid cells *per se* were not variable in number, i.e. no aneuploid cells occurred, at least within the sample cells analyzed.

A certain frequency of laggards and anaphase bridges in both divisions was observed in all hybrid plants (Table 3). However, differences amongst them were noted: (E+C) #07 and (E+C) #25 had a more irregular meiotic behavior than (E+C) #14 and (E+C) #26, but in the second division, the configuration changed, and the hybrids #25 and #26 had more irregular behavior. The frequency of bivalents at diakinesis in the hybrids varied from 55% to 78.5%. These percentages should be associated with the presence of a quadrivalent, which was found in 30.0% and 32.5% of the cells of the hybrids #07 and #26, respectively, and in 17.1% and 13.3% of the cells of the hybrids #14 and #25, respectively. The disassociation of those quadrivalents should explain the laggards and the chromosome bridges observed at division II. The hybrids #07 and #25 showed higher percentages of univalents, which were also the plants that had cells with higher proportions of laggards at meiosis I and II. The occurrence of univalents varied from 4.2% in (E+C) #14 up to 15% in (E+C) #07 and #25, being correlated with other meiotic alterations and as well bivalent frequency at diakinesis. In the plant (E+C) #25 with 15% of diakinesis cells showing univalents, they appeared as laggards at metaphase I (12.3%) and underwent centromere misdivision (2.3%). Moreover, the results suggest that chromosomes that suffered longitudinal division at meiosis I underwent centromere misdivision at meiosis II (Table 3).

Plant material	Diakinesis configurations		Laggards[1]			Bridges	Laggards[1]		Chromatid bridges	No. cells analyzed
	17II +2I	16II +1IV	18II	MI	AI		MII	AII		
(E)	—	—	—	1.0	0.0	0.6	0.4	0.0	0.2	40
(C)	—	—	—	1.3	0.0	0.2	0.0	0.8	0.6	40
(E+C) #07	15.0	30.0	55.0	19.9	28.8	1.2	15.8	16.0	1.8	40
(E+C) #14	4.2	17.1	78.5	7.1	28.2	5.0	13.0	4.0	10.6	70
(E+C) #25	15.0	13.3	71.6	12.3	26.9	2.3	15.5	7.6	16.1	60
(E+C) #26	7.5	32.5	60.0	15.3	15.5	2.1	13.4	13.4	13.3	40

[1]MI and MII, first and second metaphase, respectively, AI and AII, first and second anaphase, respectively.

Table 3. Meiotic behavior (in percentages) in *P. edulis* f. *flavicarpa*(E), *P. cincinnata*(C) and their four somatic hybrids (E+C). Modified from Barbosa et al. (2007).

The frequency of microspores with micronuclei was relatively low (6.4% up to 23.2%) considering the abnormalities found at all the meiotic stages. Probably, many laggards forming bridges at anaphase I and II were included in telophase nuclei. Certainly, a fraction of these laggards was included as entire chromosomes. The frequency of laggards in metaphase II cells was somewhat the same in metaphase I (3). However, misdivision was five times more frequent at metaphase II, although laggards were found in a reduced frequency (three times less), suggesting that almost all the chromosomes that underwent longitudinal division at first division did undergo misdivision at meiosis II. Laggards in (E+C) #25 were observed in 7.6% of the anaphase II cells and did result in a micronucleus in 8.7% of the tetrad cells. Those chromosomes that went through longitudinal division at first meiosis underwent misdivision because they were attached to the centromeres, while laggards at anaphase II were eliminated.

Parental pollen viability values were 96%. In the hybrid plants, on average, all values were high, ranging from 71.6% up to 82.1%. For the meiotic analyses, one can first state that the two bivalents attached to the nucleolus of the parental cells correspond to the smallest chromosomes, i.e. 8 and 9. Our study on *Passiflora* using fluorescent *in situ* hybridization detected positive signals corresponding to 45S rDNA sites on the secondary constrictions of chromosome 8 and satellites of chromosome 9 of *P. edulis* f. *flavicarpa* and *P. cincinnata* (Cuco et al., 2005). In addition, chromomycin A3 bandings were found in regions corresponding to 45S rDNA sites, therefore, in two chromosome pairs. No other preferentially GC-rich regions were observed in these species. The nucleolar activity was also investigated by silver staining and four silver-positive signals were detected on the smallest chromosomes pairs.

Minimal differences are also reported between the karyotypes of *P.edulis* f. *flavicarpa* and *P. cincinnata* (Cuco et al., 2005). In addition, the DNA content amongst the *Passiflora* species (at least into the subgenus *Passiflora*) is slightly variable (Souza et al. 2004; Souza et al., 2008), and *P.edulis* f. *flavicarpa* and *P. cincinnata* are very close in terms of phylogenetics (Muschner et al. 2003; Cuco et al., 2005; Padua, 2004). Based on all those features, high chromosome stability is not expected for the somatic hybrids studied here. The presence of just a few univalents in the hybrid plants is expected considering the classical taxonomy (Killip, 1938) and the modern classification proposed for the genus *Passiflora* (Ulmer & Macdougal, 2004) that indicate the proximity of these species. Univalents and the various ways they behave throughout meiosis probably contributed to the formation of the micronuclei observed in the somatic hybrid microspores. The presence of univalents as micronuclei in tetrads was frequently associated with a reduction of vigor and fertility in hybrid plants (e.g. Cao et al., 2003) as above reported for the somatic hybrids between *P. edulis* f. *flavicarpa* and *P. amethystina*. Finally, the more irregular the meiotic behavior, the lower was the pollen viability. This correlation was also abovementioned for the previous set of hybrids (E+Am).

All species used as parents are auto-incompatible and when they were artificially crossed with the somatic hybrids (theoretically, 2n=2x=18 × 2n=4x=36 → 2n=3x=27), few seeds were obtained. The hybrids were female-sterile and pollen producers. In practical terms, the levels of pollen viability and multivalent pairing in the two types of somatic hybrids suggest that these materials can be used as rootstocks due to their compatibility to *P. edulis* f. *flavicarpa* promoting resistance to soil diseases. The three *Passiflora* species investigated are closely related as they belong to the same subgenus (*Passiflora*), although *P. amethystine* belongs to the series *Lobatae*, and *P. edulis* and *P. cincinnata* are members of the series *Passiflora*. Most passionflowers are diploids, though aneuploidy and polyploidy have been reported as evolutionary mechanisms, and have symmetrical karyotypes that could be distinguished from each other by minor differences (Melo et al., 2001; Cuco et al., 2005; Souza et al., 2008).

Intriguingly, newly formed polyploids show genetic and cytological alterations. Consequently, an interest in studies aiming at understanding these genetic events involved in artificial allopolyploid formation increased in recent years, also for *Citrus*, another important fruit species (Chen et al., 2004). In the case of the interspecific allotetraploid somatic hybrids, 'Hamlin' sweet orange + 'Rough Lemon' and 'Key'lime + 'Valencia' sweet orange, meiotic analysis revealed that <20% of the meiocytes exhibited normal tetrad formation in somatic hybrid plants, and irregular chromosome behavior with univalent or multivalent pairing also occurred. Moreover, meiotic abnormalities such as chromosome bridges, chromosomes orientated away from the equatorial plate, especially lagging chromosomes, which resulted in different sizes of pollen grains, were frequently observed in both somatic hybrids. Comparing to passionflowers hybrids, pollen germinability was very low (<15%), intermediate between their corresponding *Citrus* fusion parents. Concluding, the meiotic behavior of *Citrus* somatic hybrids provided valuable information for their practical utilization, particularly for intergenomic recombination, and increased the germplasm available for interploid crosses in *Citrus* triploid seedless breeding programs.

8. Conclusion

The meiotic analysis of polyploidy hybrids is an important element in taxonomic and evolutionary studies of plants. It also provides an important starting point in alien introgression programs and permits an estimation of the frequency of recombination that will occur in interspecific hybrids. Chromosome pairing is the most used method of assessing genomic relationships between species. This information allows predictions to be made of the likelihood of transferring specific target genes from one species to another. From accumulated results obtained through cytological studies in hybrids, it is evident that cytogenetic analyses are of prime importance in determining which genotypes can continue in the process of cultivar development and which can be successfully used in breeding programs.

9. References

Adamowski FV, Pagliarini MS, Valle CB. (2008). Meiotic behavior in three interspecific three way hybrids between *Brachiaria ruziziensis* and *B. brizantha* (Poaceae: Paniceae). *Journal of Genetics*, 87, 33-38

Alexander MP. (1980). A versatile stain for pollen from fungi, yeast and bacteria. *Stain Technology*, 55, 13-18

Argel PJ, Keller-Grein G. (1996). Regional experience with *Brachiaria*: Tropical America – Humid lowlands, In: *Brachiaria: biology, agronomy, and improvement*, Miles JW, Maass BL, Valle CB (eds), 205-224, CIAT/Embrapa, Cali

Barbosa LV, Vieira MLC. (1997). Meiotic behavior of passion fruit somatic hybrids, *Passiflora edulis* f. *flavicarpa* Degener + *P. amethystina* Mikan. *Euphytica*, 98, 121-127

Barbosa LV, Mondin M, Oliveira CA, Souza AP, Vieira MLC. (2007). Cytological behavior of the somatic hybrids *Passiflora edulis* f. *flavicarpa* + *P. cincinnata*. *Plant Breeding*, 126, 323-328

Bernini C, Marin-Morales MA. (2001). Karyotype analysis in *Brachiaria* (Poaceae) species. *Cytobios*, 104, 157-171

Boddey RM, Macedo R, Torres RM, Ferreira E, Oliveira O, Cantaruti RB, Perreira JM, Alves BJR, Urquiaga S. (2004): Nitrogen cycling in *Brachiaria* pastures: the key to understanding the process of pasture decline. Agriculture and Ecosyst.em Environment ,103, 389-403.

Cao M., Bughrarab SS, Sleper DA. (2003). Cytogenetic analysis of *Festuca* species and amphiploids between *Festuca mairei* and *Lolium perenne*. *Crop Science* 43, 1659-1662

Chen CL, Guo WW, Yi HL, Deng XX. (2004). Cytogenetic analysis of two interspecific *Citrus* allotetraploid somatic hybrids and their diploid fusion parents. *Plant Breeding*, 123, 332-337

Choudhary BR, Joshi P, Ramarao S. (2000). Interspecific hybridization between *Brassica carinata* and *Brassica rapa*. *Plant Breeding*, 119, 417-420

Cuco SM, Vieira MLC, Mondin M, Aguiar-Perecin MLR. (2005). Comparative karyotype analysis of three *Passiflora* L. species and cytogenetic characterization of somatic hybrids. *Caryologia*, 58, 220-228

Dornelas MC, Tavares FCA, Oliveira JC, Vieira MLC. (1995). Plant regeneration from protoplast fusion in *Passiflora* spp. *Plant Cell Reports*, 15, 106-110

Felismino MF, Pagliarini MS, Valle CB. (2010). Meiotic behavior of interspecific hybrids between artificially tetraploidized sexual *Brachiaria ruziziensis* and tetraploid apomictic *B. brizantha* (Poaceae). *Scientia Agricola*, 67, 1-8

Felismino MF. (2011). Microsporogenesis in interspecific *Brachiaria* hybrids. Ph. D Thesis, University of Maringá, Brazil.

Ferguson JE, Crowder LV. (1974). Cytology and breeding behaviour of *Brachiaria ruziziensis*. *Crop Science*, 14, 893-895

Fuzinatto VA, Pagliarini MS, Valle CB. (2007). Microsporogenesis in sexual *Brachiaria* hybrids (Poaceae). *Genetics and Molecular Research*, 6, 1107-1117

Goodman RM, Hauptti H, Crossway A, Knauf VC. (1987). Gene transfer in crop improvement. *Science*, 236, 48-54

Keller-Grein G, Mass BL, Hanson F. (1996). Natural variation in *Brachiaria* and existing germplasm collection, In: *Brachiaria: biology, agronomy, and improvement*, Miles JW, Maass BL, Valle CB (eds), 16-42, CIAT/Embrapa, Cali

Killip EP. (1938). The American species of Passifloraceae. *Field Museum Natural History, Botanical Series*, 19, 1-613

Luckens LN, Pires JC, Leon G, Vogelzang R, Oslack L, Osborn T. (2006). Patterns of sequence loss and cytosine methylation within a population of newly resynthesized *Brassica napus* allopolyploids. *Plant Physiology*, 140, 336-348

Lutts S, Nidikumana J, Louant BP. (1991). Fertility of *Brachiaria ruziziensis* in interspecific crosses with *Brachiaria decumbens* and *Brachiaria brizantha*: meiotic behavior, pollen viability and seed set. *Euphytica*, 57, 267-274

Lutts S, Ndikumana J, Louant BP. (1994). Male and female sporogenesis and gametogenesis in apomictics *Brachiaria brizantha*, *Brachiaria decumbens* and F_1 hybrids with sexual colchicine induced tetraploid *Brachiaria ruziziensis*. *Euphytica*, 78, 19-25.

Marfil CF, Masuelli RW, Davison J, Comai L. (2006). Genomic instability in *Solanum tuberosum* x *Solanum kurtzianum* interspecific hybrids. *Genome*, 49, 104-113.

Melo NF, Cervi AC, Guerra M. (2001). Karyology and cytotaxonomy of the genus *Passiflora* L. (Passifloraceae). *Plant Systematics Evolution*, 226, 69-84

Mendes-Bonato AB, Pagliarini MS, Forli F, Valle CB, Penteado MIO. (2002). Chromosome number and microsporogenesis in *Brachiaria brizantha* (Gramineae). *Euphytica*, 125, 419-425

Mendes-Bonato AB, Pagliarini MS, Risso-Pascotto C, Valle CB. (2006a). Chromosome number and meiotic behavior in *Brachiaria jubata* (Gramineae). *Journal of Genetics*, 85, 83-88

Mendes-Bonato AB, Risso-Pascotto C, Pagliarini MS, Valle CB. (2006b). Cytogenetic evidence for genome elimination during microsporogenesis in interspecific hybrid between *Brachiaria ruziziensis* and *B. brizantha* (Gramineae). *Genetics and Molecular Biology*, 29, 711-714

Miles, J. W., B. L. Maass, and C. B. Valle, 1996: *Brachiaria*: Biology, Agronomy, and Improvement. CIAT/Embrapa, Cali

Muschner V C, Lorenz AP, Cervi AC, Bonatto SL, Souza-Chiez TT, Salzano FM, Freitas LB. (2003). A first molecular phylogenetic analysis of *Passiflora* (Passifloraceae). *American Journal of Botany*, 90, 1229-1238

Ndikumana J. (1985). Etude de l'hybridation entre espèces apomictiques et sexuées dans le genre *Brachiaria*. Ph.D. dissertation, Université Catholique de Louvain. Belgium

Padua JG. (2004). Análise genética de espécies do gênero *Passiflora* L. com base em abordagens filogenéticas, morfométricas e em marcadores microssatélites, *PhD Thesis*, Universidade de São Paulo, Brazil

Pagliarini MS, Risso-Pascotto C, Souza-Kaneshima AM, Valle CB. (2008). Analysis of meiotic behavior in selecting potential genitors among diploid and artificially induced tetraploid accessions of *Brachiaria ruziziensis* (Poaceae). *Euphytica*, 164, 181-187

Renvoize SA, Clayton WD, Kabuye CHS. (1996). Morphology, taxonomy, and natural distribution of *Brachiaria* (Trin.) Griseb, In: *Brachiaria: biology, agronomy, and improvement*, Miles JW, Maass BL, Valle CB (eds), 1-15, CIAT/Embrapa, Cali

Rieseberg LH, Baird SJE, Gardner KA. (2000). Hybridization, introgression, and linkage evolution. *Plant Molecular Biology*, 42, 205-224

Risso-Pascotto C, Pagliarini MS, Valle CB, Jank L. (2004) Asynchronous meiotic rhythm as the cause of selective chromosome elimination in an interspecific *Brachiaria* hybrid. *Plant Cell Reports*, 22, 945-950

Risso-Pascotto C, Pagliarini MS, Valle CB. (2005). Meiotic behavior in interspecific hybrids between *Brachiaria ruziziensis* and *Brachiaria brizantha* (Poaceae). *Euphytica*, 145, 155-159

Risso-Pascotto C, Pagliarini MS, Valle CB. (2006). Microsporogenesis in *Brachiaria dictyoneura* (Fig. & De Not.) Stapf (Poaceae: Paniceae). *Genetics and Molecular Research*, 5, 837-845

Sain RS, Joshi P, Satry EVD. (2002). Cytogenetic analysis of interspecific hybrids in genus *Citrullus* (Cucurbitaceae). *Euphytica*, 128, 205-210

Schwarzacher T, Heslop-Harrison JS, Anamthawat-Jónsson K. (1992). Parental genome separation in reconstructions of somatic and premeiotic metaphases of *Hordeum vulgare* x *H. bulbosum. Journal of Cell Science*, 101, 13-24

Simioni C, Valle CB do. (2009). Chromosome duplication in *Brachiaria* (A.Rich.) Stapf allows intraspecific crosses. *Crop Breeding and Applied Biotechnology*, 9, 328 - 334.

Souza MM, Palomino G, Pereira TNS, Pereira MG, Viana PA. (2004). Flow cytometric analysis of genome size variation in some *Passiflora* species. *Hereditas*, 141, 31-38

Souza MM, Pereira TNS, Vieira MLC. (2008). Cytogenetic studies in some species of *Passiflora* L. (Passifloraceae): A review emphasizing Brazilian species. *Brazilian Archives of Biology and Technology*, 51, 247-258

Souza-Kaneshima AM, Simioni C, Felismino MF, Mendes-Bonato AB, Risso-Pascotto C, Pessim C, Pagliarini MS, Valle CB. (2010). Meiotic behavior in the first interspecific hybrid between *B. brizantha* and *B. decumbens. Plant Breeding*, 129, 186-191

Stebbins GL. (1950). Variation and evolution in plants. Columbia University Press, New York

Swenne A, Louant BP, Dujardin M. (1981). Induction par la colchicine de formes autotétraploïdes chez *Brachiaria ruziziensis* Germain et Evrard (Graminée). *Agronomia Tropicale*, 36, 134-141

Ulmer T, Macdougal JM. (2004). *Passiflora* Passionflowers of the World, Timber Press, Portland

Utsunomiya KS, Pagliarini MS, Valle CB. (2005) Microsporogenesis in tetraploid accessions of *Brachiaria nigropedata* (Ficalho & Hiern) Stapf (Gramineae). *Biocell*, 29, 295-301

Valle CB, Glienke C. (1991). New sexual accession in *Brachiaria*. *Apomixis Newsletter*, 7, 42-43

Valle CB, Savidan Y. (1996). Genetics, cytogenetics, and reproductive biology of *Brachiaria*, In In: *Brachiaria: biology, agronomy, and improvement*, Miles JW, Maass BL, Valle CB (eds), 147-163, CIAT/Embrapa, Cali

Valle CB, Pagliarini MS. (2009). Biology, cytogenetics, and breeding of *Brachiaria* In: *Genetic resources, chromosome engineering, and crop improvement*, Singh RJ (ed) 103-151, CRC Press, Boca Raton

Vanderplank J. (1996). *Passion flowers*, Cambridge Press, London

Vieira MLC, Dornelas MC. (1996). Regeneration of plants from protoplasts of *Passiflora* species (Passion Fruit). In: *Plant Protoplasts and Genetic Engineering VII, Biotechnology in Agriculture and Forestry*, Bajaj YPS (ed), 38, 108-119, Springer Verlag, Berlin

Vieira MLC, Carneiro MS. (2004). Passifloraceae *Passiflora* spp. Passionfruit. In: *Biotechnology of fruit and nut crops*, Litz R (ed), 436-453, CAB International, Oxfordshire

Zerbini FM, Otoni, WC, Vieira MLC. (2008). Passionfruit. In: *Compendium of transgenic crop plants: transgenic tropical and subtropical fruits and nuts*, Kole C, Hall T. (org.), Blackwell Publishing, Oxford, 5, 213-234

Zwierzykowski Z, Lukaszewski AJ, Naganowska B, Lesniewska A. (1999). The pattern of homologous recombination in triploid hybrids of *Lolium multiflorum* with *Festuca pratensis*. *Genome*, 40, 720-726

Quantifying Meiosis: Use of the Fractal Dimension, D_f, to Describe and Predict Prophase I Substages and Metaphase I

Cynthia Ross Friedman[1] and Hua-Feng Wang[2]
[1]*Department of Biological Sciences, Thompson Rivers University,*
[2]*Beijing Urban Ecosystem Research Station,*
[2]*State Key Laboratory of Urban and Regional Ecology,*
[2]*Research Center for Eco-Environmental Sciences, Chinese Academy of Sciences,*
[1]*Canada*
[2]*China*

1. Introduction

1.1 Meiosis – Plants do it!

Plants undergo meiosis as a part of their life cycle, a fact that is sometimes forgotten and often misunderstood. Plants have a sporic life cycle wherein the living organism alternates between haploid and diploid states. Consequently, this cycle is also known as the "alternation of generations" (Gilbert, 2000). The generalization that states, "meiosis produces haploid gametes" is not technically true with respect to the intrepid members of Kingdom Plantae. Unlike the case in animal cells, the diploid plant (the sporophyte or "spore-producing plant") possesses "germ-line" cells (sporocytes) that undergo the reductive division of meiosis I followed by meiosis II to produce haploid spores. The spores then undergo mitosis, growing into a haploid "organism", which, in the flowering plants, is a small "gametophyte" (gamete-producing plant) housed within the confines of the original spore. In flowering plants, the female gametophyte (megagametophyte) held within the female spore (megaspore) inside the flower produces the egg, while the male gametophyte (microgametophyte) develops within the male spore (microspore). The microspore is better known as an immature pollen grain, and upon maturity, formation of the microgametophyte within the pollen effectively generates sperm. Ultimately, the pollen containing the sperm is released to find or "pollinate" the flower containing the egg, and two sperm are released from the pollen, one of which fertilizes the egg, forming a zygote and thus the next diploid sporophyte (Gilbert, 2000). The intricacies of flowering plant reproduction will not be described here: suffice it to say that plant sex is not quite as simple as the preceding story (for example, in flowering plants, a second fertilization event occurs, which forms a nutritive tissue called the endosperm). While leading to spores rather than gametes, meiosis the process in plant cells is *de facto* similar to that in animals.

1.2 Meiosis – A discrete continuum?

Meiosis, regardless of the organism in which it is occurring, is most often described as a series of discrete stages, even though it is really a continuum of events that flow from one to the next in a typically seamless and apparently effortless *pas de deux*. As will be described elsewhere in this book in more detail, the two meiotic divisions, meiosis I and meiosis II, are each conveniently subdivided into the familiar stages of prophase, metaphase, anaphase, and telophase, I and II, respectively. The most prolonged and intricate stage of meiosis is prophase I, which is highly regulated and concludes with the regimented alignment of homologous pairs along the equator at metaphase I (Alberts *et al.*, 2002).

During prophase I, replicated homologous pairs find each other in a process called "synapsis"; the synapsed homologues are often referred to as bivalents or tetrads, as a bivalent/tetrad has two chromosomes and four chromatids (Snustad & Simmons, 2008). DNA is exchanged between homologous chromosomes, usually resulting in chromosomal crossover. As prophase I is so protracted and complicated, it is usually described in five substages: leptotene, zygotene, pachytene, diplotene, and diakinesis.

The first stage of prophase I is the leptotene stage, also known as leptonema, from Greek words meaning "thin threads" (Snustad & Simmons, 2008), and is typically very short. Individual chromosomes—each consisting of two sister chromatids—change from the diffuse state they exist in during the cell's period of growth and gene expression, and condense into visible strands within the nucleus. However, the two sister chromatids are tightly bound to and hence indistinguishable from one another.

The zygotene stage, also known as zygonema, from Greek words meaning "paired threads", occurs as the chromosomes approximately line up with each other into homologous chromosome pairs in the initiation of synapsis (Snustad & Simmons, 2008). This is also called the "bouquet" stage because of the way the telomeres (chromosome "ends") cluster at one end of the nucleus. Pairing is brought about by a zipper like fashion and may start at any point along the chromosomes. While the initial alignment is approximate, the pairing is highly specific and exact, and complete by the end of zygotene.

The pachytene stage, also known as pachynema, from Greek words meaning "thick threads", is the stage when chromosomal crossover ("crossing over") occurs (Snustad & Simmons, 2008). In this process, nonsister chromatids of homologous chromosomes randomly exchange segments over regions of homology. The actual act of crossing over is not perceivable through the microscope, however, and pachytene is physically similar to zygotene, although the chromosomes are more condensed than in the previous stages.

During the diplotene stage, also known as diplonema from Greek words meaning "two threads", the homologous chromosomes separate from one another a little (Snustad & Simmons, 2008). The chromosomes themselves also uncoil a bit, allowing some transcription of DNA. However, the homologous chromosomes of each bivalent remain tightly bound at chiasmata, the regions where crossing-over occurred. The chiasmata remain on the chromosomes until they are severed in anaphase I.

In diakinesis, named from Greek words meaning "moving through", chromosomes condense further, and reaching maximum condensation (Snustad & Simmons, 2008). This is the first point in meiosis where the four chromatids of the tetrads are actually visible. Sites of crossing over entangle together, effectively overlapping, making chiasmata clearly visible. Diakinesis closely resembles prometaphase of mitosis: the nucleoli disappear, the nuclear membrane disintegrates into vesicles, and the meiotic spindle begins to form. Bivalents begin to migrate to the equator, guided by the spindle. When the bivalents reach the equator, the cell is said to be in metaphase I.

Thus, the chromosomes of any organism become increasingly condensed as they proceed through the substages of prophase I (Page & Hawley, 2003). Essentially, the chromosomes develop into "patches" in a "nucleoscape" before the metaphase I stage is reached. Could a method used to quantify the patchiness of chromosomal condensation provide a numerical predictor for metaphase I?   Could the continuum of meiosis, or at least aspects of the continuum, be quantified?

1.3 Quantifying patchiness: Traditional methods versus the fractal dimension, D_f

The detection and quantification of spatial pattern, including the degree of patchiness of a landscape, is frequently undertaken in ecological studies. Methods to detect and measure special heterogeneity and patchiness calculate various multidimensional configurational indices that consider size, shape, patch density, and connectivity, to name a few variables (Gustafson, 1998). However, these typical "go-to" methods for detecting patchiness in landscape ecology are scale-dependent: simply stated, magnification of a patch results in the loss of the patchy pattern over the landscape, as one essentially "enters" the patch (or non-patch region), which then becomes the new prevalent, homogeneous environment (Kenkel & Walker, 1993). Likewise, what seem like large patches at one scale might appear like random bits of noise with a more distant perspective. This problem of scale is relevant to the idea of quantifying chromosomal patchiness: magnification with a microscope would also obscure patterns of patchiness. And what magnification is ideal for patch detection and measurement? Any measure of chromosomal patchiness must not depend on the scale.

Conveniently, the fractal dimension, D_f, is scale-independent (Julien & Botet, 1987). A fractal is an object or quantity that displays self-similarity on all scales. The object need not exhibit exactly the same structure at all scales, but the same "type" of structures must appear on all scales. The prototypical example for a fractal is the length of a coastline measured with different rulers of different lengths (Mandelbrot, 1967). A shorter ruler measures more of the sinuosity of bays and inlets than a larger one, so the estimated length continues to increase as the ruler length decreases. Plotting the length of the ruler versus the measured length of the coastline on a log-log plot gives a straight line, the slope of which is the fractal dimension of the coastline, D_f. In familiar Euclidean space, a line has 1 dimension, a plane has 2, and a cube, 3. However, the dimension of a fractal "line" in a two-dimensional surface has value of $1 \leq D_f \leq 2$ (Julien & Botet, 1987; Kenkel & Walker, 1993). A D_f that approaches 1 implies that the particular object of interest (be it a landscape feature or chromosomal material) is found in patches and is less space-filling, whereas a D_f that approaches 2 suggests the measured feature is space-filling and dispersed, so that the overall landscape (or nucleoscape!) is not patchy (Sugihara & May, 1990).

The fractal dimension D_f has been used to quantify patchiness in landscapes, and often in digital landscape maps such as Landsat images (e.g., Lam, 1990; Milne, 1992), which, as will become evident, is relevant to the work described in this chapter. In analyzing digital landscape maps, a useful calculation is the "mass dimension", which describes the number of grains or digital pixels of a given type occurring within a square window of size LxL (Julien & Botet, 1987; Milne, 1992; Kenkel & Walker, 1993). A series of windows of size LxL can be "slid" across the map, centring one on each pixel of the map that represents the object of interest. Within each window, the number of pixels of the object of interest (n) is then counted. The procedure is repeated across a range of length scales. Because large windows will be limited by the extent of the map, only the data sampled by all window sizes can be included. These measures can then be used to quantify the statistical behaviour of the pattern. From the geometric measures described, a probability density function, $P(n,L)$, can be defined, which describes the probability of finding n pixels in a window of size (length) L. This probability function is analogous to a standard statistical distribution. The first moment of the probability distribution, $M(L)$, equivalent to mass dimension, is determined for increasing values of L; the slope of the plot of log $M(L)$ vs. log (L) represents D_f.

Could D_f be used to measure patchiness in a nucleoscape? A role for D_f in the quantification of meiosis seems possible: the D_f of nuclei in early prophase I, in leptotene, should approach 2, as the chromosomal material would be relatively dispersed. The D_f of nuclei progressing through zygotene, pachytene, diplotene, and diakinesis should decrease and approach 1as the chromosomal material becomes more condensed, reaching maximal condensation at diakinesis. Furthermore, some value of D_f near 1would likely be a quantitative indicator for the onset of metaphase I, especially as the alignment of chromosomes along the equator would be more organized, and essentially represent one giant cluster or "super-patch". One of the authors of this chapter has done preliminary work toward this end (Ross 2005), and the results were promising enough to warrant the larger-scale study presented here.

1.4 Megasporogenesis and microsporogenesis in three flowering plants as the systems for study

Megasporogenesis in flowering plants, the process by which a megasporocyte undergoes meiosis to produce spores -- typically of which only one goes on to produce the megagametophyte and egg -- provides a useful system for studying meiosis, particularly the prolonged and complicated prophase I. *Arceuthobium americanum* Nutt. ex Englm., the lodgepole pine dwarf mistletoe, is a particularly interesting organism and one that will provide a system in which to examine D_f as a measure of megasporocyte meiosis. The genus *Arceuthobium* comprises 42 species of economically-significant flowering plants that parasitize members of Pinaceae and Cupressaceae throughout North America, Central America, Asia, and Africa (Hawksworth *et al.*, 2002; Jerome & Ford, 2002). *Arceuthobium americanum* is found in North America, where it has the widest geographical range of any North American *Arceuthobium* species (Jerome & Ford, 2002), and can be found growing on two principal hosts, lodgepole pine (*Pinus contorta* var. *latifolia*) and jack pine (*Pinus banksiana*). The plant is dioecious, having female (Fig. 1A) and male flowers (Fig. 1B) on separate individuals (Hawksworth & Wiens, 1996), and a diploid number of 28 (2n = 28) chromosomes (Wiens, 1964). Notably, prophase I in the megasporocyte *A. americanum* female flower is especially protracted: the megasporocyte enters prophase I (leptotene) at the end of one summer, goes into a resting period over the winter, and resumes prophase I (zygotene) in the next year's growing season before proceeding into metaphase I and the rest of meiosis (Hawksworth & Wiens, 1996; Ross & Sumner, 2004; Ross, 2005). As prophase I is so prolonged, the ability to predict metaphase I within days is a considerable challenge, and hence presents the perfect opportunity to test the ability of D_f to do so.

Two other species, similar to each other but decidedly different from *Arceuthobium americanum* should also be studied in order to capture more variability. The two species ideal for study are *Decaisnea insignis* (Fig. 1C) and *Sargentodoxa cuneata* (Fig. 1D), members of the Lardizabalaceae found in China and other Southeast Asian countries. Both are "winter bud" species that undergo megasporogenesis over three weeks (typically the last three in November) in the late fall (Wang *et al.*, 2009a, 2009b, & unpubl. data).

Decaisnea is a monotypic genus, with the species, *Decaisnea insignis* (Griffith) Hook. f. & Thomson, widely distributed from central to south-western China, extending into Bhutan, Myanmar, Nepal, Sikkim and north-eastern India (Chen & Tatemi, 2001). The plant is nicknamed 'dead man's fingers', as it possesses racemes of striking deep purplish-blue elongated fruits (follicles). The plant is economically important, as it is readily cultivated as an ornamental, and its fruits are deemed a delicacy. Plants are polygamo-monoecious, as individuals possess female flowers, male flowers and bisexual flowers (Wang *et al.*, 2009b). *Decaisnea insignis* has a diploid number of 30 (2n = 30) chromosomes (Wu, 1995).

Like *Decaisnea*, *Sargentodoxa*, is also generally agreed to be a monotypic genus of the Lardizabalaceae (Rehder & Wilson, 1913; Chen & Tatemi, 2001; Soltis *et al.*, 2000; Angiosperm Phylogeny Group [APG], 2003), consisting of the single species, *Sargentodoxa cuneata* (Oliver) Rehder and E. H. Wilson. It has a wide distribution in China, and can also be found in Laos and Vietnam. The plant has ethnobotanical significance, having been used in folklore and medicine as a treatment for anemia and numerous human parasites, among other ailments (Huang et al., 2004). Both male and bisexual flowers are found on the same individual, and thus the species is considered both dioecious and monoecious (Shi *et al.*, 1994). The functionally unisexual flowers are morphologically bisexual, at least developmentally, and the diploid number is 22 (2n = 22) chromosomes (Wu, 1995).

Microsporogenesis, the process by which immature haploid pollen grains are formed by meiosis of the microsporocyte, is another useful process in which to study meiotic events, including the substages of prophase I. In fact, karyotypes for many plants are derived from studies of microsporocyte meiosis (*e.g.*, Wiens, 1964). Microsporocytes are technically much easier to harvest from flowers than megasporocytes, and as such, D_f can be determined for microsporocytes from several plant species undergoing prophase I and reaching metaphase I. *Arceuthobium americanum* also presents itself as a useful organism for study in this capacity, as microsporocytes in the male flower undergo meiosis rather rapidly over about three weeks at the end of the summer (Sereda, 2003). The resultant microspores (immature pollen) initiate microgametogenesis, but then overwinter prior to being released as mature pollen in the spring of the following year. The rapidity of meiosis in the microsporocytes should provide interesting fodder for comparison with the sluggish process in the *A. americanum* megasporocytes. Due to relative ease of obtaining microsporocytes, *Decaisnea insignis* (Fig. 1C) and *Sargentodoxa cuneata* (Fig. 1D) can also be mined for microsporocytes and D_f calculation. As with megasporogenesis, these species undergo microsporogenesis in the late fall, in the last three weeks in November (Wang *et al.*, 2009a, 2009b, & unpubl. data).

1.5 Aims and concerns

The objective of this work is to (1) calculate D_f for megasporocytes and microsporocytes from *Arceuthobium americanum*, *Decaisnea insignis*, and *Sargentodoxa cuneata* throughout prophase I and upon the onset metaphase I; (2) to assess the ability of D_f to characterize the subphases of prophase I; (3) to evaluate the ability of D_f to predict the onset of metaphase I; and (4) to compare D_f between microsporogenesis and megasporogenesis amongst the three species, where informative. The ultimate goal will be to see if there is a universal value for D_f for either or both processes at some point, or across the species.

Fig. 1. Species used in this study. A) *Arceuthobium americanum* (female plant). B) *Arceuthobium americanum* (male plant). C) Portion of *Decaisnea insignis* shrub with inflorescences. D) *Sargentodoxa cuneata* inflorescence showing female flowers borne above the male. Scale bar = 15 mm in A, 5mm in B, 15 cm in C, and 9mm in D.

2. Materials and methods

2.1 Collection and preparation for microscopy

Arceuthobium americanum female flowers were collected daily from 1 March (prior to resumption of prophase I) until 1 May (when metaphase I has been reached or surpassed) in 2000 and 2001 from a stand of heavily infested jack pine in Belair, Manitoba, Canada as described in Ross (2005). Male *A, americanum* flowers were collected daily from August 14 until August 31 from infested lodgepole pine in Stake Lake, British Columbia, Kamloops.

Male and female flowers of *Decaisnea insignis* were collected daily from Taibai Mountain in Shaanxi Province, China 1 March 2005 to 1 May 2006 in order to obtain a range of developmental stages, although further work pinpointed the time of both megasporogenesis and microsporogenesis to about a three-week period at the end of November. Special attention was paid to when prophase I and metaphase I occurred. Only unisexual flowers were examined, even though the species sports bisexual flowers as well. Likewise, male and bisexual flower of *Sargentodoxa cuneata* were collected daily from the Nanchuan district of Chongqing city, China from 1 March 2006 to 1 May 2007, and as in *D. insignis*, further work pinpointed the time of both megasporogenesis and microsporogenesis to a three-week period at the end of November. Again, the occurrence of prophase I and metaphase I was noted. Only female components of the bisexual flowers were examined.

Flowers/buds for studies of megasporogenesis were fixed in formaldehyde/acetic acid/ethyl alcohol (FAA). Ovules were dissected from the ovaries, cleared in "4½" clearing solution (Herr, 1971) and mounted directly in Hoyer's medium on glass slides (Alexopoulos & Benke, 1952). Flowers/buds for the study of microsporogenesis were fixed in 1:3 acetic acid/ethanol, and the microsporocytes were spread directly on to glass slides with a dissecting needle. No staining was used, and the specimens were viewed with phase-contrast microscopy and consistent illumination. At least 5 megasporocytes and 5 microsporocytes were examined from each species per day.

2.2 Calculation of D_f

Whole megasporocytes or microsporocytes were digitized as binary (pure black and white) raster images with Adobe Photoshop® (resolution of 2400 dots per inch). Each image was cropped to include only the nucleus (nucleolus, if present, was ignored) and exported through NIH Image (Ross 2005) or Image J (Version 1.38) as a text file (black pixel=1, white pixel=0). This effectively converted chromosomes into black pixels, and any other nuclear material into white ones. Methods for the calculation of D_f were as described in Ross (2005).

2.3 Data management and statistical analyses

All statistical work was done with Minitab®. The mean value of $D_f \pm$ standard error was calculated for the ~5 megasporocyte nucleoscapes for each day, considering each species individually, and, in the case of *A. americanum*, separately for its two sampling years (first reported in Ross, 2005). Likewise, the mean value of $D_f \pm$ standard error was calculated for the ~5 microsporocyte nucleoscapes for each day, also individually for each of the three species. A two-tailed student's *t*-test (was used to compare the mean value of D_f from one day to the next, separately for each cell type (megasporocyte or microsporocyte) and species, to determine if the change in D_f was significant or not. A two-tailed student's *t*-test was also used to compare the values of *A. americanum* megasporocyte D_f between the two sampling years for days where the D_f values appeared similar but needed statistical verification.

To compare amongst all datasets, a one-way analysis of variance ANOVA was performed across the datasets for the start and end mean values of D_f. Then, to fairly assess how similar the D_f value was during the time period in which D_f was changing significantly across the species and cell types, that time period was standardized (day number/total days of significant change), and a one-way ANOVA was performed for values of D_f across all datasets for each standardized time.

3. Results

3.1 D_f for megasporogenesis and microsporogenesis in each species

As was described in Ross (2005) but reinterpreted here, megasporocyte nucleoscapes from *A. americanum* were examined daily from 1 March to 1 May in both 2000 and 2001; thus the examination period was 66 days for each year. The mean D_f for the 5 *A. americanum* megasporocyte nucleoscapes from the first date of sampling (1 March) was 1.903 (±0.004) in 2000 and 1.900 (±0.003) in 2001 (Fig. 2); these values are not significantly different (p>0.95). Visually, the chromosomal material appeared to be in leptotene and was relatively dispersed, (Fig. 3A), as would be anticipated for a mean D_f that approached 2. Mean D_f remained constant (±0.003, 2000 and 2001 values equivalent, p>95%) over the next 9 days in 2000 and over the next 13 days in 2001 (Fig. 2). However, mean D_f then made a significant drop (p<5%) to 1.614 (±0.002) on 11 March 2000 and 1.606 (±0.003) on 15 March 2001 (Fig. 2); the "drop" values for 2000 and 2001 are equivalent (p>95%), and the drop corresponded with the chromosomal condensation associated with the resumption of prophase I, likely zygotene (Fig. 3B). The mean D_f continued to decline significantly (p<5%) each day over the next 38 days in 2000 (until 18 April) and over the next 32 days in 2001 (until 16 April), reaching statistically-similar (p>99%) values of 1.352 (±0.002) in 2000 and 1.354 (±0.002) in 2001. At that time, the chromosomes were highly condensed but not yet at the equator, and were likely at diakinesis (Fig. 3B). The mean D_f continued to decline significantly (p<5%) for the next 3 days until 21 April in 2000 (a total of 42 days of change) and for the next 3 days until 19 April in 2001 (total 36 days of change) to reach what would be a low of 1.332 (±0.001) in 2000 and 1.331 (±0.001) in 2001 (Fig. 2); values equivalent (p>95%). At the end of this period of change, the chromosomes were at the equator and cells were in metaphase I. The average period of change between the two years was thus 39 days. After metaphase I was attained, the mean D_f did not change significantly (p>99%) over the remainder of the examination period. Therefore, a D_f (both years considered) of about 1.353(±0.003) corresponded with diakinesis, which predicted metaphase I in *A. americanum* megasporocytes by 3 days in both study years, and D_f at metaphase I was the lowest stable value of about 1.332 (±0.002), both years considered.

Mean D_f for 5 megasporocyte nucleoscapes from *D. insignis* and *S. cuneata* (Fig. 4) as well as 5 microsporocyte nucleoscapes from these two species (Fig. 5) were determined each day from a period 13 November to 30 November (2005 for *D. insignis*, 2006 for *S. cuneata*); a total examination period of 18 days for each species. Similarly, mean D_f for 5 *A. americanum* microsporocyte nucleoscapes were determined each day from 14 August to 31 August 2010 (Fig. 6), also 18 days of study. The results were startlingly uniform across the abovementioned species and the cell types. The mean D_f at the beginning of the examination period ranged between 1.902 and 1.909 (±0.008) for these species and cells, and reached a low of 1.330 to 1.332 (±0.003) when the cells were in metaphase I (Figs. 4 to 6). The period in which mean D_f was changing was 9 days. In all cases, when the mean D_f had reached 1.355 to 1.360 (±0.006), the cells were apparently in diakinesis, and metaphase I occurred the day after diakinesis was reached. Notably, the high values, diakinesis values, and low values were all statistically similar to each other, respectively, and also to the values for *A. americanum* megasporocytes (Fig. 2) described previously (p>90%).

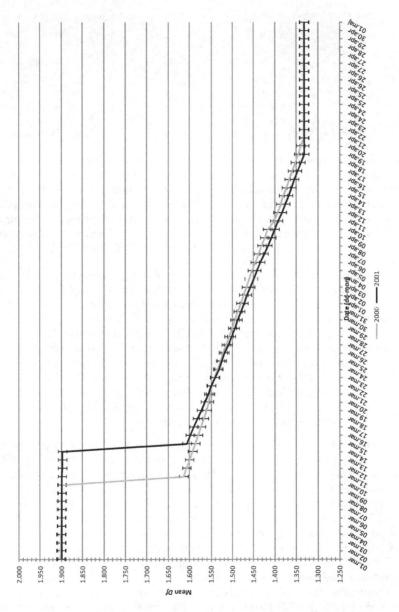

Fig. 2. Daily value of mean D_f from 1 March to 1 May in 2000 and 2001 for *Arceuthobium americanum* megasporocytes undergoing prophase I. Mean D_f makes a significant drop (p<5%) to 1.614 (±0.002) on 11 March (2000) and 1.606 (±0.003) on 15 March (2001), continues to decline significantly (p<5%) each day over the next 41 days in 2000 (until 21April) and next 35 days in 2001 (until 19April), to reach a stable low value of 1.332 (±0.001) in 2000 and 1.331 (±0.001) in 2001. There were thus 39 average total days of change.

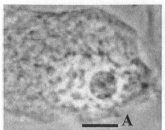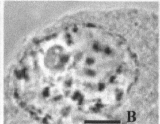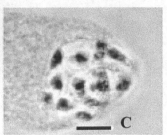

Fig. 3. Phase-contrast light micrographs of meiotic nuclei in megasporocytes of the dwarf mistletoe, *Arceuthobium americanum*. A) Nucleus sampled prior to the resumption of prophase I (near leptotene). B) Nucleus sampled when prophase I resumes (zygotene), chromosomal condensation commences, and the value of the fractal dimension, D_f, begins to drop. C) Chromosomes are maximally condensed, cell is in diakinesis. Metaphase I will occur in 3 days, and D_f will reach its lowest value. Scale bar = 10 μm in A, B, C. (Used with permission from Ross (2005) and the *International Journal of Biological Sciences* ISSN 1449-2288).

Specifically, mean D_f in both *D. insignis* and *S. cuneata* megasporocytes made a significant drop (p<5%) to 1.592 (±0.002) on 19 November (Day 7) and to 1.594 (±0.001) on 16 November (Day 4), respectively, and then continued to decrease significantly over the next 7 days, at which time diakinesis was achieved, with the mean D_f becoming 1.357 (±0.002) on 26 November (Day 14) in *D. insignis* and 1.355 (±0.002) on 23 November (Day 11) in *S. cuneata* (Fig. 4). Chromosomal condensation from leptotene (Fig. 7A) to diakinesis (Fig. 7B) was accordingly evident in the megasporocytes (Fig. 7 shows an example from *D. insignis*). The next day, 27 November (Day 15) in *D. insignis* and 24 November (Day 12) in *S. cuneata*, the lowest mean D_f was reached in both species (1.331 ±0.002 for *D. insignis* and 1.330±0.001 for *S. cuneata*), and the megasporocytes had reached metaphase I (Fig. 4). Thus, the period of change for mean D_f was, as mentioned 9 days (inclusive).

Similarly, mean D_f in *A. americanum*, *D. insignis*, and *S. cuneata* microsporocytes made a significant drop (p<5%) to 1.595 (±0.003) on 18 August (Day 5), to 1.597 (±0.002) on 17 November (also Day 5), and to 1.595 (±0.001) on 18 November (Day 6), respectively, and then continued to decrease significantly for the next 7 days, reaching diakinesis on 25 August (Day 12), 24 November (also Day 12), and 25 November (Day 13) with mean D_f values of 1.358 (±0.003), 1.360 (±0.002), and 1.357 (±0.003), respectively (Figs. 5 and 6). As in the megasporocytes, according chromosomal condensation from leptotene (Fig. 8A) to diakinesis (Fig. 8B) was evident in the microsporocytes (Fig. 8 shows an example from *S. cuneata*). The day after diakinesis occurred, microsporocytes in *A. americanum*, *D. insignis*, and *S. cuneata* reached metaphase I, specifically 26 August (Day 13), 25 November (also Day 13) and 26 November (Day 14), respectively, with respective lowest mean D_f values (Figs. 5 and 6) of 1.330 (±0.001), 1.329 (±0.002), and 1.332 (±0.002). Again, as was the case for *D. insignis* and *S. cuneata* megasporocytes, D_f decreased over a 9-day period in the microsporocytes for all three species examined.

3.2 Comparing D_f over a standard time frame: Have we some universal values?

The time frame for the progression of prophase I into metaphase I in *Arceuthobium americanum* megasporocytes was much longer than in its microsporocytes and longer than the period in both the megasporocytes and microsporocytes of *Decaisnea insignis* and *Sargentodoxa cuneata*. Specifically, the period for which mean D_f experiences significant daily drops in *A. americanum* megasporocytes averaged 39 days, but was only 9 days for the others. However, it is patently clear that D_f has statistically-significant similar value and meaning for meiosis across the cells and species studied here. Furthermore, when the time frame of mean D_f is standardized for each cell type and species, the values are all statistically equivalent (p>95%). This similarity is also evident in Fig. 9, which depicts the change in D_f for both cells and all species studied. Most notably, the slope of the regression line for the averaged values is -0.289, and the Y intercept for the regression line (the value at which mean *Df* begins to drop) is 1.602. It is this slope which could likely be used to predict substages of prophase I and metaphase I. Other key predictive values of D_f are given in Table 1.

Mean D_f ± Standard Error	Event
1.904 ±0.009	Prophase I (leptotene)
1.599 ±0.008	Prophase I (zygotene)
1.356 ±0.003	Prophase I (diakinesis)
1.331 ±0.004	Metaphase I

Table 1. Mean D_f averaged across megasporocytes and microsporocytes in *Arceuthobium americanum*, *Decaisnea insignis*, and *Sargentodoxa cuneata* as predictors (all were statistically similar, p>95%).

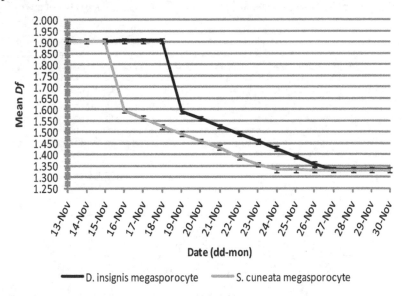

Fig. 4. Daily value of mean D_f from 13 November to 30 November for *Decaisnea insignis* (2005) and *Sargentodoxa cuneata* (2006) megasporocytes undergoing prophase I. Mean D_f makes a significant drop (p<5%) to 1.592 (±0.002) on 19 November (*D. insignis*) and to 1.594 (±0.001) on 16 November (*S. cuneata*), continues to decline significantly (p<5%) each day over the next 8 days in both species (until 27 November and 24 November, respectively), to reach a stable low value of 1.331 (±0.002) and 1.330 (±0.001), respectively (total 9 days of change).

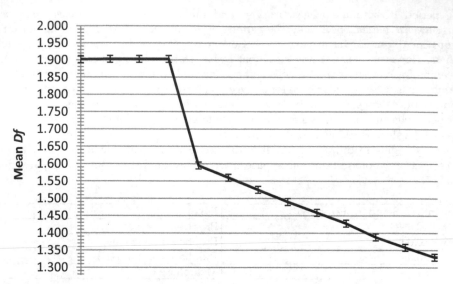

Fig. 5. Daily value of mean D_f from 14 August to 31 August in 2010 for *Arceuthobium americanum* microsporocytes undergoing prophase I. Mean D_f makes a significant drop ($p<5\%$) to 1.595 (±0.003) on 18 August, continues to decline significantly ($p<5\%$) each day over the next 9 days (until 26 August), to reach a stable low value of 1.330 (±0.001), 9 days of changing D_f.

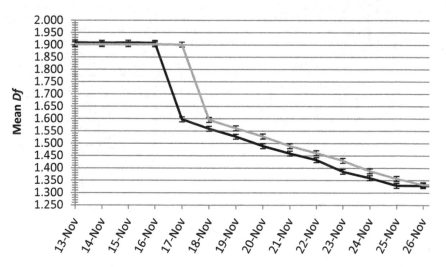

Fig. 6. Daily value of mean D_f from 13 November to 30 November for *Decaisnea insignis* (2005) and *Sargentodoxa cuneata* (2006) microsporocytes undergoing prophase I. Mean D_f makes a significant drop ($p<5\%$) to 1.597 (±0.002) on 17 November (*D. insignis*) and to 1.595 (±0.001) on 18 November (*S. cuneata*), continues to decline significantly ($p<5\%$) each day over the next 8 days in both species (until 25 November and 26 November, respectively), to reach a stable low value of 1.329 (±0.002) and 1.332 (±0.002), respectively (total 9 days of change).

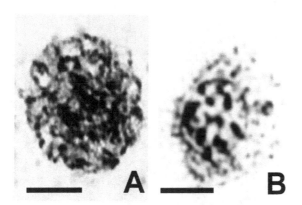

Fig. 7. Phase-contrast light micrographs of meiotic nuclei in the megasporocytes *Decaisnea insignis*. A) Nucleus sampled prior to the resumption of prophase I (near leptotene). B) Nucleus sampled when prophase I resumes (zygotene), chromosomal condensation commences, and the value of the fractal dimension, D_f, begins to drop. Metaphase I will occur in 1 day, and D_f will reach its lowest value. Scale bar = 15 μm in A, 30 μm in B.

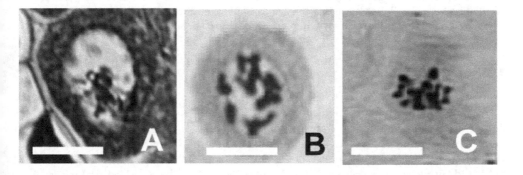

Fig. 8. Phase-contrast light micrographs of meiotic nuclei in the microsporocytes of *Sargentodoxa cuneata*. A) Nucleus sampled prior to the resumption of prophase I (near leptotene). B) Chromosomes are maximally condensed, cell is in diakinesis, and the value of the fractal dimension, D_f, has reached a low value that predicts metaphase I by one day. C) Nucleus is in metaphase I, one day after diakinesis was reached, and D_f has reached its stable, lowest value. While chromosomes are maximally condensed in B, their position along the equator drove D_f to its lowest value. Scale bar = 30 μm in A, B, C.

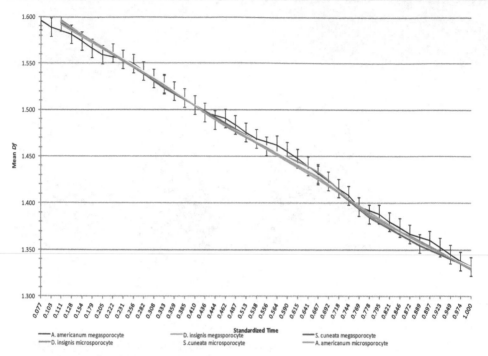

Fig. 9. Daily value of mean D_f over the standardized period of change (day number/total days of significant change) for *Arceuthobium americanum*, *Decaisnea insignis*, and *Sargentodoxa cuneata* megasporocytes and microsporocytes undergoing prophase I. Values for *A. americanum* megasporocytes were averaged over the two years of study (2000 and 2001). The values follow a statistically-similar linear trendline. Slope of the regression line = -0.289; Y intercept = value at which mean D_f begins to drop = 1.602 for the regression line.4.

4. Discussion

In the work done here with *Arceuthobium americanum*, *Decaisnea insignis*, and *Sargentodoxa cuneata* megasporocytes and microsporocytes undergoing meiosis, the fractal dimension D_f described the relative degree of chromosomal condensation in prophase I to metaphase I more sensitively and quantitatively than microscopic observation alone. Across all species and cell types, a D_f of 1.904 ±0.009 described leptotene of prophase I, D_f of 1.599 ±0.008 delimited zygotene of prophase I, and Df of 1.356 ±0.003 characterized diakinesis. A D_f approaching 1 is consistent with a nucleoscape made patchy with highly condensed chromosomal material, as would be the case for patches in a landscape (Sugihara & May, 1990). Even though the timing was different with respect to the *A. americanum* megasporocytes, the values across all species and both cell types were statistically similar. Interestingly, the timing of prophase I and the onset of metaphase I in microsporocytes across the three species as well as the timing of the same stages in the megasporocytes in *D. insignis* and *S. cuneata* was astonishingly similar: prophase seemed to occur in 9 days in all of these. One might have thought that the disparities amongst the species, particularly regarding microsporogenesis between *A. americanum* and the two members of Lardizabalaceae, would have resulted in different timing. However, it might be that these three species have similar amounts of DNA, as nuclear DNA content is known to be important factors determining or affecting the duration of male meiosis (Bennett *et al.*, 1973). These workers found that the timing of microsporogenesis and, to a certain extent, megasporogenesis, was similar in *Triticum aestivum* L. (wheat), *Hordeum vulgare* L. (barley), and *Secale cereale* L. (rye).

A mean D_f 1.331 ±0.004 quantified metaphase I, which was also predicted by the D_f measured for diakinesis. In *A. americanum* megasporocytes, metaphase I occurred 3 days after reaching diakinesis, and in the remaining cells and species examined, it occurred one day after diakinesis. "Fractally speaking", a low value of 1.331 ±0.004 make sense as a descriptor of metaphase I: while the chromosomes are as about as condensed as they are in diakinesis, their clustering at the equator contributes to the patchiness, which in turn is captured in the calculation of D_f by the probability-density function. Ross (2005) described slightly different results for the same data, but had erroneously identified metaphase I as diakinesis. In other species not studied here, the timing might be different, but perhaps the the values of D_f calculated for these stages (Table 1) can be predictive.

One might argue that the mere occurrence of diakinesis could just as effectively predict metaphase I without the calculation of D_f, but is should be stressed that the rate of change of mean D_f as determined by the slope and determined within a few days during the decrease, one would be able to predict diakinesis as well as metaphase I! In other words, once D_f begins to drop (i.e., the Y-intercept value), one could easily calculate the slope of the line (which this study showed to be consistent across all species and both cell types at around -0.289) with a few days' worth of data, then estimate when the consistent metaphase I low of 1.331 ±0.004 will be reached. This is extremely powerful.

This study has shown that meiotic timing and events amongst different species can be compared quantitative fashion that will eliminate the subjectivity inherent in qualitative descriptions. The method for harvesting nuclei and extracting D_f is relatively simple, much more so than embedding and staining tissues, so time may be saved if many metaphase I slides are required. As prophase I is always prolonged relative to metaphase I, regardless of the organism, some value of D_f should be predictive in any species examined. The method as described here will work particularly well in other species with relatively long meiotic divisions, such as peonies, genus *Paeonia* (Shamrov, 1997). In commercially important plants such as these, predicting when meiotic events occur can be particularly important and could streamline the culture of zygotic embryos. The examination and calculation of D_f in other species and other stages of meiosis should be tackled.

Use of this technique should also allow for the quantification of *mitosis* (the equational nuclear division) along with the prediction of mitotic metaphase. As for meiosis, quantification of mitosis will facilitate objective comparisons of mitotic events amongst cells, tissues, individuals, and species. Notably, a quantification of mitosis could prove considerably valuable for another reason; if D_f values can be determined for normal tissues, it is likely that deviations in D_f, either in absolute value or rate of change, could signal abnormalities in the cell cycle that might be correlative to precancer events in animals.

Fractal analysis and chaos theory have been used to describe the histological texture of tumours (Mattfeldt, 1997), and so the work described in our study seems like it would be applicable to systems beyond the study of meiosis.

5. Conclusion

Key values of the fractal dimension (D_f) is *Arceuthobium americanum, Decaisnea insignis,* and *Sargentodoxa cuneata* megasporocytes and microsporocytes undergoing meiosis were able to describe leptotene, zygotene, diakinesis, and metaphase I. When D_f began to drop, the rate of change across all species and both cell types was consistent (slope = - 0.289), and this slope could be used as a predictor for the abovementioned stages. The results are novel and exciting, and we hope they will be re-examined and extrapolated to other cells, species, processes, and systems.

6. Acknowledgements

Cynthia is grateful to Dr. M.J. Sumner for his supervision during her Ph.D., from which some of this work was derived. She fully acknowledges Dr. Norman C. Kenkel and Dr. David J. Walker for their guidance and suggestions with the initial research approach. Finally, she thanks Dr. D.R. McQueen for preparation of some of the samples, and Dr. D. Punter for his advice and expertise with the dwarf mistletoes. Hua-Feng is grateful to Mr. Cheng Guo, Mr. Zhengyu Liu, and Mr. Mingbo Ren for field work assistance and also thanks Professor Yi Ren at the Shaanxi Normal University for providing us with some material. We thank Dr Qing Cai, Ms. Jie Wen, Mr. Yinhou Xiao, and Dr. Shengxiang Yu for technical assistance in the laboratory. This work was supported by a Natural Sciences and Engineering Research Council of Canada Discovery Grant (grant number 164375 provided to C.R.F.) and by the National Basic Research Program of China (grant 2006CB403207).

7. References

Alberts, B.; Johnson, A.; Lewis, J.; Raff, M.; Roberts, K. & Walter, P. (2002). *Molecular Biology of the Cell* (4th edition), Garland Science, ISBN-10: 0-8153-3218-1, New York, USA

Alexopoulos, C.J. & Benke, E.S. (1952). *Laboratory Manual for Introductory Mycology,* Burgess Publishing Co., Minneapolis, USA

Angiosperm Phylogeny Group, APG (2003). An update of the angiosperm phylogeny group classification for the orders and families of flowering plants: APG II. *Botanical Journal of the Linnean Society,* Vol. 141, No. 4 (November 2003), pp. 399-436, ISSN 0024-4074

Bennett, M.D.; Finch, R.A.; Smith, J.B. & Rao, M.K. (1973). The time and duration of female meiosis in wheat, rye and barley. *Proceeding of the Royal Society of London Series B,* Vol. 183, No. 1072 (May 1973), pp. 301-319, ISSN 0962-8452

Chen, D.-Z. & Tatemi, S. (2001). Lardizabalaceae. In: *Flora of China,* C.-Y. Wu, (Ed.), Vol. 6., 440-454, Missouri Botanical Garden Press, ISBN 1-930723-25-3, St. Louis, USA

Gilbert, S. F. (2000). *Developmental Biology* (6th edition), Sinauer Associates, ISBN-10: 0-87893-243-7, Sunderland, USA.

Gustafson, E.J. (1998). Quantifying landscape spatial pattern: what is the state of the art? *Ecosystems,* Vol. 1, No. 2 (February 1998), pp. 143-156, ISSN 14329840

Hawksworth, F.G & Wiens, D. (1996). *Dwarf Mistletoes: Biology, Pathology, and Systematics*, *United States Department of Agriculture Forest Service Handbook Number 709*, US Government Printing Office, ISBN 0788142011, Washington, USA

Hawksworth, F.G.; Wiens, D. & Geils B.W. (2002). *Arceuthobium* in North America. In: *Mistletoes of North American Conifers, United States Department of Agriculture Rocky Mountain Research Station Forest Service Center General Technical Report 98*, B.W. Geils, J. Cibrián Tovar & B. Moody, (Tech. Coords.), Chapter 4, 29-56, Government Printing Office, OCLC 51521345, Washington, USA

Herr, J.M. Jr. (1971). A new clearing-squash technique for the study of ovule development in angiosperms. *American Journal of Botany*, Vol. 58, No. 8 (August 1971), pp. 785-790, ISSN 0002-9122

Huang, J.; Pei, S.-J. & Long, C.-L. (2004). An ethnobotanical study of medicinal plants used by the Lisu people in Nujiang, northwest Yunnan, China. *Economic Botany*, Vol. 58, Supplement (Winter 2004), pp. S253-264, ISSN 0013-0001

Jerome, C.A. & Ford, B.A. (2002). The discovery of three genetic races of the dwarf mistletoe *Arceuthobium americanum* (Viscaceae) provides insight into the evolution of parasitic angiosperms. *Molecular Ecology*, Vol. 11, No. 3 (March 2002), pp. 387-405, ISSN 0962-1083

Julien, R. & Botet, R. (1987). *Aggregation and Fractal Aggregates*, World Scientific Publishing Co., ISBN 9971502488, Singapore

Kenkel, N.C. & Walker, D.J. (1993). Fractals and ecology. *Abstracta Botanica*, Vol. 17, No. 1, (1993), pp. 53-30, ISSN 0133-6215

Lam, N. S.-N. (1990). Description and measurement of Landsat TM images using fractals. *Photogrammetric Engineering and Remote Sensing*, Vol. 56, No. 2 (February 1990), pp. 187-95, ISSN 0099-1112

Mandelbrot, B. (1967). How long is the coast of Britain? Statistical self-similarity and fractional dimension. *Science*, New Series, Vol. 156, No. 3775 (May 1967), pp. 636-638 doi:10.1126/science.156.3775.636, ISSN 0036-8075

Mattfeldt, T. (1997). Spatial Pattern Analysis using Chaos Theory: A Nonlinear Deterministic Approach to the Histological Texture of Tumours, In: *Fractals in Biology and Medicine, Volume II*, G.A. Losa, D. Merlini, T.F Nonnenmacher & E.R. Weibel (Eds.), 50-72, Birkhäuser, ISBN 3-7643-5715-0, Basel, Boston, Berlin

Milne, B.T. (1992). Spatial aggregation and neutral models in fractal landscapes. *The American Naturalist*, Vol. 139, No. 1 (January 1992), pp. 32-57, ISSN 00030147

Page, S.L. & Hawley, R.S. (2003). Chromosome choreography: the meiotic ballet. *Science*, Vol. 301, No. 5634 (August 2003), pp. 785-789, ISSN 0036-8075

Rehder, A. & Wilson, E.H. (1913). Lardizabalaceae. In: *Plantae Wilsonianae*, C.S. Sarge, (Ed.), 334-352, Arnold Arboretum of Harvard University, ISBN 9781171493068, Cambridge, USA

Ross, C.M. & Sumner, M.J. (2004). Development of the unfertilized embryo sac and pollen tubes in the dwarf mistletoe *Arceuthobium americanum* (Viscaceae). *Canadian Journal of Botany*, Vol. 82, No. 11 (November 2004), pp. 1566-1575, ISSN 0008-4026

Ross, C.M. (2005). A new way of thinking about meiosis: using the fractal dimension to predict the onset of metaphase I. *International Journal of Biological Sciences*, Vol. 1, No. 3 (August 2005), pp. 123-125, ISSN 1449-2288

Sereda, K.Y. (2003). *Phenology of Anther Development of Dwarf Mistletoe (Arceuthobium americanum Nutt. Ex Englem.): An Anatomical Investigation of Microsporogenesis and Microgametogenesis.* Department of Botany, University of Manitoba. Winnipeg, Manitoba, M.Sc. Thesis

Shamrov, I.I. (1997). Ovule and seed development in *Paeonia lactiflora* (Paeoniaceae). *Botanicheskii Zhurnal,* Vol. 82, No. 6, (June 1997), pp. 24-46, ISSN 0006-8136

Shi, J.-X.; Ren, Y. & Di, W.-Z. (1994). Taxonomic studies on Sargentodoxaceae. *Acta Botanica Boreali-Occidentalia Sinica,* Vol. 14, No. 5 (May 1994), pp. 99-103, ISSN 1000-4025

Snustad, D.P. & Simmons, M.J. (2008). *Principles of Genetics* (5th edition), John Wiley & Sons, ISBN 9780470388259, New York, USA

Soltis, D.E.; Soltis, P.S.; Chase, M.W; *et al.* (2000). Angiosperm phylogeny inferred from 18 s rDNA and atpB sequence. *Botanical Journal of the Linnean Society,* Vol. 133, No. 4 (August 2000), pp. 99-118, ISSN 0024-4074

Sugihara, G. & May, R.M. (1990). Applications of fractals in ecology. *Trends in Ecology and Evolution,* Vol. 5, No. 3 (March 1990), pp. 79-86, ISSN 0169-5347

Wang, H.-F.; Kirchoff, B.K.; Qin, H.-N. & Zhu, Z.-X. (2009a). Reproductive morphology of *Sargentodoxa cuneata* (Lardizabalaceae) and its systematic implications. *Plant Systematics and Evolution,* Vol. 280, No., 4, (November 2009), pp. 207-217, DOI: 10.1007/s00606-009-0179-3, ISSN 0378-2697

Wang, H.-F.; Ross Friedman, C.M.; Zhang, Z-X. & Qin, H-N. (2009b). Early reproductive developmental anatomy in *Decaisnea* (Lardizabalaceae) and its systematic implications. *Annals of Botany,* Vol. 104, No. 6 (September 2009), pp. 1243-1253, doi:10.1093/aob/mcp232, ISSN 0003-4754

Wiens, D. (1964). Chromosome numbers in North American Loranthaceae (*Arceuthobium, Phoradendron, Psittacanthus, Struthanthus*). *American Journal of Botany,* Vol. 51, No. 1 (January 1), pp.1-6, ISSN 0002-9122

Wu, Z.-M. (1995). Cytological studies on some plants of woody flora in Huangshan, Anhui Province. *Journal of Wuhan Botanical Research,* Vol. 13, No. 2 (May 1995), pp. 107-112, ISSN 1000-470X (in Chinese).

11

Haploid Independent Unreductional Meiosis in Hexaploid Wheat

Filipe Ressurreição, Augusta Barão, Wanda Viegas and Margarida Delgado

CBAA, Instituto Superior de Agronomia,
Technical University of Lisbon Tapada da Ajuda,
Portugal

1. Introduction

Polyploidy correspond to the presence of more than two complete sets of chromosomes and is considered a major plant evolutionary force connected with adaptive plasticity (review in Comai, 2005). Within angiosperms it is estimated that at least 50% of its members have suffered one or more rounds of polyploidization (Wendel, 2000). Formation of functional unreduced gametes (2n) due to meiotic abnormalities is considered the key event associated with chromosome doubling and has been described in several dicot and monocot plant species (Ramanna and Jacobsen, 2003, Fawcett and Van de Peer, 2010). While autopolyploids are a direct result of duplication thus having multiple chromosome sets of the same origin, allopolyploids result from interspecific or intergeneric hybridization associated or followed by chromosome doubling, allowing for the emergence of new sexually reproduced species. Among sexual polyploids are many of the most important crops worldwide as it is the case of bread wheat (*Triticum aestivum* L.).

Triticum aestivum L. is an allohexaploid with 3 genomes designated A, B and D (2n = 6x = 42, AABBDD). It is believed that the formation of hexaploid wheat has occurred about 8000 years ago (Huang et al., 2002) through the cross between *Aegilops tauschii* Coss. (2n = 2x = 14) donor of genome D, and *Triticum turgidum* L. (2n = 4x = 28; AABB) (McFadden and Sears, 1946). By its turn *Triticum turgidum* L. (durum or macaroni wheat) is an allotetraploid resulting from the hybridization between *Triticum urartu* Tumanian (2n = 2x = 14) and *Aegilops speltoides* Tausch (2n = 2x = 14) the diploid donors of the genomes A and B, respectively (Dvorak et al., 1993). Since wheat progenitor species are close related, pairing can occur between corresponding (homoeologous) chromosomes of the distinct genomes. This phenomenon is however suppressed by the *Pairing homoeologous* gene (*Ph 1*) mapped in the long arm of chromosome 5B (Riley and Chapman, 1958). The presence of *Ph 1* results in a diploid-like meiosis both in tetra- and hexaploid wheat and is believed to have arisen by mutation at the time of the tetraploid formation (Riley and Chapman, 1958; Jauhar, 2007).

It is accepted that formation of unreduced gametes is the most important mechanism for fertile allopolyploid formation. Two main pathways are associated with meiotic restitution according with the composition of the duplicated meiotic products: First Division Restitution (FDR) and Second Division Restitution (SDR). FDR and SDR correspond in *sensu lato* to the omission of the first division or the second meiotic division respectively, this is failure of either homologous chromosome segregation or sister chromatid segregation (Ramanna and Jacobsen, 2003). Meiotic restitution corresponding FDR has been described in *Triticeae* as a haploid-dependent process and repeatedly addressed through the use of intergeneric hybrids (between wheat and related species) and wheat polyhaploids (recovered from hybridization with maize) (Islam and Shepherd, 1980; Balatero and Darvey, 1993; Jauhar, 2007; Matsuoka and Nasuda, 2004; Shamina, 2011).

In summary it has been shown that the haploid dependent meiotic restitution trait is genetically controlled and is only present in some wheat genotypes. When meiotic restitution trait is present, unreduced gametes result from lack of univalent segregation at meiosis I and subsequent sister chromatid disjunction at anaphase although the level of penetrance is generally very low. Analysis of microtubule cytoskeleton revealed that haploid unreductional meiosis depends on the bipolar attachment of univalents (Cai et al., 2010). In addition, it was also recently shown that even in genotypes with meiotic restitution, formation of unreduced gametes is obstructed by the occurrence of some level of homologous pairing (Wang et al., 2010).

So far, in wheat meiotic restitution was only described for haploid genotypes in which asynapsis results from the absence of homologous chromosome pairs. Here we investigated the influence of induced asynapsis in the progression of meiosis in a diploid context. For this purpose we took advantage of aneuploid hexaploid wheat line nulisomic for 5D and tetrassomic for 5B (N5DT5B) derived from cv Chinese Spring that display inducible asynapsis. This phenotype is associated with the lack of *Low temperature pairing* (*Ltp*) gene mapped in the long arm of chromosome 5D. N5DT5B plants are fertile with regular diploid-like meiosis when grown at temperatures ranging from 19 to 29°C, however for temperatures lower than 15°C or higher than 30°C meiocytes are asynaptic (Bayliss and Riley, 1972a). Moreover it has been established that the sensitive state for asynapis is the last interphase prior to meiosis and not the meiotic division itself (Bayliss and Riley, 1972b).

2. Material and methods

2.1 Plant material and growth conditions

We analyzed the chromosome and cytoskeleton behaviour during male meiosis in aneuploid lines N5DT5B and N5DT5A derived from hexaploid wheat (*Triticum aestivum* L.) cv Chinese Spring. Plants were continuously grown at 22°C or exposed to 10°C for 15-20 days prior to meiosis (Queiroz et al, 1991). Under the low temperature regime N5DT5B line is completely asynaptic while N5DT5A line displays a low level of asynapsis (Riley et al., 1966, Bayliss and Riley, 1972a). In these lines all chromosmes are present as homologues pairs with exception of chromosome 5B or 5A that are present in four copies in N5DT5B and N5DT5A, respectively. Wheat x rye F1 hybrid between hexaploid wheat cv. Chinese Spring and Portuguese rye (*Secale cereale* L.) landrace Centeio do Alto continuously grown at 22°C was also analyzed corresponding to an haploid-dependent asynaptic genotype.

2.2 Cytological analysis of meiocytes

Anthers selected for all meiotic stages were fixed in 4% (w/v) formaldehyde in MTSB (50 mM piperazine-N,N-bis(2-ethanesulfonic acid), 5 mM MgSO4.7H2O, 5 mM EGTA, pH 6.9) for 45 min at room temperature, and then rinsed twice in MTSB. Meiocytes were processed for subcellular localization of microtubules (MT) by indirect immunofluorescence. Briefly, fixed anthers were dissected permitting meiocyte dispersal in multiwall slides coated with aminopropyltriethoxysilane (Sigma) and left to air dry. The cells were then permeabilized in 0.5% Triton X-100 in MTSB for 15 min and rinsed prior to labeling. MT were localized with mouse monoclonal antibody against alpha-tubulin DM1A (Serotec) diluted 1 : 100 in MTSB, which recognizes alpha-tubulin a component of MT. Indirect detection of DM1A was performed with a secondary antibody conjugated with fluorescein isothiocyanate, diluted 1 : 300 in MTSB. DNA was counterstained 4_,6-diamidino-2-phenylindole hydrochloride (DAPI) in Citifluor antifade buffer (AF1; Agar Scientific, Stansted, U.K.).

Immunofluorescence was recorded using an epifluorescence microscope Zeiss Axioskop2 equipped with a Zeiss AxioCam MRc 5 digital camera. Images were captured using the appropriate excitation and emission filters and composited using Adobe Photoshop 7.0 (Adobe Systems Inc.) software. For N5DT5B and N5DT5A in both growth conditions and for wheat x rye F1 hybrid several plans were analyzed and at least 100 meiocytes were scored at each stage (prophase I, metaphase I, anaphase/telophase I, prophase II and anaphase/telophase II).

3. Results

N5DT5B and N5DT5A lines grown t 22°C have a regular meiotic chromosome behaviour as previously described (Bayliss and Riley, 1972a). Also the microtubule dynamics follows the typical *Triticeae* meiotic pattern (Shamina, 2005, Cai et al., 2010). During the first meiotic division prophase a perinuclear ring is formed, from which microtubules emanate into the nuclear area with progression of prophase (Figure 1 a). Metaphase I and anaphase I are characterized by the presence of a spindle where microtubules converge at two polar foci forming a clear fusiform spindle responsible for homologous chromosome segregation (Figure 1 b and c, respectively). At telophase I a prominent fragmoplast is formed resulting in cytokinesis and formation of two reduced dyads (Figure 1 d). In the second meiotic division in each dyad following a brief prophase a spindle is observed at metaphase/anaphase leading to sister chromatid segregation and a fragmoplast is formed at telophase producing a tetrad with four haploid microspores (Figure 1 e and f).

As previously described (Bayliss and Riley, 1972a), in N5DT5B plants exposed to 10°C prior to meiosis asynapsis is induced. Under these asynaptic conditions at first meiotic division 81.6% of the meiocytes present a low level of bivalent formation (3-6 bivalents per cell). In these cases a spindle is formed although major disturbances such as twisted spindles are common. Syntetelical orientation to the spindle (monopolar attachment) of sister kinetochores is observed for both bivalents and univalents (Figure 2 a). This results in the gathering of bivalents at the metaphase plate and their correct segregation while univalents are dispersed and randomly segregated. Progression into meiosis II occurs with sister chromatid segregation and cytokenesis with consequent uneven meiotic products and high level of micronuclei (Figure 2 d, left).

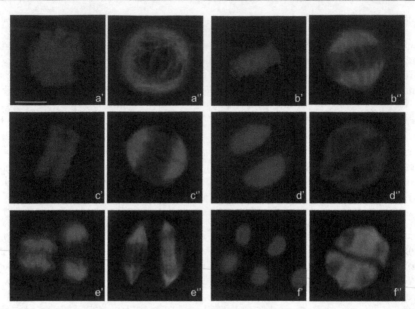

Fig. 1. Regular meiosis of the aneuploid line N5DT5B derived from hexaploid wheat cv Chinese Spring grown at 22°C. a. prophase I, b. metaphase I, c. anaphase I; d. telophase I, e. anaphase II, f. telophase II. Chromosome DAPI staining is shown in blue (a', b', c', d', e' and f') and microtubule cytoskeleton in green (a'', b'', c'', d'', e'' and f''). All images have identical magnification, bar = 10μm.

However, a significant proportion of the meiocytes (18.4%) present complete asynapsis. These cells undergo sister chromatid separation with tendentially equational chromatid segregation originating diads that further progress to interphase with a cytoskeleton organization characteristic of young pollen grain (Figure 2 d, right). Interestingly, the level of micronuclei in dyads is much low than in tetrads indicating high probability of balanced division. This results in the formation of unreduced gametes what is also supported by the fact that dyads nuclei are considerably larger than those of tetrads although the compaction level of chromatin is similar (Figure 2 d). At this stage, a high proportion (27%) of dyads are observed although simultaneously with tetrads, triads and other highly unbalanced meiotic products since unreductional division only occurs in some meiocytes within the same anther.

In the N5DT5B line under asynaptic condition we found that unreductional division can occur by two distinct pathways (Figure 2 b and c). (i) Segregation of chromatids during the first division and blockage of second division. In this case sister kinetochores orient amphitelically to the spindle (bipolar attachment) and chromosome arm cohesion is lost. At anaphase sister chromatids are pulled to opposite poles although one ore two lagging chromosomes can occur (Figure 2 b). (ii) Failure of first division with formation of a restitution nucleus encaging all univalents coincident with chromosome decondensation (Figure 2 c). The monads formed undergo to second division where sister chromatids are segregated.

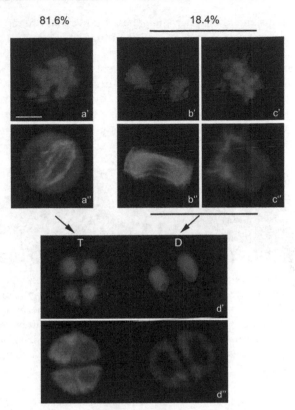

Fig. 2. Aspects of asynaptic meiosis in the aneuploid line N5DT5B due to exposure to 10°C prior to meiosis. a. Univalents at metaphase I (a') with bipolar spindle (a'') corresponding to 81.6% of the meiocytes at this stage b. and c. The remaining 18.4% of meiocytes undergo unreductional meiosis either through segregation of sister chromatids (b') with formation of bipolar spindle (b'') or through nuclear restitution and (c') absence of spindle (c'').
d. Distinct meiotic products at the end of meiosis, a tretrad on the left (T) and a dyad on the right (D) are simultaneously observed in a single microscopic field. All nuclei show identical chromatin condensation but dyad nuclei are larger than tetrad nuclei, a micronulei is present in the tetrad (d'). Microtubule organization characteristic of young pollen grain stage is present in both cell sets (d''). All images have identical magnification, bar = 10µm.

The meiotic behaviour of N5DT5A line under asynaptic conditions differs from that of N5DT5B line in the level of asynapsis that is much lower. Only a small fraction of meiocytes display complete or nearly complete asynapsis (10% of meiocytes with less than 6 bivalents). Meiocytes with a low number of bivalents behave as described for N5DT5B, while the majority of meiocytes undergo regular meiotic division although with a high frequency of micronuclei.

In the hexaploid wheat x rye F1 hybrid asynapsis at 22°C occurs due to haploid condition. Meiosis progress with a high level of irregularities as previously described for other wheat hybrid genotypes. Metaphase I form irregular spindles that result from monopolar attachement of univalents and at anaphase/telophase I univalents are randomly segregated. Meiocytes progress into meiosis II with formation of highly unbalanced gametes (Figure 3 b). Importantly, the hybrid does not display any of the characteristic features of unreductional meiosis, although it has as female parent Chinese Spring wheat from which N5DT5B and N5DT5A are derived. Neither first division nuclear restitution or sister chromatid segregation are ever observed. Occasionally meiocytes (8,6% at metaphase I) present premature sister chromatid separation but in contrast with N5DT5B this is always associated with spindled collapse and consequently impairment of sister chromatids segregation (Figure 3 a).

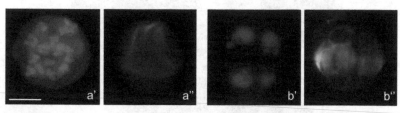

Fig. 3. Aspects of asynaptic meiosis in the hexaploid wheat cv Chinese Spring x rye Centeio do Alto F1 hybrid. a. Permature sister chromatid separation at first division (a') associated with spindle collapse (a''). b. Unbalance telophase II with several micronuclei. DAPI staining (b') and microtubule cytoskeleton (b'') reflects the random univalent segregation at previous anaphase I since both nuclei on the left have very distinct size from those on the right and dyad cleavage divides left and right nuclei. All images have identical magnification, bar = 10µm.

4. Discussion

Our results show that unreductional meiosis occurs in the aneuploid wheat line N5DT5B with formation of 2n pollen grains. To our knowledge this is the first time that unreductional meiosis is described in a wheat line with homologous pairs for all chromosomes present. Formation of unreduced gametes has recurrently been reported as a characteristic of some wheat genotypes described as having meiotic restitution trait. However, this characteristic was always considered as a haploid dependent process occurring in either interspecific F1 hybrids (for review see Silkova at al., 2011) or in the polyhaploid plants (Jauhar, 2007, Cai et al., 2010). In the case of N5DT5B this phenomena is induced under low temperature conditions as the meiotic behaviour of this line is completely regular at 22°C. N5DT5B was characterized by inducible asynapsis by low temperature exposure prior to meiosis due to the absence of the *Ltp* gene mapped in the in the long arm of chromosome 5D (Bayliss and Riley, 1972a).). This implies that in wheat asynapsis rather than haploid condition is the key feature for unreduced pollen grain formation. Supporting this hypothesis is the recent demonstration that the meiotic restitution phenotype observed in F1 hybrids between *T. turgidum* with diploid *Ae. tauschii* is completely abolished when tetraploid *Ae. tauschii* is used due to the formation of bivalents between the homologues chromosomes of *Ae. tauschii* (Wang et al 2010).

Distinct cytological processes leading to unreductional meiosis were described in distinct *Triticeae* genotypes using distinct terminology. First division restitution (FDR) was used to describe the lack of chromosome segregation at anaphase I followed by nuclear restitution and second meiotic division in hybrids between *T. turgidum* L. ssp. *durum* cultivars Langdon and Golden Ball and *S. cereale* L. cultivar Gazelle or *Ae. tauschii* Coss. (Xu and Joppa, 2000). More recently the designation unreductional meiotic cell division (UMCD) was used to describe this meiotic behavior (Cai and al., 2010). On the other hand, in very close related hybrids between *T. turgidum* L. ssp. *turgidum* and *Ae. tauschii* Coss., other authors described a single-division meiosis (SDM) characterized by a mitosis-like equational division with univalent sister chromatid segregation and blockage of second division (Zhang et al., 2007). The divergence in terminology reflects the lack of knowledge regarding the mechanisms that govern meiotic division in haploid genotypes. It should be emphasized that either nuclear restitution at meiosis I followed by regular meiosis II or sister chromatid segregation at anaphase I and absence of the second division have exactly the same genetic outcome, i.e. formation of two genetically identical unreduced microspores since cell division only involves sister chromatids segregation. Additionally, it has been shown that both types of division coexist with various frequencies depending on the genotype analysed (Xu and Joppa, 2000, Zhang et al., 2007, Silkova et al., 2011). In fact this is also the case observed here for the N5DT5B line under asynaptic conditions. This suggests that processes leading to the previously described haploid dependent formation of unreduced gametes in F1 hybrids or polyhaploids also occur when two homologous chromosomes are present if synapsis is inhibited.

Interestingly, when the wheat cv Chinese Spring, the genotype from which N5DT5B line is derived, is crossed with Portuguese rye Centeio do Alto, the F1 hybrid does not present meiotic restitution features since at meiosis I univalents are either randomly segregated or if sister chromatid cohesion is lost microtubule spindle collapses and therefore no segregation of chromatid is observed. Similarly, Zhang et al. (2007) found no evidence of meiotic restitution in the F1 hybrid between Chinese Spring wheat and a Chinese rye landrace or *Ae. variabilis*. On the other hand, analysis of polyhaploid lines derived from hexaploid wheat cv Chinise Spring showed that although meiotic restitution can occur with seed production (Jauhar, 2007) the level of seed set is much lower than that observed in polyhaploid lines derived from the tetraploid wheat Langdon, one of the most studied genotypes for its haploid dependent meiotic restitution trait (Cai et al., 2010). Together these observations indicate that hexaploid wheat cv Chinese Spring genotypes cannot be characterized as having haploid dependent meiotic restitution trait as previously assumed by Zhang et al. (2007). This does not exclude however that a high level of meiotic restitution can occur in some in wheat cv Chinese Spring interspecific hybrids as has been observed with *Hordeum vulgare* cv. Betzes (Islam and Shepherd, 1980) since this phenomenon is influenced by both parental genomes.

The present results show that meiotic restitution and unreduced gamete formation can occur in hexaploid wheat genotypes without meiotic restitution trait in a haploid independent manner in conditions of induced asynapsis. These observations raise the hypothesis that in some step of hexaploid wheat evolution interspecific hybridization could result from unreduced gamete fertilization.

Taken together, this challenges the widely accepted notion that interspecific hybridization has taken place prior to chromosome duplication in wheat evolution. Considering that formation of 2n gametes in sexual species is rare and highly dependent on environment

(Ramanna and Jacobsen, 2003), it is plausible that allopolyploidization results from fusion of unreduced gametes which in turn are the result of environmental conditions in established species sharing the same geographical niche. In fact, it has been proposed that this is the primarily process in the formation of neopolyploids in flowering plants (Ramsey and Schemske, 2002).

From the extensive data on wheat haploid-dependent meiotic restitution it is clear that this genetically controlled trait occurs in some genotypes of tetraploid wheat but not in others. If we consider that haploid dependent restitution is the genesis of wheat species it must be assumed that at least in some genotypes controlling gene(s) were conserved throughout wheat evolution, although only effective in the haploid condition. Based on our observations, we are proposing the alternative hypothesis that the haploid-dependent meiotic restitution trait has been acquired in some genotypes after allopolyploidization and that this acquired trait resulted from a loss or reduction in function of genes involved in promoting meiotic pairing.

Several genes promoting meiotic pairing have been identified in wheat (Feldman, 1966, Riley, 1974, Queiroz et al., 1991). The most well studied genes involved in wheat meiotic chromosome pairing are *Ph* (*Pairing homoeologous*) genes, which ensure that pairing is limited to homologous rather than between homoeologous chromosomes. We suggest that in genotypes without haploid-dependent meiotic restitution, promoting pairing genes allow transient chromosome synapsis and thus result in a dysfunctional meiosis. On the other hand, in genotypes with haploid-dependent meiotic restitution, reduced pairing capacity could lead to a situation similar to that observed in the N5DT5B line at low temperatures. In this case, absence of meiotic synapsis can be attributable to defective initiation of meiotic chromosome pairing therefore becoming reminiscent of a mitotic division.

Our hypothesis that haploid-dependent meiotic restitution in wheat evolved more recently is supported by phylogenetic evidence. Chromosome structure of Chinese Spring wheat, which does not have the haploid-dependent meiotic restitution trait, is more similar to the primitive tetraploid wheat than other wild tetraploids (Kawahara, 1988). Also significantly, a recent report showed that tetraploid wheat genotypes without haploid-dependent meiotic restitution have less meiotic abnormalities in conditions of heat stress than tetraploid genotypes with this trait, suggesting that haploid-dependent meiotic restitution is directly associated with less efficient meiotic control (Rezaei and Sayed-Tabatabaei, 2010). Taken together, we believe our results provide a novel and comprehensive view of wheat evolution.

5. Conclusion

Unreductional meiosis and formation of 2n gametes is the main mechanism in the emergence of sexual polyploids. Bread wheat (*Triticum aestivum* L.) is a natural allohexaploid with regular diploid-like meiosis resulting from two sequential events of hybridization associated with chromosome doubling. Several studies have addressed the question of chromosome duplication considering that this is a haploid-dependent process. We show here that unreductional meiosis occurs in hexaploid wheat genotypes where two homologues are present for each chromosome as an asynapsis dependent process controlled by a genotype-temperature mechanism. This is the first time that formation of 2n microspores is observed in wheat as a haploid independent process. The present results raise the hypothesis that wheat evolution could result from unreduced gamete fertilization challenging the generalized idea that wheat intergeneric hybridization occurred prior to chromosome duplication.

6. Acknowledgements

This work was funded by Pest-OE/AGR/UIO240/2011, FCT-Portugal

7. References

Balatero CH, Darvey NL. 1993. Influence of selected wheat and rye genotypes on the direct synthesis of hexaploid triticale. *Euphytica* 66: 179-185

Bayliss MW, Riley R. 1972a. An analysis of temperature-dependent asynapsis in *Triticum aestivum*. *Genetical Research* 20: 93-200

Bayliss MW, Riley R. 1972b. Evidence of premeiotic control of chromosome pairing in *Triticum aestivum*. *Genetical Research* 20:201-212

Cai X, Xu SS, Zhu X. 2010. Mechanism of haploidy-dependent unreductional meiotic cell division in polyploid wheat. *Chromosoma* 119:275–285

Comai L. 2005. The advantages and disadvantages of being polyploid. *Nat Rev Genet.* 6:836-846.

Dvorak J, DiTerlizzi P, Zhang H-B, Resta P. 1993. The evolution of poly- ploid wheats: identification of the A genome donor species. *Genome.* 36:21–31.

Fawcett JA, Van de Peer Y. 2010. Angiosperm polyploids and their road to evolutionary success. *Trends in Evolutionary Biology* 2:e3.

Feldman M. 1966 The effect of chromosomes 5B, 5D and 5A on chromosomal pairing in Triticum aestiuum. *Proc. Natl. Acad. Sci.* US. 55: 1447-1453.

Huang S, Sirikhachornkit A, Su X, Faris J, Gill B, Haselkorn R, Gornicki P. 2002. Genes encoding plastid acetyl-CoA carboxylase and 3-phosphoglycerate kinase of the Triticum/Aegilops complex and the evolutionary history of poly- ploid wheat. *Proc Natl Acad Sci* USA. 99:8133–8138.

Islam AKMR, Shepherd KW. 1980. Meiotic Restitution in Wheat-Barley Hybrids. *Chromosoma* 79:363-372

Jauhar P. 2007. Meiotic Restitution in Wheat Polyhaploids (Amphihaploids): A Potent Evolutionary Force. *Journal of Heredity* 98:188–193

Kawahara T. 1988. Confirmation of primitive chromosome structure in the hexaploid wheats. *Theor Appl Genet* 75: 717-719.

Matsuoka Y, Nasuda S. 2004. Durum wheat as a candidate for the unknown female progenitor of bread wheat: an empirical study with a highly fertile F1 hybrid with *Aegilops tauschii* Coss. *Theor Appl Genet* 109:1710–1717

McFadden ES, Sears ER. 1946. The origin of *Triticum spelta* and its free threshing hexaploid relatives. *J Hered.* 37:81–89.

Queiroz A, Mello-Sampayo T, Viegas W. 1991. Identification of low temperature stabilizing genes, controlling chromosome synapsis or recombination, in short arms of chromosomes from the homoeologous group 5 of *Triticum aestivum*. *Hereditas* 115: 37-41

Ramanna MS, Jacobsen E. 2003. Relevance of sexual polyploidization for crop improvement – A review. *Euphytica* 133: 3–18

Ramsey J, Schemske DW. 2002. Neopolyploidy in flowering plants. *Annu. Rev. Ecol. Syst.* 33:589–639

Rezaei M, Arzani A, Sayed-Tabatabaei B. E. 2010 Meiotic behaviour of tetraploid wheats (*Triticum turgidum* L.) and their synthetic hexaploid wheat derivates influenced by meiotic restitution and heat stress. *J. Genet.* 89, 401–407

Riley R. 1974. Cytogenetics of chromosome pairing in wheat. *Genetics* 78: 193-203

Riley R, Chapman V. 1958. Genetic control of the cytologically diploid behaviour of hexaploid wheat. *Nature.* 182:713–715.

Riley R, Chapman V., Young RB, Belfield AM. 1966. Control of meiotic chromosome pairing by the chromosomes of the homoeologous group 5 of *Triticum aestivum. Nature.* 212:1475–1477.

Silkova OG, Shchapova AI Shumny VK. 2011 Meiotic restitution in amphihaploids in the tribe *Triticeae. Russian Journal of Genetics.* 47:383–39.

Shamina NV. 2005. Formation of division spindles in higher plant meiosis. *Cell Biol Int.* 29:307-318.

Shamina NV. 2011."Bouquet arrest", monopolar chromosomes segregation, and correction of the abnormal spindle. *Protoplasma.* 2011 Jan 28. [Epub ahead of print]

Wang CJ, Zhang LQ, Dai SF, Zheng YL, Zhang HG, Liu DC. 2010. Formation of unreduced gametes is impeded by homologous chromosome pairing in tetraploid *Triticum turgidum* x *Aegilops tauschii* hybrids. *Euphytica* 175:323–329

Wendel JF. 2000. Genome evolution in polyploids. *Plant Mol Biol.* 42:225-249.

Xu SJ, Joppa LR. 1995.Mechanisms and inheritance of first division restitution in hybrids of wheat, rye, and Aegilops squarrosa. *Genome.* 38:607-15.

ZhangL-Q, Yen Y, Zheng Y-L, Liu D-C. 2007.Meiotic restriction in emmer wheat is controlled by one or more nuclear genes that continue to function in derived lines. *Sexual Plant Reproduction.* 20: 159-166

12

Embryology of Flowering Plants Applied to Cytogenetic Studies on Meiosis

Jorge E. A. Mariath et al*

*Plant Anatomy Laboratory, Biosciences Institute,
Electron Microscopy Centre, Federal University of Rio Grande do Sul-RS,
Brazil*

1. Introduction

The present chapter has three specific goals: (1) to discuss the incongruencies in terminology applied to meiosis and to pollen grains, (2) to describe cellular and subcellular aspects of plant meiosis and pollen, and (3) to review cytogenetic studies in plant meiosis involving embryological approaches.

The life cycle of plants is constituted by two generations: sporophytic and gametophytic. Gametophytic generation is the sexual generation. Differently from meiosis in animals, which gives rise to gametes, meiosis in plants originates spores. Heterospory occurs in some pterydophytes and in seed plants. This consists of the formation of two types of spores in separate sporangia (androsporangium and gynosporangium). In angiosperms, when meiosis occurs in anther sporangia, the spores are called androspores. When it occurs in seminal rudiment sporangia, they are called gynospores. The sporogenesis develops in a complete endosporic manner. In the case of gynospores formed, generally only one is viable in each sporangium. The viable spore germinates and, after three mitotic divisions, forms the female gametophyte, which develops in the sporangium tissue - the nucellus. During the development of the male gametophyte, the first mitosis occurs inside the sporangium, the other ones may occur after male gametophyte release. The androgametophyte is called pollen grain and the gynogametophyte is called embryo sac. The two sperm cells formed in the second mitosis of the male gametophyte, are the male gametes. They are present in the tricellular pollen grain or after mitosis of the generative cells during the pollen tube germination. The female gametes are called egg cell and central cell. In this way, sexual

* André L. L. Vanzela[2], Eliane Kaltchuk-Santos[3], Karen L. G. De Toni[4], Célia G. T. J. Andrade[5], Adriano Silvério[1], Erica Duarte-Silva[1], Carlos R. M. da Silva[2], Juca A. B. San Martin[1], Fernanda Nogueira[5] and Simone P. Mendes[6]
[1]*Plant Anatomy Laboratory, Biosciences Institute,*
 Electron Microscopy Centre, Federal University of Rio Grande do Sul-RS, Brazil
[2]*Biodiversity and Ecosystems Restorations Laboratory, State University of Londrina-PR, Brazil*
[3]*Cytogenetic Laboratory, Biosciences Institute, Brazil*
 Federal University of Rio Grande do Sul-RS, Brazil
[4]*Botanical Garden of Rio de Janeiro- RJ, Brazil*
[5]*Electron Microscopy and Microanalyses Laboratory, State University of Londrina-PR, Brazil*
[6]*National Museum- Federal University of Rio de Janeiro-RJ, Brazil*

generation in flowering plants is reduced to few cells and nutritionally dependent on the sporophyte. All the structures of the plant body, except the pollen grain and the embryo sac belong to the sporophytic generation.

We must not confuse or mix the different plant generations based on Hofmeister's alternation of generations (1851). When we analyze meiosis we are analyzing the reproductive results of the sporophyte, with the genetic recombinations present in the spores. No relation with sexuality was needed to obtain the expected result from a sporophytic generation – only SPORITY – androspores or gynospores, maintaining the reference to its localization in the androceum or gynoeceum in the word etymology, but without a sexuated connotation.

On the other hand SEXUALITY is present in the gametophytes, a generation that produces the male and female gametes, in individuals that are separate and thus unisexuated. They form, through fertilization, the new sporophytic generation (embryo) and the xenophytic generation (endosperm). The xenophyte is an accessory generation responsible for nourishing the embryo that is being formed, not directly connected to the reproductive cycle, since it never manages to produce its own reproductive structures.

In this chapter we will discuss the terminology applied to meiosis and to pollen grains, in accordance with the sporities and sexuality of generations, the cellular and subcellular aspects of these structures, and cases of cytogenetic studies involving embryological approaches.

2. Terminology applied to meiosis in plants and pollen grains

The scientific investigation of plant sexuality began in the 17th century with Rudolph Camerarius (1694) in his work De Sexu Plantarum Epistola. Camerarius was the first to prove the existence of sex in plants. His discovery was a Copernicus-like event for Botany, and for several fields of biology (Zàrsky & Tupy 1995). According to Cocucci (1969), Camerarius work was preceded by Nehemiah Grew (1682) that described the pollen, and Marcello Malpighi (1687) that described ovary and ovules, nowadays, known as seminal rudiment. Camerarius mistakenly identified sexuality in sporophytic structures, since he established that the stamina and pistils were male and female "organs". Carolus Linnaeus established a classification system that includes two basic principles: the use of Latinized or Latin words to name groups of organisms, as well as the use of categories that distribute the organisms from large to limited groupings. Linnaeus (1810) created the "Sexual System" positioning the plants with flowers in twenty-four classes based on the number of stamina and pistils. According to Quammen (2007) descriptions of flowers were compared to sexual relations among human beings, causing polemic and scandal in the society of his time. The *Fuchsia* flower, for instance, was classified in this system containing eight male stamina around a female pistil, belonging to class 8, described as 'eight men in the nuptial chamber of only one woman'.

In 18th century, Hedwig (1784) thought the antheridia of mosses to be equivalent to anthers (Wagenitz 1999). In the 19th century, Hofmeister (1851) published the theory of alternation of generations, accepting that there could be two generations for briophytes and pterydophytes, one of them sporophytic and the other gametophytic.

Darwin (1877) was the first to discuss the presence of different floral types in plants of the same species, and called the flower hermaphrodites, female and male. The terminologies adopted by Darwin, attributing sexuality to flowers, are used in biology even today, in a large part of scientific production. (Richards 1997; Ainsworth 1999; Barrett 2002; Mitchell & Diggle 2005; Karasawa 2009).

The Hofmeister's work, considered as genial as that of Mendel or Darwin, was not widely disseminated in scientific circles (Kaplan & Cooke 1996). Scholars in the field of biology, taking Darwin as their primordial, guiding assumption (Cohen 2010), commonly observe a flower from the perspective of Camerarius and Linnaeus, as a sex organ, attributing sexuality to the sporophyte. The terms female flower and male flower are wrong, since they include both sporophytically originated structures, to which no sex is ascribed (calyx, corolla, androecium and gynoecium), and gametophytic structures that express sexuality such as the female gametophyte (embryo sac) and the male one (pollen grain). Further criticism of the term hermaphrodict is that every gametophyte in angiosperms and gymnosperms is unisexual, and in turn, hermophroditism is a condition present until isospore pterydophytes along the evolutionary scale of plants, in which a single spore gives rise to a prothallus with male and female gametangia. (Cocucci 1969; 1973; 1980; Cocucci & Hunziker 1994; Cruden & Lloyd 1995; Cocucci & Mariath 1995).

The terms 'female', 'male' and 'hermaphrodite' (Camerarius 1694; Linnaeus 1754, 1810; Darwin 1877) should be replaced by "imperfect pistillate", "imperfect staminate" and "perfect", respectively (Cocucci & Mariath 1995; Cocucci 1980; Greyson 1994; Cruden & Lloyd 1995). These terms are utilized by the APG (2007). This is due to the fact that the flower constitutes a sporophytic structure holding one or two gametophytes that depend nutritionally from the sporophyte (Coccuci 1973).

Based on these assumptions, Cocucci & Mariath (1995) propose to name flowers as: "monosporic" when only one kind of sporangium developed, producing androspores or gynospores; and "bisporic", when the flower presents the two types of sporangia, producing androspores and gynospores. The present study adopts the terms perfect flower, imperfect pistillate and imperfect staminate, due to their frequent use in current scientific literature, avoiding an inappropriate sexual connotation.

Despite the conceptual issue of considering flowers as a structure that includes tissues of the sporophytic and gametophytic generation, studies on their evolution have developed greatly, and are currently one of the topics discussed in the fields of evolution, ecology and genetics. Charles Darwin (1877) recognized that plants with seeds have an incredible diversity of reproductive systems, a 'sexual diversification' that is determined by ecological and genetic factors that are one of the core problems of evolutionary biology. The integration of phylogeny, ecology and studies on population genetics has supplied new ideas regarding the selection mechanisms that are responsible for the greatest evolutionary transitions between the modes of reproduction (Barrett 2002).

However, in our view, this so-called 'sexual diversity' is actually the expression of the 'spority' of the antophytes, a diversification that includes flowers with different morphologies, and may carry both functional sporangia, androgynosporangiates (perfect flower), or only one of these two types, androsporangiate (androic) or gynosporangiate (ginoic), or other combinations of these sporangia (monoic, dioic, trimonoic, andromonoic and gynomonoic).

As regards the gametophytic generation, based on the heterospored pterydophytes, the gametophytes are always unisexuate, and they are never hermaphrodite or bisexuate. Cocucci & Mariath (1995) and Cocucci (2006), within the sphere of studies on plant reproduction and improvement, propose to limit the sexual terminology to the gametophytes, erradicating the sexual terminology applied to sporophytes.

3. Meiosis and pollen grain: Cellular and subcellular aspects

The life cycle of plants consists of two generations: a sporophytic, diploid, and a gametophytic, haploid (Hofmeister 1851; Cocucci 1969; Cocucci & Mariath 1995). In seeds plants, gametophytic and thus sexuated generation begins from a spore that develops a multicellular structure called gametophyte through mitotic divisions. If this process occurs in the androecium, the male gametophyte is called androgametophyte, microgametophyte, androphyte or pollen grain. If it occurs in the gynoecium, the female gametophyte is called gynogametophyte, megagametophyte, gynophyte or embryo sac. Therefore, in the seed plants, the gametophytic sexuated generation is small and depends nutritionally on the sporophyte, and the morphological structures, except the embryo sac and the pollen grains, belongs to the sporophytic generation (Cocucci 1969; Cocucci & Hunziker 1994; Cocucci & Mariath 1995).

As a case study we shall analyze *Passiflora elegans* Master (Passifloraceae), a species of passion flower native to Brazil, regarding aspects of the development of androsporogenesis and androgametogenesis, divided into four stages described below.

Stage I – Sporangium and archesporial cells. The completely formed young anther has four sporangia, two of them in the ventral-lateral regions and the other ones in the dorsal-lateral regions (Figure 1 A). The epidermal cells, as well as the endothecium and middle layers, present vacuolated cells with conspicuous nuclei (Figure 1B). The epidermis and endothecium are indifferentiated at this stage. The middle layers present nuclei with portions of condensed chromatin and small vacuoles in the cytoplasm. The tapetal cells are radially elongated, multinucleated, their nuclei have conspicuous nucleoli, the cytoplasm is dense, with the presence of small vacuoles (Figure 1B). The archesporial cells are the largest cells of the sporangium, the nuclei are hyaline, with conspicuous nucleoli, and few portions of condensed chromatin, the cytoplasm presents storage lipids granules (Figure 1B) which, through their oxidation, ensure the energy needed to trigger the meiotic process.

Stage II – Meiosis and the end of sporophytic generation. Concomitantly, when meiosis begins, deposition of a callose wall begins on the inner side of the primary wall of the androspore mother cell. Individual isolation of the young androspores occurs through the formation of a callose wall, only after the tetranucleated phase, thus characterizing cytokinesis as being of the simultaneous type (Figure 2A-H). The tetrads present a tetrahedric spatial arrangement (Figure 2E,G-H). The sporoderm begins to be deposited around the future androspores, already in the tetrad phase, through the formation of the primexine that is initially internal to the callose wall. This wall is constituted by polysaccharide and protein.

The deposition of sporopollenin inside the tetrad occurs after the primexine wall is formed, and it was confirmed with Auramine O (Figure 2G). The reticulated aspect of the future exine is noted when it is submitted to analysis with differential interference contrast. (Figure 2H).

In monocotyledons, the most common cytokinesis is of the successive type, as demonstrated in *Pitcairnia encholirioides* (Bromeliaceae) (Figure 3A-C). After telophase I, the androspore mother cell gives rise to a dyad of androspores (Figure 3B). A callose wall is syntethized between the dyad of androspores. Meiosis II takes course forming the androspore tetrads, mostly with an isobilateral arrangement (Figure 3C).

Stage III – Beginning of the gametophytic phase (unicellular gametophyte). The young androspores are released from the tetrads inside the anther loculus and take on a spherical shape (Figure 2I). The androspores present a large hyaline nucleus, with a conspicuous

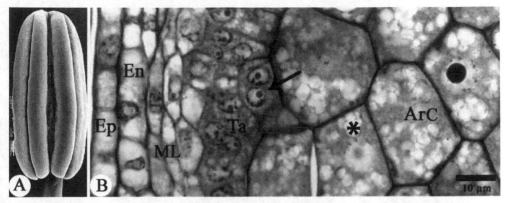

Fig. 1. Young anther of *Passiflora elegans* at the pre-meiotic stage. (A) Tetrasporangiated anther under scanning electron microscopy. (B) Cross-section of anther in pre-meiosis with archesporial cells. (Ep) epidermis; (En) Endothecium; (ML) middle layer; (ArC) Archesporial cell; (*) storage granules with lipidic nature.

nucleolus, and a large vacuole inside them (Figure 2I), signaling the beginning of gametogenesis. Sporoderm formation begins during the tetrad phase and in the free androspore phase the exine presents two stracta (ectexine and endexine) (Figure 4A). The ectexine, in *Passiflora elegans*, does not present a basal layer, so that the collumelae are plunged directly into the endexine. The intine, as well as the exine, are thick in *P. elegans* and are constituted by three chemically distinct pectic-proteic-cellulosic strata (Figure 4a). The cytoplasm presents a dense aspect and concentrates its largest volume at the polarized nucleus. After polarization, the androspore nucleus undergoes mitotic division forming the vegetative cell (VC), and the generative cell (GC) (Figure 4B). The tapetal cells, which previously were elongated and organized around the loculus, present irregular contours and degrade at the end of meiosis.

Stage IV – Bicellular pollen grain (bicellular gametophyte). Once the mitotic division has occurred, the central vacuole becomes smaller and the vegetative nucleus returns to the median portion of the cytoplasm, while the generative cell is kept in a parietal position. The wall separating the two cells is continuous with the intine. The vegetative cell includes the generative cell, acquiring a shape that ranges from lenticular to sickle-like (Figure 4B). From this phase on, the vegetative cells accumulates a large quantity of starch in their cytoplasm. This starch is distinctly hydrolyzed during the differentiation of pollen, so that different species can present pollen with or without starch during anthesis. In cases in which the starch is hydrolyzed, the cytoplasm is reactive to the PAS reaction while in the others, without hydrolysis, there is a weaker PAS reaction of the cytoplasm. The bicellular mature pollen grain, in cross-section, presents a hyaline vegetative nucleus, with a rounded shape and a conspicuous nucleolus. The generative cell is found immediately next to it and has an elongated shape (Figure 4B), in longitudinal section and a roundish cross-section, and a nucleus with portions of condensed chromatin (Figure 4B). Depending on the plant analyzed, the pollen grain is released in this bicellular form, completing the formation of gamete cells along the germinated pollen tube, as a result fo the mitosis of the generative cell (Figure 4D).

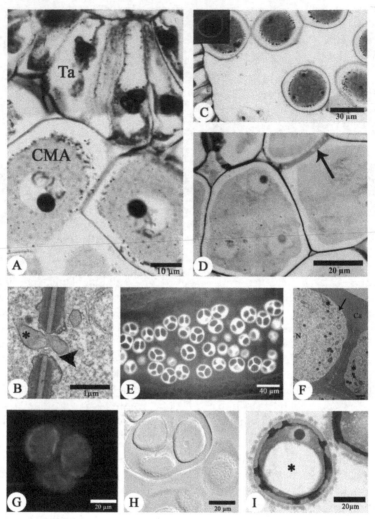

Fig. 2. Young anther of *Passiflora elegans* in meiotic and post-meiotic stage. (A) Androspores mother cells in prophase I. (B) Electron micrography of a cytomitic channel between androspores mother cells with the transit of plastids during meiosis. * means plastid, and arrowhead means cytomitic channel. (C) Meiocytes in telophase II. Insert reveals positive reaction to callosis with aniline Blue under fluorescence microscopy. (D) Cytomitic channels between meiocytes (arrow). (E) Tetrad after simultaneous cytokinesis, with callose between the androspores, marked with Aniline Blue under fluorescence microscopy. (F) Electron micrography of the callose wall involving the androspores in the tetrad. * means plastid, and arrow means mitochondria. (G) Androspores tetrad with exine deposited in the primexine, below the callose walls. Tetrad stained with Auramine O and observed under fluorescence microscopy. (H) Androspores tetrad in differential interference contrast, showing the reticulated aspect of the exine. (I) Androspore released from the tetrad in a vacuolated state (asterisk) stained with Toluidine Blue O.

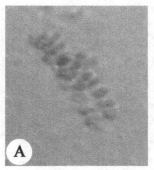
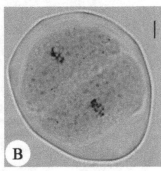
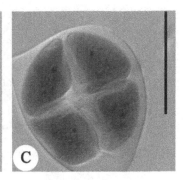

Fig. 3. Successive cytokinesis in P*itcairnia encholirioides*. (A) Androspore mother cell, metaphase I (B) Androspores dyad in metaphase II . (C) Androspores tetrad at the end of meiosis. Scale: 3A-B =20μm; 3C=50μm.

4. Cytogenetic studies involving embryological approaches

4.1 Cytogenetic and embryological analyses of sporophytic sterility and gametophytic sterility, in flowering plants

Staminal sterility, also known in agronomic and biotechnological spheres as "male-sterility", consists in the lack of some stage of androsporogenesis and androgemetogenesis, which leads to the non-formation of a viable androphyte. A viable androphyte is considered to be a pollen grain morphologically typical and metabolically active that can emit a pollen tube (Dafni & Firmage 2000; Duarte-Silva et al. 2011). "Male sterility" would be a more appropriate term to designate only sterility events related to failures in androgametogenesis, without attributing sex to sporophytic phases (Cocucci & Mariath 1995). Since the term is used in the literature equally for any failure in the development of the anther and of the pollen grain, in the present study we will adopt the term 'staminal sterility' for failures in the initial stages of androsporogenesis (sporophytic sterility) and 'pollen sterility' for failures in stages after sporoderm formation (gametophytic sterility). The detection and investigation of staminal sterility in edible and medicinal plants are highly important, since they help improve the plant, because they are naturally free of self-pollination and therefore more easily manipulated in reproductive system experiments to produce strains with a given character of interest (Bhat et al. 2005). The study of androsporogenesis and androgametogenesis in cases of staminal sterility can determine the stage when spore development (in the sporophytic phase) or the pollen grain (in the gametophytic phase) cease. Besides, together with the tools of cytogenetics, immunocytochemistry, and transmission electron microscopy, it allows inferring the cause of this failure. Once the critical stage for the occurrence of staminal sterility has been identified, molecular biology studies can be performed to detect the genes involved in the expression of sterility. Embryological studies to investigate staminal sterility are common in widely sold edible plants, such as soy, wheat, rice, beans and maize, and little studied in medicinal plants used in the phytotherapeutic industry (Duarte-Silva et al. 2010).

Staminode formation is caused by different embryological events, motivated by equally diverse environmental and genetic factors. Different genes act on each stage of embryo development, and their failures may lead to the same result: stamen and pollen sterility.

Mutants *bam1* and *bam2*, of *Arabidopsis* do not develop the endothecium, middle layer and tapetum cells, and the mother cells of androspores, in turn, degenerate before meiosis (Hord et al., 2006).

Many cases of staminal sterility are related to tapetum problems. The function of the tapetum is related to the the synthesis of enzymes needed to separate the androspores from the tetrad and also to nourish the pollen grains (Goldberg et al., 1993). Tapetum degeneration occurs normally at the late stages of pollen grain development and it has been considered a process of programmed cell death or apoptosis (Papini et al., 1999; Li *et al.* 2006). However, apoptosis, early or late, leads to staminal sterility events, as in rice (*Oriza sativa* L.) where tapetum degradation occurs late, obliterating the anther locule and causing the androspores to collapse (Li et al., 2006). Already in mutant BR 97-17971, of *Glycine max*,

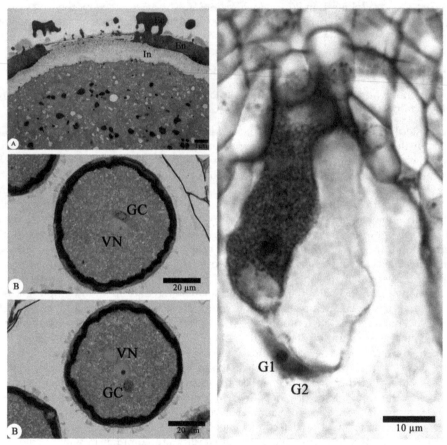

Fig. 4. Pollen grain of *Passiflora elegans*. (A) Eletron micrography of sporoderm (B) Slice of mature pollen grain with degenerative cell in a longitudinal section encompassed by the vegetative cell. (C) Slice of mature pollen cell at cross section. Generative cell enveloped by the vegetative cell. (D) Gametic cells (G1 and G2) after discharge of the pollen tube. (Ec) ectexine; (En) endexine; (In) Intine; (NV) nucleus of vegetative cell; (GC) generative cell, (G1 and G2) male gametes., (Si) Synergid , (EC) Egg cell.

early degeneration of the tapetum occurs in telophase II of meiosis, with the absence or failure of cytokinesis of the androspore mother cells, forming a cenocyte, or, sometimes, a tetrad with some degenerated androspores (Bione et al., 2002). Another tapetum-related event is the abnormal vacuolation of tapetal cells in *Bidens cervicata* Sherff. (Asteraceae) which leads to the disintegration of the sporogenic tissue (Sun & Ganders 1987). In the case of maize (*Zea maiz* L., Poaceae), mutants for the *ms 23* and *ms 32* genes presented periclinal divisions in the tapetum layer followed by their non-differentiation, causing the cell death of the sporogenic tissue in prophase I of meiosis (Chaubal et al., 2000). In *Helianthus annus* L. (Asteraceae) disordered periclinal divisions occur, showing signs of disorganization in the organelles and in the cell wall. Thus, the androspore tetrads disintegrate (Horner, 1977).

Problems in the formation of the androspore tetrad also lead to sterility. In bean plants (*Phaseolus vulgaris* L., Leguminosae) abortions occur in the tetrad stage due to cytoplasmatic connections, which are maintained between the androspores, indicating incomplete or aberrant cytokinesis (Johns et al., 1998). A similar event occurs in pistillated flowers of *Valeriana scandens* (Caprifoliaceae). The androsporogenesis of pistillated flowers is the same as in perfect flowers until the androspore tetrad phase. In this stage the exine is deposited in the androspores, both in the primexine wall and in the aperture regions, forming connections between the androspores; as the callose degrades through the callase enzyme, the androspores are released in the locule in a tetrad shape; an exine continuum is formed among them; they undergo vacuolation and begin androgametogenesis, but become senescent, and only the exine of the sporoderm remains collapsed in the locule of the anther in anthesis (Figure 5A-B) (Duarte-Silva et al. 2010).

A mutant of strain 6492 of Arabidopsis (Brassicaceae), in the tetrad stage presents eight microspores enveloped by callose (Peirson et al., 1996). Two other mutants (strains 7219 and 7593) present tetrads with other sizes of microspores, with failures in callose production and in vacuole development, and also multinucleate cenocytes (Peirson et al., 1996). Disorders in the synthesis and degradation of callose, and in the timing at which these processes occur, lead to sterility. Mutant *ms32* of *Arabidopsis* presents early degradation of callose right after meiosis, because of the formation, in the same stage, as a large amount of rough

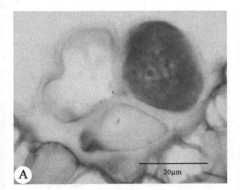
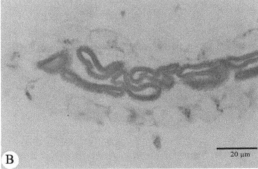

Fig. 5. (A) Malformed tetrad of *Valeriana scandens* (Caprifoliaceae). Observe the continuum of the exine between the tetrad androspores. (B) Sterile pollen in the anther in anthesis.

endoplasmic reticulum in the tapetum cells, probably responsible for callase synthesis and/or secretion in the anther locule (Fei & Sawhney 1999). On the other hand, in the mutant for gene *cals5* of the same species, callose deposition practically does not occur in the mother cell of androspores and in the tetrad, causing androspore degeneration (Dong et al., 2005). Once the androspore tetrads have been formed, there are cases in which the androspores interrupt their development at this stage. In *Allium schoenoprasum*, two mutants for staminal sterility (*wi* e *st1*) do not present dissolution of the callose after microspore tetrad formation, causing the non-deposition of sporopollenin (Engelke et al. 2002).

Pollen grain abortion can also occur in the sporoderm development phase. The mutant *nef1* of *Arabidopsis thaliana* presents pollen grains with primexin and sporoderm without sporopollenin, synthetized and accumulated in the locule of the anther (Ariizumi et al. 2004). In mutants *cesa1* and *-6* of *Arabidopsis*, cellulose synthesis deficiency in the sporoderm leads to abortion of the pollen grain (Persson et al. 2007). Mutants *rip1* of *Oryza sativa* developed sterile pollen grains in consequence of no intine formation (Han et al. 2006). Also, in *Oryza sativa*, the presence of empty pollen grains is the result of allele interactions of loci *S-a*, *S-b* and *S-c* which lead to abnormalities in the migration of the androspore nucleus to the cell periphery, where asymmetric mitose occurs and the consequent formation of vegetative and generative cells (Zhang et al. 2006). In the same study, stained, but non viable pollen grains are listed among the failures of generative cell migration and of the vegetative nucleus to the center of pollen, as well as failures in sperm cell formation (Zhang et al. 2006). In *Arabidopsis*, the bicellular androphyte goes into programmed cell death after a deficiency in the division of the generative cells to form the gametic cells, due to the lack of the key enzyme 'SerinePalmitoyltransferase' (Teng et al. 2008).

The synthesis of reserve substances is a critical event in androphyte development. The synthesis and accumulation of starch were investigated in androphytes of *O. sativa* and six genes that act on the sucrose-starch metabolic pathway were identified (*Rsus, OSINV2, OsPGM, OsUGP, OsAGPL3* and *OsSSI*). Mutants for these genes are deficient in the production of amyloplasts and constitute a sterile class called HL type (Kong et al. 2008). *Arabidopsis* sporophytes recessive for the gene (*atatg6/atatg6*) develop morphologically typical androphytes that are incapable of emitting a pollen tube (Fujiki et al. 2007). On the other hand the transgenics of *Brassica campestris*, that do not have gene *BcMF2*, presented, in the sporoderm, the pectic proportion of the overdeveloped exintine and the underdeveloped endintine, causing in vitro development of balloon-shaped pollen tubes (Huang et al. 2009).

4.2 Cyperaceae and their uncommon microsporogenesis

4.2.1 Ultrastructural and structural studies of pseudomonads in cyperaceae

Androsporogenesis in the Cyperaceae family follows a distinct pattern from that found in other Angiosperm groups. After meiosis, three of the four nuclei formed take up a defined region of the cell. They are isolated, degenerate and only one becomes the functional androspore. The four nuclei present a polarized distribution in an asymmetric cell known as pseudomonad (Figure 6A-B) (Selling 1947; Erdtman 1971; Strandhede 1973). In the development of the pseudomonad, all the cells come into contact with the tapetum inside the sporangium, in an arrangement known as peripheral (Kirpes et al. 1996). The pear-shaped pseudomonad has the broader region oriented toward the tapetum (abaxial region) and the narrower one oriented toward the center of the anther locule (adaxial region) (Figure 6b) (Strandhede, 1973). This spatial distribution, referring to the adaxial and abaxial aspects of a plant organ, may lead to mistaken interpretations, considering a entire sporangium, half of it oriented toward the adaxial aspect of the growth axis (rachila), while the other half is oriented toward its abaxial aspect. We suggest the orientation of these structures to the contiguous region, referring to the tapetum or locule.

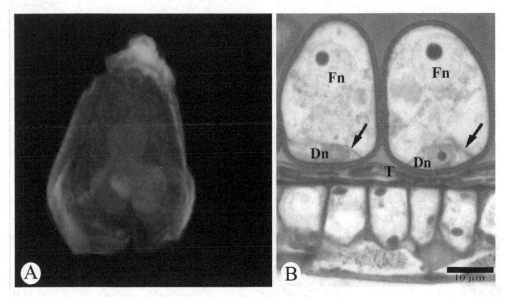

Fig. 6. (A) *Rhynchospora pubera* pseudomonad in confocal scanning microscopy showing the four nuclei in tetrahedral arrangement. The degenerative nuclei are located in a broad region of the cell, while the functional ones is placed in the central region. (B) Pseudomonad in contact with tapetum. The functional nucleus oriented toward the anther locule and the degenerative nucleus isolated by a thin inner cell wall (arrow). (Dn) degenerative nuclei; (Fn) functional nucleus; (T) tapetum.

In most genera of the Cyperaceae family, polarization follows a single pattern. The degenerative nuclei of the pseudomonad are oriented toward the locule, while the functional nucleus is oriented toward the tapetum region (Furness & Rudall 1999). This pattern of nuclear placement led to propose the "tapetum-pore" hypothesis. According to this hypothesis, the determining factor of the functional domain of the pseudomonad is due to the presence of the pollen pore close to the tapetum region (Strandhede 1973; Kirpes et al. 1996). This hypothesis can be applied to species of genus *Eleocharis* as shown in this chapter (see below). However, it cannot be applied to all Cyperaceae, since in *Rhynchospora* species the degenerative nuclei are found to be tapetum oriented and the functional nucleus is locule oriented (Figure 6B).

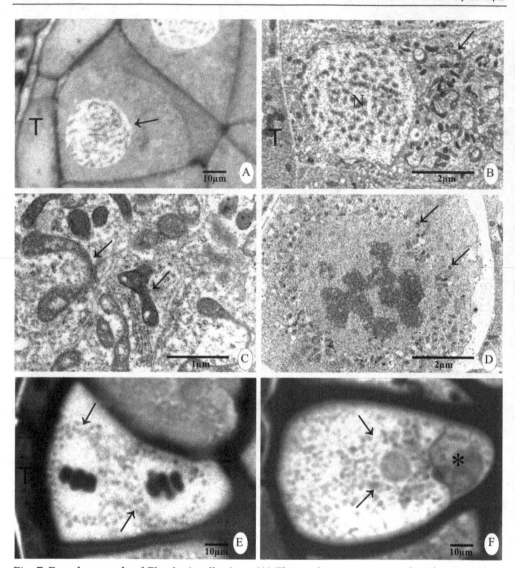

Fig. 7. Pseudomonads of *Eleocharis sellowiana*. (A) The nucleus is present in basal region of the androspore mother cell (arrow), during initial prophase. Toluidine blue 100X. (B) Electron micrograph of prophase I showing the organelles (arrow) located in the apical cytoplasm. (C) Detail of the figure 8B. Note organelles (arrow) in process of division, most of them are proplastids, mitochondria and rough endoplasmic reticulum. (D) Electron micrograph showing metaphase I. Arrows indicate organelles organized surrounding chromosomes. (E) Androspore mother cell during metaphase II exhibiting a cord of organelles (arrows) surrounding chromosomes. Toluidine blue. (F) Photomicrograph showing pseudomonad during final meiosis. The degenerative nuclei are at the cell apex (*). The arrows indicate remained organelles involving the functional nucleus. Toluidine blue. Tapetal cells (T). Nucleus (N).

In *Rhynchospora pubera* the androspores mother cell presents organelles with a polarized distribution, oriented towards the locule region, and the nucleus oriented towards the tapetal region (San Martin et al. 2009). These authors suggest that the pseudomonad undergoes a polarization process even before meiosis occurs. However, in *Carex blanda*, meiosis occurs with the nucleus of the androspore mother cell in the center of the cell and the three degenerative nuclei migrate to the pseudomonad region oriented toward the locule (Brown & Lemmon 2000). The reason why there is more than one type of polarization in the pseudomonads of Cyperaceae is still unknow. However, some cytoplasmic evidences indicate that this organization may be influenced by the behavior of organelles and cytoskeleton.

In *Eleocharis sellowiana*, as well as in other Cyperaceae, the androspores appears asymmetrically organized, culminating in pseudomonads polarized as found in *Carex*. However, in the early stages of meiosis the organelles develop an atypical arrangement around the chromosome complement. At prophase I, the nucleus occupy the basal region of the cell (Figure 7A-B) and the organelles increase in size and in number (Figure 7C). At metaphase I the organelles surround the chromosomes establishing a dense "Organelles Cord" (Figure 7D). This arrangement has not been documented for other Cyperaceae species, although it was reported in *Malva silvestris* (Kudlicka and Rodkiewicz 1990) and *Lavatera thuringiaca* (Tchórzewska *et al.* 2008). This "organelles cord" is held up sourrounding the chromosomes set until telophase I, when the phragmoplast is established. As meiosis progresses, the regions occupied by the phragmoplast are occupied by organelles cord, being one for each chromosome complement. These are held up until metaphase II / anaphase II (Figure 7E). There is evidence of organelles cord being maintained in one of the four nuclei after telophase II, and it is possibly important to select the functional nucleus. The functional one appears surrounded by an organelles cord in the pseudomonads in *E. sellowiana* (Figure 7F). The biological role of the organelles cord found in the *E. sellowiana* androspore mother cell remains uncertain, but the data suggest that it could influence in the delimitation of functional and degenerative nuclei at the transition from anaphase II to telophase II.

In *Rhynchospora pubera*, polarization occurs independent of the presence of organelles cord, and in the pseudomonads the functional nucleus appears directed to the anther locule. Toward the asymmetric and simultaneous cytokinesis, the degenerative nuclei are isolated after phragmoplast fusion and formation of an electron dense pectic wall (Figure 8A-B) (San Martin et al. 2011 in press). The degenerative nuclei are thus isolated with portions of cytoplasm containing organelles (Figure 8A-B). This shows the presence of two cell domains: one of them functional and the other degenerative. In this way, four androspores are formed, but three are eliminated (Figure 6A-B, 8A). This entire ensemble is contained in a single sporoderm with a maturating exine, constituted by tectum and columelle. (Figure 8A). In the cytoplasm of the functional domain, there are lipid droplets, which is not seen in the degenerative domain (Figure 8A). Vacuoles, mitochondria and other organelles are found both in the functional and in the degenerative domain (Figure 8A-B). Folded of rough endoplasmicc reticulum membranes are present in the functional cytoplasm, in some cases associated with lipid droplets (Figure 8C). The cytoplasm portions that envelop the degenerative androspores show signs of shrinkage, which does not occur in the cytoplasm of the functional androspore (Figure 8A-B). The degenerative androspore nuclei are small with more packed chromatin, seen both at light microscopy and at transmission electron microscopy. The chromatin in the functional nucleus is unpacked, indicating high metabolic activity (Figure 6B, 8A-B).

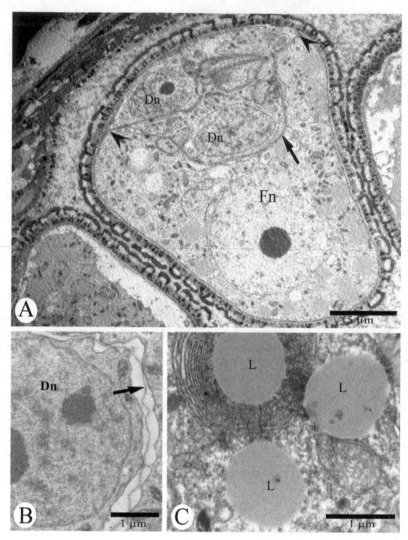

Fig. 8. Transmission electron micrographs of the pseudomonad of *Rhynchospora pubera*. (A) General view. Degenerative nuclei with chromatin packed and isolated by a wall (arrow), which establishes contact with intine (arrowhead). In the functional nucleus chromatin is unpacked. (B) Detail of inner cell wall isolating the degenerative androspores, arrows show the electrondense string in the inner of phragmoplast. (C) Free lipid droplets associated with folded membranes in the cytoplasm of functional androspore. (Dn) degenerative nucleus; (Fn) functional nucleus; (L) lipid droplet.

The degenerative androspores present ultrastructural characteristics of programmed cell death, highlighted by the packed chromatin, cytoplasmic shrinkage and vacuolation (Coimbra et al., 2004). On the other hand, the functional androspore shows characteristics of a cell during a phase of high molecule synthesis, evidenced by the unpacked chromatin, lipid droplets accumulation and presence of a well-developed rough endoplasmic reticulum. A crucial aspect in pseudomonad development in Cyperaceae is the occurrence of an asymmetrical cytokinesis after meiosis. This enables cell polarization and the isolation of the three degenerative androspores. In the same way, it allows the signalization to the programmed cell death, in degenerative domain, does not influences in the development of the functional domain.

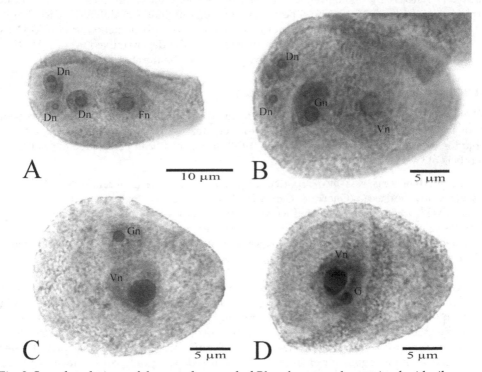

Fig. 9. Squash technique of the pseudomonad of *Rhynchospora pubera* stained with silver impregnation (AgNO3) to evidence ribonucleoproteins. (A) Four pseudomonad nuclei, three degenerative and one functional. (B) Gametophyte after first mitosis, showing the vegetative nucleus and the nucleus of the generative cell, in the presence of degenerative sporophytic nuclei. (C) After the complete PCD of the non-functional microspores, the generative cell and the vegetative nucleus are placed together in the central region of the pseudomonad. (D) Mature androphyte with vegetative cell and one gametic cell after generative cell mitosis. (Dn) Degenerative nucleus; (Fn) functional nucleus; (Gn) generative nucleus; (G) gametic cell; (Vn) vegetative nucleus.

Even in the presence of the degenerative androspores, the functional androspore undergoes mitosis giving rise to the vegetative and degenerative cell (Figure 9A-B), forming the young male gametophyte. After the complety elimination of the degenerative androspores , the generative cells gives rise to the gametic cells, through mitosis. Thus, the pollen grain (mature male gametophyte) is released from the anther in its tricellular form (Figure 9C-D).

The evolutionary history and adaptive advantage of developing a single functional androspore, instead of the four usual ones, remain unknown. A hypothesis would be that the increased volume of the cytoplasm would also increase the adaptive character of the cell (Ranganath & Nagashree 2000). This idea can be supported by the similarity of gynospore formation, where three of the four products of meiosis degenerate, resulting in an increased volume of the functional gynospore that will give rise to the female gametophyte.

Even so, some questions remain unanswered about androsporogenesis in the Cyperaceae family: 1) What is the biological reason for the degeneration of three of the four nuclei formed after meiosis? 2) Why does polarization occur differently in different genera? 3) Is the selection of functional and degenerate nuclei a chance phenomenon or not? 4) Does selection occur in favor of the functional nucleus or of the degenerative ones? 5) At what time during meiosis does the cellular signalling system act on the choice of nuclei?

4.2.2 Post-reductional meiosis in cyperaceae

Another poorly understood cytological feature about representatives of Cyperaceae, as well as, of holocentric chromosomes, is the post-reductional or inverted meiosis. This meiotic behavior was reported in *Carex* (Heilborn 1928) and it has been considered typical from members of Cyperaceae.

The inverted meiosis was also reported to *Luzula*, Juncaceae (Nordenskiöld 1951), and *Cuscuta*, Convolvulaceae (Guerra & Garcia 2004), as well as for some insect groups, but in the last case it was always associated with sex chromosomes and not with autosomes (Solari 1979; Pérez et al. 1997; Bongiorni et al. 2004). Besides holocentric chromosomes presented kinetic activity diffused along its major axis at mitosis (Nagaki et al. 2005), there is evidence of kinetic activity only in the terminal region of each chromatid, with spindle fibers attached directly to each one of the four chromatids (Vieira et al. 2009; Guerra et al. 2010). This arrangement could allow the formation of "box structures" of the bivalents (see Pazy & Plitmann 1987, Vanzela et al. 2000 and Guerra et al. 2010). However, there is no convincing image in the literature that shows a "box structure" directed to inverted meiosis, with the four chromatids completely individualized. Thus, we do not consider this enough to define the existence of inverted meiosis. In fact, it is very difficult to show if segregation of sister chromatids or homologous chromosomes has occurred using only conventional staining, unless there is a chromosome marker that allow to visualize each one of the chromatids. It is possible to univalents and sex chromosomes of insects (Hughes-Schrader & Schrader 1961; Vieira et al. 2009). The last author questioned the wide use of inverted meiosis for holocentric autosomes because reduction of part of the chromatid always occurs when there is chiasmate in bivalents, independent of orientation. This has been well documented in insects (Bongiorni et al. 2004; Nokkala et al. 2006), but not mentioned in plants (Nordenskiöld 1951, Strandhede 1965; Vanzela et al. 2000).

There are at least two excellent examples of inverted meiosis in plants. In the first case, Pazy (1997) showed the separation of sister chromatids in B-chromosomes in anaphase I of *Cuscuta babylonica*. In the second case, Da Silva et al. (2005) reported the separation of sister chromatids for all homologous of *Eleocharis subarticulata* (Cyperaceae). However, the two cases can be considered special situations due to the presence of a univalent B-chromosome and of a karyotype with multivalents rearranged by disploidy, respectively. But independent of these arguments, this atypical meiotic behavior has been referred only to organisms with holocentric chromosomes (Viera et al. 2009). Here we show evidence of reductional, or normal meiosis, in at least one chromosome pair of *Rhynchospora pubera*.

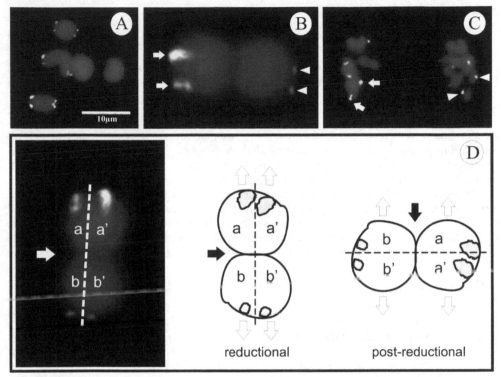

Fig. 10. Meiotic cells of *Rhynchospora pubera* hybridized with biotiniled 45S rDNA probe and detected with avidin-FITC. (A) Metaphase I showing pairs 1, 2 and 4 with terminal hybridizations signals. (B) Pair 2 isolated. Observe sister chromatids laterally joined while homologous are joined end-to-end. This bivalent shows a heteromorphism in the hybridization signals size (arrows and arrowshead). (C) Anaphase I and (D) Scheme with the second pair. Both images show segregation of homologous chromosomes and not sister chromatids. Arrows in (C) point out homologous of pair 2 with major signals and the arrowheads indicate the other homologous with minor hybridization signals. Thus, at least to pair 2 the meiosis was reductional and not inverted. The scheme in (D) was elaborated from pair 2 (on the left). The normal sense is the segregation of the homologues (center image), and not the segregation of sister chromatids in the first anaphase (on the right).

This event is enough to question the occurrence of inverted meiosis in all chromosomes and species of Cyperaceae. *R. pubera* exhibits six chromosomes with terminal 45S rDNA sites (Vanzela et al. 1998). Fortunately, we found one individual with a polymorphism in pair 2, which presents one of the homologous with a major hybridization signal after FISH with 45S rDNA probe. This event allowed accompanying the bivalent segregation in anaphase I and found the separation of homologous chromosomes, and not of sister chromatids (Figure 10).

This evidence raises doubts about some published images on bivalents in "box", which were interpreted as a conclusive diagnosis of post-reductional meiosis of all bivalents (Vanzela et al. 2000). At this time, there are no certainties as to whether all bivalents or part of them are oriented together or re-oriented independently, i. e., the post-reductional meiosis may occur but, is it necessary that all autosomes always behave the same way?

5. Acknowledgment

This work is part of doctoral thesis from Adriano Silvério and Érica Duarte-Silva, carried out at Programa de Pós-Graduação em Botânica da UFRGS. We thank CAPES, CNPq and FAPERGS for finnacial support, and the laboratories of Porto Alegre and Londrina for the use of the structure and tecnologies.

6. Conclusion

Firstly, based on Hofmeister's theory (1851), the life cycle of plants consists of two generations: sporophytic and gametophytic. In flowering plants, gametophytic generation is reduced to pollen (male gametophyte) and embryo sac (female gametophyte). In the sphere of studies on plant reproduction, is necessary limit the sexual terminology to the gametophytes, erradicating the sexual terminology applied to sporophytes, specially, the terms: male, female and hermophroditic flowers. In the second, the study of cellular and sub-cellular aspects of meiosis can improve the comprehension of plant meiosis process, particularly in cases of staminal or pollen sterility, as well as, in studies of uncommon types of meiosis, illustrated by pseudomonads in Cyperaceae.

7. References

Ainsworth, C. C. (1999) *Sex determination in plants.* BIOS Scientific Publishers Limited. ISBN:1859960421.USA.

APG III. (2009). Un update of the Angiosperm Phylogeny Group classification for the orders and families of flowering plants: APG III. *Botanical Journal of Linnean Society.* Vo. 161, No. 2, (October 2009), pp. 122-127, ISSN: 0024-4074.

Ariizumi, T.; Hatakeyama, K.; Hinata, K.; Inatsugi, R.; Nishida, I.; Sato, S.; Kato, T.; Tabata, S.; Toriyama, K. (2004). Disruption of the novel plant protein NEF1 affects lipid acumulation in the plastids of the tapetum and exine formation of pollen resulting in male sterility in Arabidopsis thaliana. *The Plant Journal.* Vo. 39, No. 2, (June 2004): pp. 170-181, ISSN: 0960-7412

Barrett, S. C. H. (2002). The evolution of plant sexual diversity. *Nature Genetics.* Vo. 3, (April 2001), pp. 274-284, ISSN: 1061-4036.

Bhat, V.; Dwivedi, K. K.; Khurana, J. P.; Sopory, S. K. (2005). Apomixis: An enigma with potencial applications. Special section: Embriology of flowering plants. *Current Science.* Vo. 89, No. 10, (December 2005), pp. 1879-1893, ISSN: 0011-3891.

Bione, N. C. P.; Pagliarini, M. S.; Almeida, L. A.; Seifert, A. L. (2002). An ms2 male sterile, female-fertile soybean sharing phenotypic expression with other ms mutant. *Plant Breeding.* Vo. 121, No. 4, (August 2002), pp. 307-313, ISSN: 0179-9541.

Blanvillain, R.; Boavida, L. C.; McCormick, S.; Ow., D. W. (2008). *EXPORTIN1* genes are essential for development and function of the gametophytes in *Arabidopsis thaliana.* *Genetics.* Vo. 180, No. 3, (November 2008), pp. 1493-1500. ISSN: 0016-6731.

Boavida, L. C.; Vieria, A. M.; Becker, J. D.; Feijó, J. A. (2005). Gametophyte interaction and sexual reproduction: how plants make a zygote. *The International Journal of Developmental Biology.* Vo. 49, No. 5/6, pp. 615-632. ISSN: 0214-6282.

Bongiorni, S.; Fiorenzo, P.; Pippoletti, D.; Prantera, G. (2004). Inverted meiosis and meiotic drive in mealybugs. *Chromosoma.* Vo.112, No.7, (March 2004), pp331–341, ISSN: 0009-5915.

Brown, R. C.; Lemmon, B. E. (2000). The cytoskeleton and polarization during pollen development in *Carex blanda* (Cyperaceae). *American Journal of Botany*. Vo. 87, No. 1, pp. 1-11, ISSN: 0002-9122.

Camerarius, R. J. (1694). *De Sexu Plantarum Epistola*. Literis Erhardtianis.Tübingen, Germany.

Chaubal, R.; Zanella, C.; Trimnel, M.R.; Fox, T.W.; Albertsen, M.C.; Bedinger, P. (2000). Two male-sterile mutants of *Zea mays* (Poaceae) with an extra cell divison in the anther wall. *American Journal of Botany*. Vo. 87, No. 8, (August 2000), pp. 1193-1201. ISSN: 0002-9122.

Cocucci, A. E. (1969). El processo sexual en angiospermas. *Kurtziana*. Vo. 5, No. 5, (June 1969), pp. 407-423, ISSN: 1852-5962.

Cocucci, A. E. (1973). Some suggestions on the evolution of gametophytes of higher plants. *Phytomorphology*. Vo. 23, pp. 109-124, ISSN: 0031-9449.

Cocucci, A. E. (1980). Precisiones sobre la terminología sexológica aplicada a angiospermas. *Boletín de la Sociedad Argentina de Botánica*. Vo.11, No.1-2, (July 1980), pp. 75-81, ISSN: 0373-580X.

Cocucci, A. E.; Hunziker A. T. (1994). *Los ciclos biológicos en el reino vegetal* (2°ed.), Academia Nacional de Ciências, ISBN: 0325-3406, Argentina.

Cocucci, A. E.; Mariath, J. E. A. (1995) Sexualidade em plantas. *Ciência Hoje* Vo. 18, No. 106, pp. 51-61, ISSN: 0101-8515.

Cocucci, A. E. (2006). La embriologia e los sistemas reproductivos de Angiospermae. In: *Os Avanços da Botânica no início do século XXI*. Mariath, J. E. A.; Santos, R. P. p. 97-102. Sociedade Botânica do Brasil, ISBN: 8560428003. Brazil.

Coen, I. J. (2010). A case to which no parallel exists: the influence of Darwin's diferente forms of flowers. *American Journal of Botany*. Vo. 97, No. 5, (April 2010), pp.701-716, ISSN: 0002-9122.

Coimbra, S.; Torrao, L.; Abreu, I. (2004). Programmed cell death induces male sterility in *Actidia deliciosa* female flowers. *Plant physiologi and biochemistry*. Vo. 42, pp. 537-541. ISSN: 0981-9428

Cruden R. W.; Lloyd R. M. (1995). Embryophytes have equivalent sexual phenotypes and breeding systems: why not a common terminology to describe them? *American Journal of Botany*. Vo. 82. No. 6, (June 1995), pp. 816-825, ISSN: 0002-9122.

Da Silva, C. R. M.; González-Elizondo, M. S.; Vanzela, A. L. L. (2005). Reduction of chromosome number in *Eleocharis subarticulata* (Cyperaceae) by multiple translocation. *Botanical Journal of the Linnean Society*. Vo. 149, No. 4, (December 2005), pp.457–464, ISSN: 1095-8339.

Dafni A.; Firmage D. (2000). Pollen viability and longevity: practical, ecological and evolutionary implications. *Plant Systematics and Evolution*. Vo. 222, No. 1, (March 2000), pp. 113-132, ISSN: 0378-2697.

Duarte-Silva, E.; Vanzela, A. L. L.; Mariath, J. E. A. (2010). Developmental and cytogenetic analysis of pollen sterility in *Valeriana scandens* L. *Sexual Plant Reproduction*, Vo. 23, No. 2, (June 2010), pp. 105-113, ISSN 0934-0882.

Duarte-Silva, E.; Rodrigues, L. R.; Mariath, J. E. A. (2011). Contradictory results in pollen viability determination of Valeriana scandens L. *Gene Conserve, in press*. ISSN 1808-1878.

Darwin, C. R. (1877). *The different forms of flowers on plants of the same species*. Murray, England.

Dong, X.; Zonglie, H.; Sivaramakrishnan, M.; Mahfouz, M.; Verma, P. S. (2005). Callose synthase (CalS5) is required for exine formation during microgametogenesis and for pollen viability in *Arabidopsis*. *The Plant Journal*. Vo. 42, No. 3, (May 2005), pp. 315-328, ISSN: 0960-7412.

Engelke, T.; Hülsmann, S.; Tatlioglu T. (2002). A comparative study of microsporogenesis and anther wall development in different types of genic and citoplasmic male sterilities in chives. *Plant Breeding* Vo. 121, No. 3, (June 2002), pp. 254-258, ISSN: 0179-9541.

Fei H.; Sawhney V. K. (1999). MS32-regulated timing of callose degradation during microsporogenesis in Arabidopsis is associated with the accumulation of staked rough ER in tapetal cells. *Sexual Plant Reproduction*.Vo. 12, No. 3, (June 2002), pp. 188-193, ISSN: 0934-0882.

Fujiki, Y.; Yoshimoto, K.; Ohsumi, Y. (2007). An *Arabidopsis* homolog of yeast ATG6/VPS30 is essential for pollen germination. *Plant Physiology*. Vo. 143, No. 3, (March 2007), pp. 1132-1139, ISSN: 0032-0889.

Furness, C. A.; Rudall, P. J. (1999). Microsporogenesis in monocotyledons. *Annals of Botany* Vo. 84, pp. 475–499, ISSN: 1095-8290.

Guerra, M.; García, M. A. (2004). Heterochromatin and rDNA sites distribution in the holocentric chromosomes of *Cuscuta approximata* Bab. (Convolvulaceae). *Genome*. Vo. 47, No. 1, (February 2004), pp. 134–140, ISSN: 0831-2796.

Guerra, M.; Cabral, G.; Cuacos, M.; González-García, M.; González-Sánchez, M.; Vega, J.; Puertas, M. J. (2010). Neocentrics and Holokinetics (Holocentrics): Chromosomes out of the Centromeric Rules. Cytogenetic and Genome Research. Vo. 129, pp. 82–96, ISSN: 1662-3797.

Goldberg, R. B.; Beals, T. P.; Sanders, P. M. (1993). Anther development: basic principles and practical applications. *The Plant Cell*. Vo. 5, No. 10, (October 1993), pp. 1217-1229, ISSN: 1040-4651.

Greyson, R.I. (1994). *The development of flowers*. Oxford University Press, ISBN: 019506688X, England.

Han, M.; Jung, K.; Yi, G.; Lee, D.; An, G. (2006). *Rice immature pollen 1 (RIP1)* is a regulator of late pollen development. *Plant and cell physiology*. Vo. 47, No. 11, (January 2007), pp. 1457-1472, ISSN: 0032-0781.

Heilborn, O. (1928). Chromosome studies in Cyperaceae. *Hereditas*. Vo. 11, pp. 182–192. ISSN: 0018-0661.

Hofmeister, W. (1862). On the germination, development and frutification of higher Cryptogamia ando n the frutification of Coniferae (English translation by F. Curry form the 1851 germany edition). London.

Hord, C.L.H.; Chen C.; DeYoung B.J.; Clark S.E.; Ma H. (2006). The *BAM1/BAM2* receptor-like kinases are importantt regulators of early *Arabidopsis* anther development. *Plant Cell*. Vo. 18, No. 7, (July 2006), pp.1667– 1680, ISSN: 1040-4651.

Horner, Jr., H. T. (1977). A Comparative Light- And Electron-Microscopic Study of Microsporogenesis in Male-Fertile and Cytoplasmic Male-Sterile Sunflower (Helianthus Annuus). *American Journal of Botany*. Vo. 64, No. 6, (July 1977), pp.745-759, ISSN: 0002-9122.

Huang, L.; Cao, J.; Zhang, A.; Ye, Y.; Liu, T. (2009). The Polygalacturonase gene BcMF2 from *Brassica campestris* is associated with intine development. *Journal of Experimental Botany*. Vo. 60, No. 1, (November 2008), pp. 301-313, ISSN: 0022-0957.

Johns, C.; Lu, M.; Lyznik, A.; Mackenzie, S. (1998). A mitochondrial DNA sequences isassociated with Abnormal Pollen Development in Cytoplasmic Male Sterile Bean Plants. *Genetics*. Vo. 150, No. 1, (September 1998), pp. 383-391, ISSN: 0016-6731.

Hughes-Schrader, S.; Schrader, F. (1961). The kinotochore of the Hemiptera. *Chromosoma*. Vo.12, pp.327 – 350. ISSN: 0009-5915.

Kaplan, D. R.; Cooke, T. J. (1996). The genius of Wilhelm Hofmeister: The origin of causal analytical research plant development. *American Journal of Botany*. Vo. 83, No. 12, (December 1996), pp. 1647-1660, ISSN: 0002-9122.

Karasawa, M.M.G. (2009). *Diversidade reprodutiva de plantas*. Sociedade Brasileira de Genética. ISBN:978-85-89265-12-6. Brazil.

Kaul, M.L.H. (1988). *Male sterility in higher plants. Monographs on the theoretical and applied genetics, 10*. Springer. ISBN: 0387179526. Germany.

Kirpes, C. C.; Clark, L. G.; Lersten, N. R. (1996). Systematic significance of pollenarrangement in microsporangia of Poaceae and Cyperaceae: review andobservations on representative taxa. *American Journal of Botany* Vo.83, pp1609–1622. ISSN: 0002-9122.

Li, N.; Da-Sheng, Z.; Hai-Sheng, L.; Xiao-xing, L.; Wan-qi, L.; Zheng, Y.; Huang-Wei, C.; Wang J.; Tie-Qiao, W.; Hai, H.; Luo, D.; Hong, M.; Da-Bing, Z. (2006). The rice tapetum degeneration retardation gene is required for tapetum degradation and anther development. *The Plant Cell*. Vo. 18, No. 2, (February 2006), pp. 2999-3014, ISSN: 1532-298X.

Linnaeus, C. (1810). *Système sexuel des végétaux: suivant les classes, les orders, les genres et les espèces, avec les caractères et les différences* (2° ed.). Arthus-Bertrand. France.

Nagaki, K.; Kashihara, K.; Murata, M. (2005). Visualization of diffuse centromeres with centromere-specific histone H3 in the holocentric plant *Luzula nivea. The Plant Cell*. Vo. 17, No. 7, (June 2005), pp. 1886–1893, ISSN: 1532-298X.

Nokkala, S.; Kuznetsova, V. G.; Maryanska-Nadachowska, A.; Nokkala, C. (2006). Holocentric chromosomes in meiosis. II. The modes of orientation and segregation of a trivalent. *Chromosome Research*. Vo. 14, No. 5, (March 2006), pp. 559-565, ISSN: 1573-6849.

Nordenskiöld, H. (1951). Cyto-taxonomical studies in the genus *Luzula* I. Somatic chromosomes and chromosomes numbers. *Hereditas*. Vo. 37, pp. 325-355, ISSN: 0018-0661.

Papini, A; Mosti, S.; Brighigna, L. (1999). Programmed-cell death events during tapetum development of angiosperms. *Protoplasma*. Vo. 207, No. 3-4, (September 1999), pp. 213-221, ISSN: 0033-183X.

Pazy, B. (1997). Supernumerary chromosomes and their behaviour in meiosis of the holocentric *Cuscuta babylonica* Choisy. *Botanical Journal of the Linnean Society* Vo.123, No. 2, (February 1997), pp. 173–17,. ISSN: 1095-8339.

Pérez, R.; Panzera, F.; Page, J.; Suja, J. A.; Rufas, J. S. (1997). Meiotic behaviour of holocentric chromosomes: orientation and segregation of autosomes in *Triatoma infestans* (Heteroptera). *Chromosome Research*. Vo. 5, No. 1, (February 1997), pp.47–56, ISSN: 1573-6849.

Peirson, B.N.; Owen H.A.; Feldmann K.A.; Makaroff, C.A. (1996). Characterization of three male sterile mutants of *Arabidopsis thaliana* exhibiting alterations in meiosis. *Sexual Plant Reproduction* Vo. 9, No. 1, pp. 1-16, ISSN: 0934-0882.

Ranganath, R. M.; Nagashee, N. R. (2000) Selective cell elimination during microsporogenesis in sedges. *Sexual Plant Reproduction*. Vo. 13, No. 1, (March 2000), pp. 53–60, ISSN: 0934-0882.

Quammen, D.; Shimitz, H. (2007). A passion for order. *National Geographic*. (June 2007). ISSN: 0027-9358.

San Martin, J. A. B.; Andrade, C. G. T. J.; Vanzela, A. L. L. (2009). Early meiosis in Rhynchospora pubera L. (Cyperaceae) is marked by uncommon ultrastructural features. *Cell Biology Interantional* Vo. 33, No. 10, (October 2009), pp. 1118-1122, ISSN: 106i5-6995.

San Martin, J. A. B.; Andrade, C. G. T. J; Mastroberti, A. A.; Mariath, J. E. A; Vanzela, A. L. L. (2011). Asymmetric tetrads and programmed cell death in the pseudomonads of Rhynchospora pubera (Cyperaceae). *Annals of Botany, in press*. ISSN: 1095-8290.

Selling, O. H. (1947). Studies in the Hawaiian pollen statistics. Part II. Thepollens of the Hawaiian phanerogams. *Bishop Museum Bulletin in Botany* Vo. 38, pp. 1–430, ISSN: 0893-3138.

Smith, M.B.; Palmer, R.G.; Horner, R.T. (2002). Microscopy of a citoplasmic male-sterile from an interspecific cross between Glycine max and G. soja (Leguminosae). *American Journal of Botany*. Vo. 89, No. 3, (March 2002), pp. 417-426, ISSN: 0002-9122.

Solari, A. J. (1979). Autosomal synaptonemal complex and sex chromosomes without axes in *Triatoma infestans* (Reduviidae, Hemiptera). *Chromosoma*. Vo. 72, pp. 225-240, ISSN: 0009-5915.

Strandhede, S. O. (1965). Chromosome studies in *Eleocharis*, subser. Palustres. *Opera Botanica*. Vo. 9, pp. 1–86. ISSN: 0078-5237.

Strandhede, S. O. (1973). Pollen development in the Eleocharis palustris group (Cyperaceae). II. Cytokinesisand microspore degeneration. *Botaniska Notiser*. Vo. 126, pp. 255–265. ISSN: 0006-8195

Sun, M.; Ganders, F.R. (1987). Microsporogenesis in male-sterile and hermaphroditic plants of nine gynodioecius taxa in Hawaiian *Bidens* (Asteraceae). *American Journal of Botany*. Vo. 74, No. 2, (February 1987), pp. 209-217, ISSN: 0002-9122.

Teng, C.; Dong, H.; Shi, L.; Deng, Y.; Mu, J.; Zhang, J.; Yang, X.; Zuo, J. (2008). Serine Palmitoyltransferase, a key enzyme for male gametophyte development in *Arabidopsis*. *Plant Physiology* Vo. 146, No. 3, (March 2008), pp. 1322- 1332, ISSN: 0032-0889.

Trelease, W. (1916). Two new terms cormophytaster and xeniophyte anxiomatically fundamental in botany. *Proccedings of the American Philosophical Society* Vo. 55, No. 3, pp. 237-242.ISSN: 0003-049X.

Vanzela, A. L. L.; Cuadrado, A.; Jouve, N.; Luceño, M.; Guerra, M. (1998) Multiple locations of the rDNA in species of *Rhynchospora* (Cyperaceae). *Chromosome Research* Vo.6, No.5, (March 1998), pp.345–349.

Vanzela A. L. L.; Luceño, M.; Guerra, M. (2000). Karyotype evolution and cytotaxonomy in Brazilian species of *Rhynchospora* Vahl (Cyperaceae). *Botanical Journal of the Linnean Society*. Vo.134, No.4 (June 2000), pp.557–566. ISSN: 1095-8339.

Viera, A.; Page, J.; Rufas, J. S. (2009). Inverted meiosis: the true bugs as a model to study. *Genome Dynamics*. Vo. 5, pp. 137–156, ISSN: 1662-3797.

Wagenitz, G. (1999). Botanical Terminology and Homology in Their Historical Context. *Plant Ecology and Evolution* Vo.68, No. 1/2 (March 1999), pp. 33-37, ISSN:1374-7886.

Zàrsky, V.; Tupy, J. (1995). A missed anniversary: 300 years after Rudolf Jacob Camerarius 'De sexu plantarum epistola'. *Sexual Plant Reproduction*. Vo. 8, No. 6, (November 1995), pp. 375-376, ISSN:0934-0882.

Zhang, Z.; Lu, Y.; Liu, X.; Feng, J.; Zhang, G. (2006). Cytological mechanism of pollen abortion resulting from allelic interaction of F1 pollen sterility locus in rice (*Oryza sativa* L.). *Genetica* Vo. 127, No. 1-3, (May 2006), pp. 295-302, ISSN: 0016-6707.

Investigating Host Induced Meiosis in a Fungal Plant Pathogen

B. J. Saville, M. E. Donaldson and C. E. Doyle

Trent University,
Canada

1. Introduction

Fungal spores lend the smut and rust plant diseases their names. Smut fungi produce massive numbers of dark, dust like, thick walled teliospores and the name, smut, is derived from the older definition meaning dark smudge from soot, smoke or dirt. The rust fungi produce diseases characterised by the production of pustules erupting from the plant surface. They contain urediniospores which are often orange or rusty in colour. Spores are essential for fungal survival, providing a means of dispersal and often a structure to protect the fungus; they are also integral to fungal meiosis. Smut and rust fungi are biotrophs, meaning they derive their nutrients from living plant hosts. This interaction is very intimate, involving fungal penetration of the plant cell walls but not the plasma membranes (e.g. Snetselaar & Mims, 1992; Voegele & Mendgen, 2003). As such, most smut and rust fungi have only evolved to infect (and become meiotically competent within) one or a limited number of host species.

The economic impact of these pathogens is well illustrated by considering two crops significantly damaged by them: corn and wheat. According to Capitol Commodity Services (2011), corn remains the largest valued crop in the United States, totalling $67 billion in 2010. World-wide, corn crops were estimated at $163 billion in 2010 (U.S. Grains Council, 2010). The comparable numbers for wheat were $13 billion, and $140 billion, respectively (Capitol Commodity Services, 2011; U.S. Department of Agriculture, 2011). Although mitigated by varieties with partial resistance, the maize crop loss resulting from common smut of corn, caused by *Ustilago maydis*, is 2% annually, equivalent to ~$1 billion (Allen et al., 2011; Martinez-Espinoza et al., 2002). Wheat crop losses due to wheat leaf rust *Puccinia triticina* Eriks, which is the most common and widely distributed wheat rust, results in trace to 10% crop losses in many countries around the world. In the US, from 2000 to 2004, the loss was $350 million/year, and it can be $100 million/year in Canada. The production in China is more than twice that of the US and commonly suffers 10-30% crop loss per year (Huerta-Espino et al., 2011). There is also an extreme threat from emerging races of stripe rust of wheat (*Puccinia striiformis* f. sp. *tritici*) and wheat stem rust (*Puccinia graminis* f. sp. *tritici*). The emerging stem rust races are referred to collectively as UG99 after their location of origin (Uganda), and year of detection (1999). These races are virulent on the vast majority of wheat varieties cultivated around the world. It is predicted that if resistant varieties are not developed and utilized that the UG99 epidemic in Africa will become global (Singh et al., 2011).

The impact of smut and rust fungi is limited by deploying resistant crop varieties; however, the fungi overcome the resistance leading to cycles in which varieties with new resistances are released and fungi with new virulence genotypes arise. While new virulence alleles ultimately result from mutation, genotypic diversity is created through recombination. Some populations of leaf rust have a genetic structure consistent with an asexual dikaryotic population "within which stepwise mutation at avirulence or virulence loci regularly occurs" (Ordoñez & Kolmer, 2009). In contrast, greatly increased genetic diversity and epidemics of stem rust have been linked to sexual reproduction (Burdon & Reolfs, 1985; Jin, 2011) and eradicating the alternate host for stem rust, common barberry and other *Berberis* spp. on which sexual reproduction occurs, has provided substantial benefit in controlling wheat stem rust (Roelfs, 1982) and, inadvertently, stripe rust of wheat (Jin, 2011). Further, the corn smut pathogen *U. maydis* exists in predominantly out-crossing populations (Barnes et al., 2004). This suggests a key role for sexual reproduction in the emergence and maintenance of virulence genotypes.

The rust fungi are obligate biotrophs and cannot be cultured outside their hosts. The wheat rusts, as typified by stem rust, have five spore stages and require two completely unrelated hosts (Schumann & D'Arcy, 2009). The primary host is wheat and the alternative host is barberry. This complex and interesting life cycle will not be discussed in detail here except to note that, in the stem rust life cycle, meiosis likely initiates *in planta* followed by teliospore maturation (see paragraph below on rust teliospore microscopy). The diploid teliospores are produced late in the season on the primary host, wheat. They germinate and complete meiosis yielding basidiospores that infect the alternate host. In contrast to the rust fungi, the model fungal biotrophic pathogen *U. maydis* (Banuett, 1995; Brefort et al., 2009) is readily cultured in the laboratory on defined media and its sexual cycle can be completed within 28 days following injection of compatible haploid cells into seedlings of the host *Zea mays* (corn). *U. maydis* is amenable to genetic analysis and molecular manipulation, including homologous gene replacement, and several vectors are available for gene expression analysis. An annotated version of the genome sequence of *U. maydis* was released in 2007 (Kämper et al., 2006) and the annotation continues to be improved (e.g. Donaldson & Saville, 2008; Doyle et al., 2011; Ho et al., 2007; Kronstad, 2008; Morrison et al., in preparation). This allows molecular manipulation of *U. maydis* outside the host, followed by molecular analysis in the host.

The *U. maydis* life cycle (Figure 1) begins with teliospore germination and the completion of meiosis to create haploid basidiospores, which divide by budding. Compatible non-pathogenic haploids fuse to form the pathogenic filamentous dikaryon, which proliferates, branches, and penetrates the plant via the formation of specialised cells called appressoria. It grows within and between plant cells eliciting the formation of a tumour. Banuett and Herskowitz (1996) describe a series of developmental events that *U. maydis* undergoes in the tumour leading to teliospore formation. These events occur within the enlarged host cells and include: 1) the formation of hyphal branches at close intervals, 2) the production of a mucilaginous matrix in which the hyphae are embedded and the hyphal tips become lobed, 3) hyphae fragmentation, 4) rounding of fragmented hyphae and 5) the deposition of a pigmented thick cell wall. The pigmented teliospores enter a dormant state, the tumours disintegrate, and the teliospores are dispersed, continuing the cycle.

An overview of how meiosis proceeds in *U. maydis* was presented by Donaldson and Saville (2008). Since the early stages of meiosis occur *in planta* and meiosis is temporally linked to the formation of thick walled dormant teliospores, direct microscopic observation of meiotic events has not been possible. Therefore, it is informative to review how meiosis precedes in the related homobasidiomycete, *Coprinopsis cinerea*. This fungus can be induced to form mushrooms (fruiting bodies) in culture and, in these fruiting bodies, meiosis proceeds in a synchronous manner with over 60% of the approximately 10 million basidia in a given cap at the same stage (Pukkila et al., 1984). Kües, (2000) reviewed meiosis in *C. cinerea* and noted that chromatid duplication in premeiotic S phase is followed by karyogamy, and the cytological events of prophase I precede with the condensation and alignment of chromosomes (leptotene), synapsis (zygotene), and recombination nodule appearance (pachytene). This process, from post karyogamy to pachytene, is completed in six hours (Celerin et al., 2000). It is followed by desynapsis (diplotene) and the transition to metaphase (diakinesis). The second meiotic division occurs fairly rapidly following interphase, with prophase II through telophase II being completed in ~1 hour. The second division occurs in the same plane as the first, across the longitudinal axis of the basidium. Then, chromatid separation is followed by the four nuclei migrating toward the basidium tip where basidiospores form and the nuclei migrate into them then complete a round of mitosis. This overview of basidiomycete meiosis provides a framework for *U. maydis* investigations.

U. maydis, like other smut fungi, does not form a fruiting body. Instead, when teliospores germinate, basidia are formed in which meiosis is completed. So while a *C. cinerea* fruiting body has millions of basidia undergoing meiosis, in *U. maydis*, millions of teliospores are dispersed and each produces a basidium. What we know of the cytological events of meiosis in *U. maydis* is that when hyphae are enveloped in the mucilaginous matrix during teliospore formation, they contain a single nucleus, indicating karyogamy has occurred (Banuett & Herskowitz, 1996). If *U. maydis* meiosis follows the pathway of *C. cinerea*, then premeiotic S phase and the duplication of chromatids would have been completed before karyogamy occurred. The next meiotic event we are aware of in *U. maydis* is the germination of the teliospore when the nucleus is in late prophase I (O'Donnell & McLaughlin, 1984). Between karyogamy and germination the teliospore is dormant with extremely limited metabolic activity. This indicates that major meiotic events cannot be occurring; this leaves the possibility that, either there is a pause after karyogamy and meiosis continues with teliospore germination, or that prophase I and recombination events begin immediately after karyogamy and pause, perhaps at the pachytene checkpoint, when the teliospore becomes dormant. Following germination, the metaphase I spindles align with the longitudinal axis of the metabasidium, and a transverse septum forms, indicating the completion of telophase I and leading to the formation of two cells (O'Donnell & McLaughlin, 1984). This is rapidly followed by meiosis II, in which the nucleus in each cell migrates to a central location, divides and septa are formed, resulting in three haploid nuclei, each in a basidium cell. The fourth nucleus migrates to the base of the teliospore (Ramberg & McLaughlin, 1980). Basidospores form by budding, each basidium cell nucleus migrates into the respective basidiospore and divides, then one nucleus remains and the other migrates back into the basidium cell (Banuett, 1995). Basidiospores continue to divide by budding.

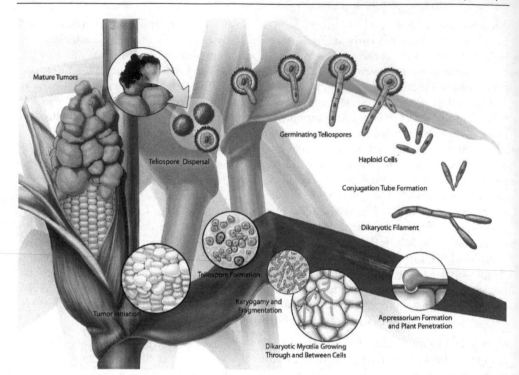

Fig. 1. The Life Cycle of *Ustilago maydis*. In this diagram, meiosis begins soon after karyogamy, pauses at pachytene during teliospore dispersal, and meiosis resumes during teliospore germination.

Support for the idea of meiosis proceeding immediately after karyogamy and pausing at pachytene comes from microscopic analysis of a number of rust fungi, notably *Coleosporium ipomoeae* (Mims & Richardson, 2005). This rust fungus, like other members of the *Coleosporium* genus, has thin cell walls which enable stained nuclei to be observed in developing, mature and germinating teliospores. Mims and Richardson (2005) observed synaptonemal complexes in a high percentage of the nuclei in unhydrated *C. ipomoeae* teliospores. This indicated that meiosis had begun soon after karyogamy and was interrupted or arrested at pachytene, where it remained until the teliospores were hydrated. Mims and Richardson (2005) also reported that synaptonemal complexes were observed in ungerminated teliospores of a number of other rust species including *Puccinia graminis* f.sp. *tritici* (Boehm et al., 1992). They report that: "arrested meiosis is common in teliospores of rust fungi and may, in fact, be the rule rather than the exception in these organisms."

The microscopically visible events of meiosis in *U. maydis,* and the model of meiosis beginning immediately after karyogamy and pausing at pachytene will be discussed further at different points throughout this review. However, the focus will switch to the discussions of environmental signals that trigger meiosis, transduction pathways that transmit these signals, control of gene expression, and an update of meiosis gene identification in *U. maydis,* with information on meiosis gene expression and evidence for post-transcriptional control mechanisms in *U. maydis*. In each section, relevant data from other fungal models, notably *Saccharomyces cerevisiae, Schizosaccharomyces pombe* and *C. cinerea,* will be reviewed for comparison and insight.

2. Signals triggering meiosis initiation in fungi: Genetic and environmental signals

The switch from mitotic division to meiosis involves a dramatic shift in cellular processes; in fact, this can be considered the most traumatic change a cell can undergo. Therefore, it is essential that entry into meiosis is tightly controlled, preventing the inappropriate execution of the meiotic program. The signals that trigger meiosis vary extensively between organisms, possibly due to a need for organisms to respond to the unique environmental niches in which they reside (Pawlowski et al., 2007). While the signals that trigger meiosis initiation are different, the timing is conserved from yeast to mice, occurring prior to the premeiotic S phase (Pawlowski et al., 2007). While this may seem obvious, the complex developmental processes that precede and accompany meiosis have obscured the timing until recently (Pawlowski et al., 2007). In this section, the nature and timing of signals leading to the initiation of meiosis in the model laboratory fungi *S. cerevisiae, S. pombe* and *C. cinerea* are reviewed and used as a reference in presenting hypotheses regarding meiotic initiation signals for the model plant pathogen *U. maydis.*

2.1 *Saccharomyces cerevisiae*

Meiosis in the ascomycete fungus *S. cerevisiae* has been extensively studied. For the purposes of this discussion, we will consider meiosis initiation as the first of three stages; the others being DNA replication, recombination, and the meiotic divisions leading to haploid products (Honigberg, 2003). This separation is interesting because it indicates an order of events different than that of *C. cinerea.*

In *S. cerevisiae,* meiosis initiation is triggered by environmental and genetic signals working in concert. To be receptive to the environmental signals, *S. cerevisiae* cells must be diploid and possess both MATa and MATα mating type alleles (reviewed in Piekarska et al., 2010). MATa and MATα encode components of the transcriptional repressor a1/α2, (Mitchell, 1994; Piekarska et al., 2010). The environmental signal involves three nutritional shifts: 1) the absence of an essential nutrient, 2) the presence of a non-fermentable carbon source, and 3) the absence of glucose (Honigberg, 2003). The essential nutrient typically eliminated in laboratory studies is nitrogen and, while there may also be a direct requirement for nitrogen sensing, limiting carbon, phosphates or sulphates can also provide the required signal to trigger meiosis initiation (Honigberg, 2003; Mitchell, 1994). The CO_2 produced through respiration, stimulated by the presence of a non-fermentable carbon source, results in the alkalization of the media, which may be a component of the 2nd shift (Honigberg, 2003). While respiration is a required signal throughout meiosis, the non-fermentable carbon source is only required prior to meiosis I (Honigberg, 2003; Piekarska et al., 2010). Finally, the presence of glucose can override the other signals and repress meiosis in *S. cerevisiae,* (Honigberg, 2003; Mitchell, 1994; Piekarska et al., 2010).

While the signal transduction pathways will be discussed later, other key aspects of meiosis initiation in *S. cerevisiae* are the timing of entry and the link between genetic and nutritional signals. A *S. cerevisiae* diploid cell commits to mitosis before DNA replication in the S phase.

It has been proposed that the commitment involves the separation of the spindle pole bodies (SPBs) during the cell cycle. In budding yeast cells where the SPBs are still together, the cells may arrest and enter the meiotic cycle, whereas after the SPBs have separated, the cell can no longer enter the meiotic cell cycle, and must complete mitosis (Simchen, 2009). Starvation for an essential nutrient results in the arrest of the cell in G1, when the SPBs are together (Honigberg, 2003; Piekarska et al., 2010), this specific arrest allows a switch to meiosis. The nutritional and genetic signals also converge to initiate meiosis through the transcriptional regulation of two main inducers of meiosis, *Ime1* (initiator of meiosis), which is a transcription factor that stimulates the expression of many early meiosis genes and *Ime2*, a serine/threonine protein kinase. The expression of *Ime1* is controlled by the a1/α2 repressor and by nutritional signals (Honigberg, 2003; Mitchell, 1994). Multiple nutritional signals converge on *Ime2* as well. The full expression of both of these genes is essential to the initiation and continuation of meiosis in *S. cerevisiae* (Honigberg, 2003; Mitchell, 1994).

2.2 *Schizosaccharomyces pombe*

Similar to *S. cerevisiae*, the initiation of meiosis in *S. pombe* requires diploid cells and nutrient starvation. However, *S. pombe* also requires pheromone signalling, whereas the *S. cerevisiae* pheromones are turned off after mating. In *S. pombe*, mating type is determined by the *mat1* locus. Each mating type allele codes two proteins, *mat1-P* codes *mat1-Pc* and *mat1-Pi*, while *mat1-M* codes *mat1-Mc* and *mat1-Mi*. *mat1-Mc* and *mat1-Pc* are essential for mating and meiosis, as they control pheromone and receptor production (Harigaya & Yamamoto, 2007; Yamamoto, 1996a). When compatible haploid cells are nitrogen starved, pheromone signalling is induced, which initiates cellular fusion and the formation of the diploids (Nielsen, 1993). These diploids grow mitotically, under rich nutrient conditions, but under nutrient starvation conditions they arrest in G1 and proceed to meiosis. As in *S. cerevisiae*, meiosis can only occur if the cells arrest in G1; beyond this point they are committed to mitosis (Harigaya & Yamamoto, 2007). However, meiosis in *S. pombe* will not proceed without the pheromone signal. Diploid cells lacking one pheromone receptor can still undergo meiosis but those lacking both cannot (Yamamoto, 1996a, 1996b).

The linkage to environmental signals in *S. pombe* comes through *Ste11*, a transcription factor expressed under nutrient starvation (Yamamoto, 1996a, 1996b). *Ste11* controls the expression of *mat1-Mc* and *mat1-Pc* along with other mating and meiosis genes. *Ste11* plays a similar role in *S. pombe* to *Ime1* in *S. cerevisiae*, as both transcription factors respond to environmental signals and lead to meiotic initiation. However, despite their functional similarities, these two proteins are not structurally similar (Burns et al., 2010a). *Mat1-Mc* and *Mat1-Pi* together stimulate the expression of *Mei3*, an inhibitor of *Pat1*, a serine/threonine protein kinase that itself inhibits meiotic initiation (Harigaya & Yamamoto, 2007; Willer et al., 1995). All of these signals converge on *Mei2*, an RNA binding protein which is essential for entry into, and continuation of, meiosis in fission yeast (Harigaya & Yamamoto, 2007).

2.3 *Coprinopsis cinerea*

C. cinerea is a filamentous, basidiomycete fungus that can be induced to form fruiting bodies (mushrooms) in the laboratory. As noted above, it is a model for the study of meiosis because the millions of basidia, the cells in which meiosis occurs, in a single cap develop synchronously. Initiation of meiosis in *C. cinerea* depends on light cues, not the nutritional cues used by *S. cerevisiae* and *S. pombe*. It has been hypothesized that linking fruiting body formation and meiosis to light/dark cycles provides a selective advantage because the fruiting bodies would be produced when animals are grazing, and *C. cinerea* depends upon animal ingestion for dispersal (Lu, 2000). This hypothesis is consistent with the concept of fungi responding to niche specific signals to initiate meiosis.

C. cinerea is well tuned to changes in lighting. Not only is light essential for karyogamy and the initiation of meiosis (Lu, 2000), but increasing the intensity of light speeds up the process, with less time being required to reach karyogamy under more intense light. It is proposed that the number of photons received is important in stimulating the progression of the cell into the premeiotic S phase and karyogamy (Lu, 2000). This timing is the same as the other fungi; therefore, although the signals are very different, the timing of commitment to meiosis is conserved, with the signal that initiates meiosis coming before the premeiotic S phase.

As may be expected, based on the different environmental triggers for initiation, *C. cinerea* has no orthologs to either *Ime1* or *Ste11*, the master meiotic regulators from *S. cerevisiae* and *S. pombe*, respectively. However, studies have shown that during meiosis, successive waves of transcription occur in *C. cinerea*, much like waves noted in both yeasts (Burns et al., 2010b). Hence, it may not be unreasonable to assume that there is a heretofore unidentified transcription factor that responds, directly or indirectly, to light signals and initiates this transcriptional program, making it similar in function, if not structure, to the regulators in budding and fission yeast.

2.4 *Ustilago maydis*

U. maydis is the model biotrophic basidiomycete plant pathogen (Banuett, 1995; Brefort et al., 2009). Like *S. cerevisiae* and *S. pombe*, there is a genetic and an environmental requirement for the initiation and completion of meiosis. *U. maydis* has two mating type loci, the multiallelic *b* locus and the diallelic *a* locus. The b locus codes a pair of homeodomain proteins that act as transcription factors when a heterodimeric protein consisting of polypeptides from different alleles is formed. The *a* locus codes for the pheromone and pheromone receptors. Alleles at each of these loci must be different in order for haploid cells to mate and for the maintenance of filamentous growth (reviewed in Banuett, 1995, 2002, 2010); however, only *b* locus heterozygosity is required for completion of meiosis (Banuett & Herskowitz, 1989). The environmental input required for meiosis is growth within the plant, and Banuett and Herskowitz (1996) suggested a peptide produced by the plant may stimulate karyogamy. In light of the earlier discussions here regarding a requirement for a signal to initiate premeiotic S phase and karyogamy, the suggestion by Banuett and Herskowitz (1996) could easily be extrapolated to suggest that the plant peptide stimulates premeiotic S phase and subsequent karyogamy in *U. maydis*. As reviewed by Banuett (2002), Kahmann and Kämper (2004), and Klosterman et al. (2007), the influence of the mating type loci is modulated by nutrition, pH, temperature, oxygen tension and plant signals, so it is possible that other factors influence meiosis in *U. maydis*.

To provide context for the possibility that nutritional conditions act as a signal influencing meiosis, Horst et al. (2010a) determined that, upon infection, *U. maydis* creates a strong nitrogen and carbon sink around the site of infection, and this stimulates the productivity of the remaining source leaves, allowing import of nutrients to the developing tumour tissue.

It is believed that the imported sucrose is used for building the tumour and for feeding *U. maydis,* and that the nitrogen may be fuelling both host defense protein synthesis and fungal growth (Horst et al., 2010b). This indicates that nutrition availability is an important aspect of the plant-pathogen interaction and that there is a competition for nutrients between the fungus and the plant. As an important part of its ecological niche, it is conceivable that changes in nutrient availability influence the progression of meiosis in *U. maydis.*

Genes that are involved in regulating the transition to meiosis have not yet been identified in *U. maydis.* Bioinformatic comparisons have determined that *U. maydis* does not possess an ortholog to *Ime1,* the master regulator of meiosis in *S. cerevisiae* (Donaldson & Saville, 2008). However, an ortholog to *Ste11,* the key transcription factor in *S. pombe,* is present in *U. maydis.* This putative ortholog is known as *Prf1* in *U. maydis* and its function has been previously characterized. Interestingly, it is a transcription factor that is involved in the sexual development of *U. maydis* in response to environmental signals, much like *Ste11.* *Prf1* is involved in regulating *a* and *b* mating type gene expression, resulting in high levels of pheromones and receptors during mating, and controlling *b* gene expression during pathogenesis (Hartmann et al., 1999). Four different environmental signals affect *Prf1*: the carbon source, pheromones, the *b* heterodimer and the cAMP pathway. These signals act to control *Prf1* transcriptionally and post-transcriptionally (Hartmann et al., 1999). This is similar to the function of *Ste11* in *S. pombe,* which controls pheromone gene expression in response to environmental signals, allowing for conjugation, the initiation of the sexual cycle and the commencement of meiosis. It is feasible that *Prf1* is also involved in initiating meiosis in *U. maydis,* possibly by stimulating the expression of a gene that controls further meiotic gene expression.

3. Signal transduction pathways and meiotic progression

The requirement for genetic and environmental signals to stimulate the entrance into meiosis implies there must be a way to transduce the environmental signals and integrate them with the genetic status of the cells. In this section we provide an overview of the signal transduction in *S. cerevisiae* and *S. pombe* and then, with this background, we link what is known about pathogenic signal transduction in *U. maydis* to its potential role in meiosis. Since research on these organisms has historically emphasized different levels of the signal transduction pathways, the focus in each section varies. In *S. cerevisiae,* major regulators are known and the focus has been on transcriptional control, and as such the overview will focus on signalling as it influences transcription. In *S. Pombe,* the major regulator is also known, but the emphasis has not been as strongly focused on transcription so this section is somewhat more pathway oriented. In *U. maydis,* the master regulators are not known so the knowledge of signal transduction in pathogenesis is reviewed and a model is presented for how this may stimulate the initiation of meiosis.

3.1 *Saccharomyces cerevisiae*

The initiation and continuation of meiosis are linked to environmental cues through multiple signal transduction pathways in *Saccharomyces cerevisiae.* The master controller of meiosis is the gene *Ime1.* The influences on this gene primarily result in changes in its transcription, which is controlled by the genetic and environmental signals. *Ime1* has an unusually large promoter region of 2,100bp, which is divided into 4 different Upstream Controlling Sequences, UCS1-4. These UCSs respond to different signals. Nutritional signals affect UCS1 and 2, with UCS2 promoting the transcription of *Ime1* and UCS1 inhibiting it. Cell-type signals affect UCS3 and 4, repressing the expression of *Ime1* in MAT-insufficient cells (Sagee et al., 1998). This section will focus on how these different signals are relayed to influence the transcription of this master controller.

3.1.1 Genetic control

The cell-type signals that control *Ime1* are transmitted through a repressor, *Rme1*, and an activator, *Ime4*. In haploid *S. cerevisiae*, the RME1 protein inhibits meiosis by repressing the expression of *Ime1*. It does so by binding to the *Rme1* Repressor Element (RRE), within UCS4 of the *Ime1* promoter (Covitz & Mitchell, 1993; Sagee et al., 1998). In diploid MATa/MATα cells, the proteins a1, a MATa product, and α2, a MATα product, form the a1/α2 heterodimer. This heterodimer binds upstream of *Rme1* and directly represses its transcription (Covitz et al., 1991). Repression of *Rme1* results in the de-repression of *Ime1*, as RME1 is no longer available to bind to the RRE, and in this way *Rme1* transmits the cell-type signal directly to *Ime1*. IME4 expression is necessary for the full expression of *Ime1* (Shah & Clancy, 1992). The a1/α2 heterodimer regulates *Ime1* expression by repressing the transcription of *Ime4* antisense and allowing the transcription of *Ime4* sense transcript (discussed further in section 6.4). The MATa/MATα cell-type signal also regulates *Ime1* expression through the UCS3 repressor region, however the protein involved has not been identified (Sagee et al., 1998). The status of the diploid cell is determined by the mating type loci, and the outlined transcriptional control pathways ensure that *Ime1* is only expressed, and meiosis can only proceed, in diploid MATa/MATα cells.

3.1.2 Nutritional control

The nutritional signals that control meiotic initiation in *S. cerevisiae* are transmitted through a signalling network composed of the RAS, cAMP and TOR pathways, all of which regulate the expression of transcription factor *Ime1* and kinase *Ime2* (reviewed in Piekarska et al., 2010).

While nitrogen limitation is often described as a requirement for meiosis in *S. cerevisiae*, starvation of any essential nutrient can stimulate meiosis. In each case, nutrient starvation may not have a direct effect; rather, it may act indirectly since nutrient limitation results in G1 arrest, and G1 arrest is required for meiosis initiation (Honigberg & Purnapatre, 2003). The response to nitrogen starvation is mediated, in part, by the TOR pathway. This pathway controls the expression of metabolism genes and, as such, its role in meiosis is also proposed to be indirect, because *Tor2* causes changes in metabolism that result in G1 arrest (Honigberg & Purnapatre, 2003). G1 arrest is essential to meiosis initiation, as we have discussed. CLN3 is a G1 cyclin that is part of the mitotic G1 to S phase transition. In nitrogen deprived cells, CLN3 is strongly down-regulated (Gallego et al., 1997). G1 cyclins, like CLN1, 2 and 3 down-regulate *Ime1* in cells grown in nutrient rich medium. This, in turn, represses the initiation of meiosis until cells are starved (Colomina et al., 1999). Starvation triggers a reduction in G1 cyclin levels resulting in the cells arresting at G1 (Colomina et al., 1999; Gallego et al., 1997). Lowered cyclin levels allow IME1 to be transferred to the nucleus, where it initiates meiosis by stimulating transcription of early meiotic genes (Zaman et al.,

2008). Apart from the indirect effects of nitrogen limitation, *Ime1* transcription may be directly influenced by nitrogen limitation, since deletion of the UCS1 upstream controlling sequence allows meiosis in the presence of nitrogen (Kassir et al., 2003). This nitrogen signal is transmitted through *Cdc25*, a positive regulator of the cAMP/PKA and MAPK pathways. The pathway involved in the transduction of *Cdc25*'s effect on *Ime1* is currently unknown (Kassir et al., 2003).

Carbon source is another essential element in the regulation of *Ime1* activity, both in repressing its function under non-favourable conditions and in activating its function under favourable conditions. Carbon source signals act on UCS2, the only controlling region that possess upstream activating sequences (UAS). In the presence of glucose, *Ime1* transcription is repressed at UCS2, preventing *Ime1* expression. UCS1 is also a target for repressing *Ime1* when glucose is present (Kassir et al., 2003). However, UCS2's promoter activity is stimulated in the presence of a non-fermentable carbon source (Sagee et al., 1998; Kassir et al., 2003). So when glucose is absent, *Ime1* expression is stimulated, but it is opposed by the constitutive repressor elements of USC2 and UCS1, unless nitrogen is also limited, which results in a high level of *Ime1* expression, inducing meiosis (Govin & Berger, 2009; Kassir et al., 2003).

The cAMP/PKA pathway is known to transmit the glucose signal to *Ime1* in many ways. Glucose is sensed by the G coupled receptor, GPR1, which activates GPA2, a component of a transmembrane heterotrimeric G protein that activates PKA through adenylyl cyclase (reviewed in Honigberg & Purnapatre, 2003). Adenylyl cyclase activity increases the level of cAMP in the cell, and increased cAMP leads to repression of *Ime1* (Kassir et al., 2003). Repression is mediated through transcription factor MSN2 which stimulates *Ime1* transcription, but with increased cAMP levels it is not transmitted to the nucleus, preventing *Ime1* activation (Kassir et al., 2003). SOK2 is another DNA binding protein that mediates the response of *Ime1* to glucose through the cAMP/PKA pathway. SOK2 functions as a repressor by associating with MSN2. When glucose is not present, SOK2 is converted to an activator (Shenhar & Kassir, 2001). As a further control, *Sok2* expression is dependent on glucose, when cells are growing in a non-fermentable carbon source, SOK2 levels drop dramatically, alleviating its repression of *Ime1* (Shenhar & Kassir, 2001). RIM15 is a serine/threonine protein kinase that is inactivated by PKA phosphorylation when cells are growing in glucose rich media and is increased in acetate media. RIM15 promotes the disassembly of the *Ume6* repressor complex, contributing to the activation of *Ime1* (Zaman et al., 2008). Intracellular acidification of yeast cells also plays into the cAMP pathway. Lowered pH inside the cell stimulates *Ras2*, which stimulates cAMP synthesis (Thevelein & De Winde, 1999). Outside of the cAMP/PKA pathway, glucose sensing also affects *Ime1* through the *Snf1* signal transduction pathway. Glucose inhibits the SNF1 protein kinase, which is necessary for full expression of *Ime1*. However, this is not the only use for SNF1; it also plays a role in *Ime2* regulation and spore formation (Honigberg & Lee, 1998).

It is clear that the regulation of *Ime1* integrates multiple factors to control meiotic initiation in response to environmental cues. However, *Ime1* is not the only target of nutritional regulation; *Ime2* is a meiosis specific protein kinase that is the second major regulator of meiosis in *S. cerevisiae*. It affects multiple stages of meiotic progression, and its transcription and activity are controlled by nitrogen and carbon source signals. *Ime2* activity is inhibited by glucose through GPA2, the α subunit of the heterotrimeric G-protein (a component of the cAMP/PKA pathway). When active GPA2 interacts with the C terminus of IME2, this interaction represses the activity of IME2, which in turn inhibits entry into meiosis (Donzeau & Bandlow, 1999). Glucose also modifies the protein stability of IME2 through the glucose sensors SNF3 and RGT2 (Rubin-Bejerano et al., 1996). UME6 binds to the *Ime2* promoter, repressing transcription in the presence of glucose and nitrogen. During vegetative growth, *Ime2* expression is repressed, like many early meiosis genes, by the UME6-SIN3-RPD3 complex. Under meiotic conditions, UME6 disassociates from SIN3 and RPD3, forming a complex with IME1, which activates the transcription of *Ime2* (Honigberg & Purnapatre, 2003; Purnapatre et al., 2005). The stabilization of this UME6-IME1 complex requires starvation for both nitrogen and glucose. The stabilization is mediated through phosphorylation by RIM11, a glycogen synthase kinase (Chung et al., 2001; Purnapatre et al., 2005) and RIM15, a protein kinase. The expression of RIM15 is repressed when glucose is present in the media and the activity of RIM15 and RIM11 are repressed through the cAMP/PKA pathway, which destabilizes the UME6-IME1 complex (Honigberg & Purnapatre, 2003; Piekarska et al., 2010; Xiao & Mitchell, 2000). Finally, media alkalization effects the expression of *Ime2* through the activation of the UME3-UME5 complex, which has been shown to be required for the full expression of *Ime2* (Cooper & Strich, 2002; Honigberg & Purnapatre, 2003). Thus it is clear that carbon and nitrogen nutritional signals converge on both *Ime1* and *Ime2* in order to control the initiation of meiosis.

3.2 Schizosaccharomyces pombe

In *S. pombe*, the master controller of meiosis is *Ste11*, a transcription factor that stimulates both mating and meiosis. It triggers the expression of both mating type loci and *Mei2*, another key meiosis control gene (Sugimoto et al., 1991). Regulation of both *Ste11* and *Mei2* integrates cell type and environmental signals that lead to initiation of meiosis in *S. pombe*. This section will focus on how these signals are conveyed to the regulators of meiosis through signal transduction pathways, and how these pathways are required for the initiation of meiosis.

3.2.1 Genetic control

A requirement for meiosis in *S. pombe* is that cells are diploid and contain mating type loci *mat1-P* which codes *mat1-Pc* and *mat1-Pi*, as well as *mat1-M*, which codes *mat1-Mc* and *mat1-Mi*. The genes *mat1-Mc* and *mat1-Pc* stimulate pheromone signalling and are essential for both mating and meiosis (Willer et al., 1995). The expression of these two genes requires STE11 (Yamamoto, 1996a) and *Ste11* is only expressed under nutrient starvation conditions (see next prargraph). The pheromones produced bind to their respective receptors (Yamamoto, 1996b). A G protein α subunit, GPA1, is coupled to the pheromone receptors, transmitting the signal downstream (Obara et al., 1991). This activates a MAP kinase cascade including: MAPKKK BYR2, MAPKK BYR1 and MAPK SPK1 (reviewed in Yamamoto, 1996b). Signals received at SPK1 are transmitted to stimulate expression of *mat1-Pi* and *mat1-Mi*. These gene products then allow for the initiation of meiosis by stimulating the expression of *Mei3* (Willer et al., 1995; Yamamoto, 1996b). Another GTP binding protein, RAS1, helps to regulate the MAPK cascade through activating BYR2. RAS1 binds to BYR2 and controls its translocation to the plasma membrane (Bauman et al., 1998). Interestingly, a Ras homolog in *S. cerevisiae*, *Ras1*, is also involved in meiotic initiation, but by repressing it through the cAMP/PKA pathway (Honigberg & Purnapatre, 2003). This is another example of how similar signals are utilized in different ways by divergent organisms. The requirement for starvation to stimulate *Ste11*, which stimulates expression at the mating type loci, provides a link between environmental signals and genetic status of the cells.

3.2.2 Nutritional control

Sexual development in *S. pombe* requires nutrient starvation; with nitrogen starvation, in particular, playing an essential role. Starvation initiates mating, which is typically immediately followed by meiosis. The nutritional signals are linked to meiosis through the cAMP/PKA pathway. In *S. pombe*, similar to *S. cerevisiae*, increased levels of intracellular cAMP inhibit meiosis progress, while lower levels lead to its initiation (reviewed in Yamamoto, 1996a). When cells are growing in nutrient rich media, cAMP levels are high, but when they are transferred to nitrogen-free media, the cAMP levels decrease by approximately 50% before meiosis occurs. The cAMP then increases to a level greater than or equal to those in nutrient rich media during sporulation (the last stage of meiosis). When the cAMP level is artificially elevated in the cells, it results in sterility (Mochizuki & Yamamoto, 1992). Carbon starvation also results in a decrease in intracellular cAMP levels and can contribute to the initiation of meiosis (Isshiki et al., 1992). The cAMP levels in *S. pombe* are controlled by GPA2, the ortholog of GPA2 in *Saccharomyces cerevisiae*, a heterotrimeric G protein which controls the activity of adenylate cyclase. *Gpa2* null mutants had low levels of intracellular cAMP and were able to mate and sporulate, even in rich media (Honigberg & Purnapatre, 2003; Isshiki et al., 1992). GPA2 is necessary for the cell to be able to increase cAMP levels upon glucose stimulation, indicating that it is directly involved in sensing carbon starvation. The ability of the GPA2 mutant to sporulate, even on nitrogen rich media, may also indicate its involvement in nitrogen sensing (Isshiki et al., 1992). Changes in cAMP levels alter the activity of Protein Kinase A (PKA). PKA controls the expression of the major meiosis control gene *Ste11* through its impact on RST2 (Kunitomo et al., 2000). RST2 binds to an upstream *cis*-element, inducing *Ste11* expression. Phosphorylation of RST2 by PKA suppresses its ability to induce transcription of *Ste11*. PKA activity also controls the nuclear localization of RST2, where high levels of PKA result in RST2 being mostly located in the cytoplasm, while low levels result in it being found in the nucleus (Higuchi et al., 2002). When nutritional starvation results in the decrease in cAMP, and thus PKA activity, this results in the activation of RST2, which in turn stimulates *Ste11* expression, leading to meiotic gene expression. In addition to control over *Ste11*, the cAMP signalling pathway also acts to control *Mei2*, another crucial regulator of meiosis initiation. As with *Ste11*, increased cAMP levels inhibit the expression of *Mei2* (Y. Watanabe et al., 1988). MEI3 inactivates PAT1, which inhibits both MEI2 and STE11 through phosphorylation. When MEI3 is expressed, its inactivation of PAT1 results in the accumulation of unphosphorylated and active *Mei2*. The active MEI2 stimulates the continuation of meiosis. In fact, the expression of *Mei3* can bypass both nutritional and genetic requirements and result in ectopic meiosis (Peng et al., 2003). MEI3, is a substrate of PKA; however, decreased phosphorylation does not affect MEI3's ability in inactivate PAT1 (Peng et al., 2003). Regardless of this remaining uncertainty, this information clearly indicates that the cAMP pathway transmits the nutritional starvation signal and influences the initiation of meiosis on multiple levels in S. pombe.

The cAMP/PKA pathway is not the only signal transduction pathway for nutritional sensing. The TOR pathway transmits signals involved in nitrogen source availability. Genes induced by TOR2, a component of the TORC1 complex, include those induced by nitrogen starvation. Tor2 is a negative regulator of meiosis, with Tor2 inhibition increasing meiosis (Matsuo et al., 2007). TOR2 forms a complex with, and inhibits the function of both Ste11 and Mei2, leading to the repression of meiosis (Álvarez & Moreno, 2006). The TOR pathway interacts with the PKA pathway; both are used as a means to drive cell growth and inhibit sporulation. They also work together to regulate Ste11 expression and localization within the cell. STE11 is located throughout the cell, but under meiosis conditions, it builds up in the nucleus. PKA appears to have a controlling role in nuclear localization, when PKA is absent, STE11 localizes to the nucleus, even in the presence of TOR2; the absence of TOR2 also results in nuclear localization of STE11 (Valbuena & Moreno, 2010). Cells with constitutively active Tor2 are impaired in mating, but they regain functional mating when PKA is deactivated, suggesting that PKA is a more potent regulator. When PKA is at a high level in the cell, RST2 represses Ste11 transcription, and mating and meiosis are inhibited (Valbuena & Moreno, 2010). This indicates that the cAMP/PKA pathway interacts with the TOR pathway to control expression and localization of STE11. This, in turn, controls the initiation of meiosis.

There is one additional pathway that transmits nutrient starvation signals to Ste11, the stress response pathway (SRP). Stress includes starvation, the typical trigger for meiosis initiation in S. pombe. The SRP includes the MAPKK WIS1 and MAPK STY1 that play a role in meiosis initiation and stress response in the cell (Kato et al., 1996; Shiozaki & Russell, 1996; Wilkinson et al., 1996). STY1 phosphorylates and modifies the activity of the transcription factor, ATF1 (Shiozaki & Russell, 1996; Wilkinson et al., 1996). ATF1 is necessary for the expression of Ste11 during nutrient starvation (Takeda et al., 1995). Therefore, nutrient starvation signals are also transmitted through the stress response pathway to control Ste11 expression and meiosis in S. pombe. This is notably different from what occurs in S. cerevisiae, where the closest homolog to STY1 is HOG1, which in budding yeast responds only to osmotic stress, not stress in general (Wilkinson et al., 1996).

3.3 Ustilago maydis

In U. maydis research, the focus has been on signals leading to pathogenesis. A look at the life cycle of this fungus (Figure 1) illustrates how closely pathogenesis is tied to the events of sexual reproduction. There are differences given U. maydis is a basidiomycete, for example, when compatible haploid cells fuse they form a filamentous dikaryon and not a diploid. This dikaryon is the pathogenic form and persists for some time before karyogamy is stimulated and meiosis ensues. Recall this is also the situation in the model basidiomycete mushroom, C. cinerea. However, like the yeasts, the proteins coded at the U. maydis mating type loci interact with the output of signal transduction pathways to influence continued development toward meiosis. In order to integrate signals from the mating type loci with environmental signals it is reasonable to expect that, in U. maydis, these signals converge on a given gene or gene(s). These genes have not yet been identified. In the following discussion, the signal transduction pathways, as they are currently understood, will be outlined. There are interesting similarities to signalling pathways in the yeasts and this enables hypotheses to be generated regarding the signalling leading to meiosis in *U. maydis*.

3.3.1 Host/environmental control

Mating, morphogenesis and pathogenicity depend on the cAMP/PKA and MAPK pathways in *U. maydis*. While each pathway transmits signals independently, there is crosstalk between them. The cAMP pathway, as it has been elucidated thus far, begins with a heterotrimeric G protein for which the α subunit and the β subunit have been identified. The α submit is GPA3, the only one of four α subunits coded by *U. maydis* that influences the cAMP pathway. GPA3 mutants are sterile and unable to respond to pheromone signalling, thus unable to mate (Regenfelder et al., 1997). GPA3 associates with the β subunit BPP1 and together they convey the signal to adenylate cyclase, UAC1, the next component of the pathway (Muller et al., 2004). Adenylate cyclase produces cAMP, which activates protein kinase A (PKA) by causing its regulatory subunit, UBC1, to dissociate from the catalytic subunit, ADR1 (Feldbrugge et al., 2004; Gold et al., 1997). cAMP signalling in *U. maydis* has several roles. Its influences: 1) alter the expression of *a* and *b* mating type genes in response to pheromone signalling (Kaffarnik et al., 2003), 2) direct the switch from budding to filamentous growth (Lee et al., 2003), and 3) control pathogenic development (Gold et al., 1997). The influence on filamentous growth is linked to cAMP levels and thus PKA activity. Lower levels of cAMP or altered PKA activity, such as is the case with a defective ADR1 subunit, results in constitutive filamentous growth, while high cAMP/PKA levels result in a budding phenotype (Lee et al., 2003). The influence of the cAMP pathway on pathogenic development was determined through mutation of *Ubc1* (the PKA regulatory subunit), which resulted in high PKA activity. *U. maydis* strains with these mutations were able to colonize the plant, but were unable to form tumours or teliospores. *Uac1* mutants, with low PKA activity, are non-pathogenic (Gold et al., 1997). This suggests that tight control of PKA is required for proper progression of pathogenesis, with low PKA being required for filamentous growth, followed by increased PKA activity needed for infection of the plant, and then lowered PKA once again for tumour and teliospore formation (Gold et al., 1997). Consistent with the requirement for tight control, *U. maydis* strains carrying a constitutively active *Gpa3* can infect corn, leading to tumour formation but not teliospore development (Krüger et al., 2000). It was suggested that the difference between the *Uac1* and the *Gpa3* mutant phenotype is due to different levels of PKA activity in the two mutants, with the *Gpa3* mutant likely representing a less active version with a less defective pathogenic cycle (Krüger et al., 2000). Thus, carefully regulated levels of cAMP appear to be required throughout sexual development, and pathogenesis.

In addition to cAMP signalling, mating and pathogenesis are also regulated by a MAPK signalling cascade. The pathway consists of MAPKKK *Ubc4/Kpp4*, MAPKK *Ubc5/Fuz7* and MAPK *Ubc3/Kpp2* and it may respond to signals transmitted through *Ras2*, a *U. maydis* homolog of the *S. pombe Ras1* (Muller et al., 2003). Evidence supports the pheromone signal being transmitted through a single MAPK pathway, which is also similar to *S. pombe* (Muller et al., 2003). In a parallel pathway to pheromone response, the MAPK cascade is necessary for appressorium formation and function, as well as filamentous growth in the plant. While *Kpp2* is required for appressorium formation, a second MAPK, *Kpp6*, is involved in plant penetration (Muller et al., 2003). Unlike *S. pombe*, no known G-protein α subunit plays a role in the *U. maydis* MAPK cascade; however, a plant signal likely influences this pathway through some means, since the maintenance of *U. maydis* filamentous growth requires the host plant. One possibility is through the link with another pathway. This is suggested because disruption of the MAPK pathway resulted in repression of the constitutive filamentous growth phenotype that is caused by *Adr1* mutation (Muller et al., 2003), recall ADR1 is the catalytic subunit of PKA which is activated by cAMP.

Exploration of the links between the cAMP/PKA and MAPK pathways revealed crosstalk through proteins that are putative orthologs to two major meiotic regulators discussed above. The first is CRK1, which is an IME2 related protein kinase. *Ime2* is a key meiotic regulator and target of environmental signals in *S. cerevisiae*. The second is *Prf1*, a putative ortholog of *Ste11*, a key regulator of meiosis in *S. pombe*. *Crk1* is a target of environmental stimuli in *U. maydis*: *Crk1* mutants are impaired in their response to environmental signals, and *Crk1* is highly expressed when cells are grown in nutrition stress conditions (Garrido & Pérez-Martín, 2003). *Crk1* also plays a role in mating and pathogenesis since *Crk1* mutants are unable to mate on plates and have attenuated pathogenesis producing few tumours and no observed black teliospores (Garrido et al., 2004). *Crk1* is also involved in cell morphogenesis. When *Crk1* is overexpressed, it causes filamentous growth. When *Crk1* is inactivated, it suppresses the constitutive filamentous growth that results from *Adr1* and *Gpa3* mutants. This indicates that it acts downstream of these cAMP pathway genes, however high levels of *Crk1* cannot repress the budding phenotype of a *Ubc1* mutant, indicating it cannot override all cAMP mediated responses (Garrido & Pérez-Martín, 2003). The expression of *Crk1* is regulated by both the cAMP and MAPK pathways, which have antagonistic effects on its transcription. *Crk1* is transcriptionally repressed by the cAMP pathway, with high PKA levels resulting in a low level of *Crk1* expression and vice versa (Garrido & Pérez-Martín, 2003). The MAPK pathway, conversely, positively regulates *Crk1* expression, with *Kpp2* (a MAPK) mutants resulting in much lower levels of *Crk1* in the cell (Garrido & Pérez-Martín, 2003). KPP2 also interacts physically with CRK1, and is required for the role of CRK1 in cell morphogenesis (Garrido et al., 2004). In addition, *Fuz7* (a MAPKK) is required for activation of *Crk1*. FUZ7 phosphorylates CRK1, activating it (Garrido et al., 2004). Thus *Crk1* is clearly involved in the integration of the cAMP and MAPK signalling pathways. However, many of the phenotypes of *Crk1* mutants appear to result from an effect on *Prf1*, since *Crk1* controls the transcription of *Prf1*.

PRF1 is an HMG protein that controls the expression of mating type genes, which regulate mating, pathogenesis and cell morphology. PRF1 binds to the pheromone response element, or PRE, upstream in the *a* and *b* loci, stimulating their expression (Hartmann et al., 1996). Therefore, a *Prf1* mutant strain is unable to mate because it is unable to produce or respond to pheromones. The receptor and pheromone are coded by the *a* mating type locus. Through its control of the *b* mating–type locus *Prf1* influences filamentous growth and pathogenesis. A solopathogenic *Prf1* mutant is unable to cause tumours when it infects the plant (Hartmann et al., 1996). It is possible that *Prf1* acts as a mediator for response to plant signals during pathogenic growth. PRF1 is controlled through the cAMP/PKA and MAPK pathways, facilitating control by pheromones and environmental signals. PRF1 has phosphorylation sites for both MAPK and PKA, and mutations in either of these sites impede mating. This indicates that both MAPK and PKA phosphorylation are required for proper function of PRF1 in sexual development (Kaffarnik et al., 2003). Interestingly, these phosphorylation sites also determine which genes are activated by PRF1, with MAPK phosphorylation being necessary for *b* gene expression, but not for *a*, while PKA phosphorylation is required for both (Kaffarnik et al., 2003). The MAPK pathway may also have a role in inducing the transcription of *Prf1*, as a constitutively active *Fuz7* can increase *Prf1* levels (Kaffarnik et al., 2003). Beyond these post-translational controls, *Prf1* has a cis-regulatory element in its promoter that is termed the UAS, an upstream activator sequence. The *Prf1* UAS appears to regulate transcription of *Prf1* in response to cAMP and carbon source signals. Glucose or sucrose stimulates *Prf1* transcription via the UAS (Hartmann et al., 1999). High cAMP levels repress *Prf1* transcription though the UAS. It is important to note that this is a separate mechanism from the post-transcriptional activation of *Prf1* by PKA, and this seemingly contradictory activity of the cAMP pathway results in increase in *a* gene expression at moderate cAMP levels, but repression through transcriptional control at higher levels (Hartmann et al., 1999). This emphasizes the fine scale control imparted by cAMP levels. The cAMP pathway could also mediate the carbon source signal, or it could be mediated by a separate pathway; this is not yet elucidated (Hartmann et al., 1999).

The link between PRF1 and CRK1 is that CRK1 is required for transcriptional activation of *Prf1* through the UAS (Garrido et al., 2004). This provides another avenue for *Prf1* control by MAPK and cAMP. Kaffarnik et al. (2003) theorized that the cAMP and MAPK paths may be required to control mating because mating typically occurs on the plant, and if sensing the plant results in increased cAMP levels, this would be sufficient to increase *a* gene expression, increasing pheromone expression and making mate detection easier, then pheromone signalling would feed back into the cAMP and MAPK pathways. The MAPK pathway would then initiate conjugation tube formation and mating (Muller et al., 2003), and the cAMP and MAPK pathways would increase the transcription and the activity of *Prf1*, triggering *b* gene expression, and pathogenesis. Thus it is clear that the integration between the two signalling pathways provides a mechanism whereby a plant signal received before penetration could lead to the subsequent events of pathogenesis; however, what triggers meiosis?

The discovery that a decrease in cAMP level is required for the completion of teliospore development suggests that the fungus must lower cAMP levels during pathogenesis to allow teliospores to form. This could be in response to a signal received from the plant. Interestingly, as we discussed above, a decrease in cAMP/PKA levels is necessary to stimulate meiosis in both *S. cerevisiae* and *S. pombe*. Since, in *U. maydis*, meiosis initiation begins around the time of teliospore formation, it is compelling to link the arrest of teliospore development, resulting from elevated cAMP levels, to meiosis. This mutation-stimulated arrest occurs sometime between when the hyphae form lobed tips, and when they fragment and begin rounding and swelling (Krüger et al., 2000). Interestingly, this is very shortly after the time that karyogamy occurs during normal pathogenic development, recall karyogamy occurs before hyphal fragmentation, but after the cells are imbedded in the mucilaginous matrix (Banuett & Herskowitz, 1996). These findings can be integrated in a model where the *U. maydis* dikaryon infects the plant and grows within and between cells, stimulating the initiation of tumour formation, and then it receives a signal from the plant which leads to the fungal cells entering premeiotic S phase and karyogamy, concomitant with the reduction of cAMP, allowing teliospore formation to proceed. The signals for karyogamy and teliospore formation must at least be interrelated, as these two processes need to proceed simultaneously to avoid crucial disruptions in both developmental pathways. As meiosis proceeds, the teliospore develops such that when it enters a dormant state, meiosis arrests at pachytene. This could be the result of reaching the end of a developmental cascade initiated by the plant signals, or a response to another plant signal.

4. Control of meiotic gene expression

The signals received from the environment are transduced through the pathways noted above and result in cascades of transcription that guide meiosis. These waves of transcription have been well studied in *S. cerevisiae* and *S. pombe*. In this section, knowledge of these transcriptional cascades is reviewed and compared. The existing data regarding transcription during meiosis in *U. maydis* is then presented and compared to that of the yeasts.

4.1 *Saccharomyces cerevisiae*

In *S. cerevisiae*, the stages of meiosis have been defined by the waves of genes expressed in a transcriptional cascade. Typically these genes are classified as early, middle and late, depending upon their time of expression during meiosis. Some researchers have found it necessary to further subdivide expression, and, as such, genes may be referred to as belonging to an intermediate expression time; for example, mid-late genes are expressed before late genes, but after the typical middle gene expression (Chu et al., 1998; Mitchell, 1994). In this section we provide an overview of the transcriptional waves, with information on the control of transcription and the relationship of expression to meiotic progression.

4.1.1 Initiation of meiosis

The master regulator of meiosis in *S. cerevisiae* is IME1, a transcription factor that initiates the transcriptional cascade. It is the point of integration of environmental signals and directly controls the expression of early meiosis genes. Under meiotic conditions, IME1 interacts with UME6. UME6 was first identified as a repressor of meiotic genes under vegetative growth conditions. It binds at the upstream repression sequence 1 (URS1) found in target genes (Mitchell, 1994). However, during meiotic growth, UME6 forms a complex with IME1, and this complex activates early meiotic genes, often through the URS1 (Chu et al., 1998; Mitchell, 1994; Rubin-Bejerano et al., 1996). URS1 is a weak upstream activator sequence, and as such, the signal to initiate transcription is often augmented by binding of activator ABF1 at a distinct recognition sequence (Vershon & Pierce, 2000). Based on their expression patterns, early genes have been subdivided into three groups: early (I) induction, early (II) induction and early-middle induction (Chu et al., 1998). These early genes are involved in controlling DNA replication and the events of prophase I: chromosome pairing, homologous recombination and spindle pole body movements (Chu et al., 1998; reviewed in Piekarska et al., 2010). Interestingly, though the early-middle phase genes grouped with other early genes, most lack the URS1, indicating that they are unlikely to be controlled by IME1/UME6. Instead, about half of these genes possess the MSE (middle sporulation element) indicating expression is controlled by NDT80 (Chu et al., 1998). *Ime2* is a key early gene whose transcription is initiated by IME1. IME2 acts through a second pathway that does not directly involve IME1. So there is an IME1 dependant pathway that does not involve IME2 and an IME2 dependant pathway (Mitchell et al., 1990; Mitchell, 1994). IME2 is a cdk-like protein kinase, which plays a key role in transitions in meiosis. It acts to amplify transcription of meiosis genes including itself, activates NDT80 to trigger middle meiosis gene expression and stabilizes Clb cyclins through its inhibition of the APC/C (anaphase promoting complex/cyclosome), which in turn controls chromosome segregation (Marston & Amon, 2004; reviewed in Piekarska et al., 2010). IME2 also targets SIC1, resulting in its degradation, triggering the initiation of the S phase and premeiotic DNA replication (Piekarska et al., 2010). Additionally, IME2 phosphorylates IME1, which then signals it for destruction by the proteasome (Guttmann-Raviv et al., 2002). In this way, IME1 initiates meiosis and IME2 reinforces this and enables it to proceed to the next stage of meiosis, middle gene expression.

4.1.2 Commitment and continuation

Fully active NDT80 is necessary for the full expression of middle meiosis genes and thus for the continuation of meiosis in *S. cerevisiae*. These middle meiosis genes are required for meiotic divisions, and include genes such as B-type cyclins and those involved in spore morphogenesis (Chu & Herskowitz, 1998). NDT80 binds to the conserved MSE element, found upstream of 70% of middle meiosis genes (Chu et al., 1998; Chu & Herskowitz, 1998). NDT80 competes for some of the MSEs with another transcription factor, SUM1, which acts as a repressor of middle meiosis genes during vegetative growth and early meiosis. NDT80 and SUM1 bind to overlapping, yet different, sequences within the MSE, resulting in MSEs that function as *Sum1* repressors, *Ndt80* activators, or both simultaneously (Pierce et al., 2003). A combination of upstream elements also controls the expression of *Ndt80*. While *Ndt80* is a middle meiosis gene, it is expressed slightly before the rest of the middle meiosis genes. This expression pattern, termed pre-middle, results from two URS1s and two MSEs, located upstream of *Ndt80* (Pak & Segall, 2002a). During vegetative growth, expression of *Ndt80* is repressed by both UME6 and SUM1 acting on URS1 and MSE respectively. After the initiation of meiosis, the UME6 repressor complex is replaced with UME6-IME1, but expression is still repressed by the MSE (Pak and Segall, 2002a). IME2 and CDK1 phosphorylate SUM1, leading to its release from the MSE and relieving its repression of *Ndt80* (Ahmed et al., 2009; Pak & Segall, 2002a; Shin et al., 2010). This allows for low level *Ndt80* expression, stimulated by IME1 at URS1. The expressed NDT80 then binds to the MSE in its own promoter region, stimulating expression, leading to full middle gene expression and progression into the first meiotic divisions. However, both *Sum1* and *Ndt80* expression are also controlled by the pachytene checkpoint, which can prevent the expression of middle meiotic genes. Middle gene expression is essential for the cell to exit from the pachytene checkpoint and enter into meiotic divisions. The pachytene checkpoint, or meiotic recombination checkpoint, is part of a surveillance system in eukaryotic cells that arrests the cell cycle in response to defects. To ensure the integrity of the events of meiosis, this checkpoint prevents the cell from exiting the pachytene stage of prophase I and entering into meiotic divisions before the completion of recombination (Roeder & Bailis, 2000). Cells arrested at pachytene are not yet committed to meiosis, meaning they can revert to mitotic growth if conditions are adjusted. This is the last point at which the cell can return to mitotic growth, as after the transition to the first meiotic divisions, the cell is committed to meiosis

(Shuster & Byers, 1989). This makes the pachytene checkpoint the "point of no return" for the cell, allowing one last chance for the cell to arrest and abort meiotic progression, and revert to mitotic growth. In *Dmc1* mutants, DSB repair is impaired, in *Zip1* mutants SC formation is impaired and in *Hop2* mutants, synapsis is defective. Each of these mutants trigger pachytene arrest (Roeder & Bailis, 2000). Checkpoint arrest is mediated through proteins that monitor synapsis and recombination and exert their effects on downstream targets of checkpoint regulation. The checkpoint targets and stabilizes SWE1, a kinase that inactivates CDC28, preventing exit from the pachytene, and SUM1, which represses the expression of *Ndt80* (Pak & Segall, 2002b; Roeder & Bailis, 2000). The checkpoint machinery also directly inhibits the activity of NDT80 by inhibiting its phosphorylation (Hepworth et al., 1998; Pak & Segall, 2002b; Tung et al., 2000). The CDC28/Clb complex allows cell cycle progression past the pachytene checkpoint and into meiotic divisions (Tung et al., 2000). Fully active NDT80 then allows for the expression of middle meiosis genes, leading to meiotic divisions and spore formation and full commitment to meiosis.

The final waves of meiotic gene expression are mid-late and late genes. These genes are involved in spore wall formation and spore maturation, but the transcription factors that initiate their expression are not currently known (Chu et al., 1998; Vershon & Pierce, 2000). In the mid-late genes, 36% have at least one MSE located upstream, indicating that these may be regulated by NDT80, SUM1 or both. Their delay in expression is theorized to be due to other negative regulatory elements (NREs) present in the promoter region that delay expression until the mid-late phase. The factor that acts on these regulatory elements is not known; however, there is evidence that it requires a co-repressor complex of SSN6 and TUP1 (Chu et al., 1998; Vershon & Pierce, 2000). The late genes do not contain either of the previously identified regulatory elements and the control of their expression is not yet understood (Vershon & Pierce, 2000). What is known, however, is that it requires two separate pathways, one involving SPS1 and SMK1, part of a MAPK cascade, and one involving SWM1, a middle meiosis gene that is part of the anaphase promoting complex (Piekarska et al., 2010; Vershon & Pierce, 2000). *Smk1* transcription is regulated through the APC, which links the completion of meiosis to spore formation and maturation, as controlled by the late genes, and through the RAS/cAMP pathway; indicating that nutritional control is still having an effect on spore formation, even after the full commitment to meiosis is made (reviewed in Piekarska et al., 2010). There is still much to learn about the control of meiotic gene expression in *Saccharomyces cerevisiae*, but is it clear that tightly controlled waves of transcription, coupled with a key meiotic checkpoint, ensure that each stage of this transcriptional cascade proceeds only when the cell is prepared to proceed.

4.2 *Schizosaccharomyces pombe*

In *Schizosaccharomyces pombe*, meiosis is controlled by the key transcription factor *Ste11* and the RNA binding protein, MEI2. As in *S. cerevisiae*, many genes are differentially expressed once meiosis is initiated. Mata et al. (2002), proposed four temporal classes; starvation/pheromone induced genes, early genes, middle genes, and late genes. While this progression is similar to *S. cerevisiae*, the control of meiosis in *S. cerevisiae* and *S. pombe* are highly divergent. There are few conserved genes among these species and the regulatory machinery differs. A transcription analysis of *S. pombe* with comparison to the core meiotic transcriptome from two strains of *S. cerevisiae* identified 75 shared genes (Mata et al., 2002). This compares to hundreds of genes with meiosis specific expression in each species (Mata et al., 2002). As such, one would expect differences in the transcriptional control of meiosis between *S. cerevisiae* and *S. pombe*. Here we provide an overview of how transcription triggered by *Ste11* initiates meiosis, how *Mei2* then controls meiotic progression, and how the transcription factors control different waves of transcription during meiosis in *S. pombe*.

4.2.1 Initiation of meiosis

The first genes induced during *S. pombe* meiosis are those that act in response to starvation, including nitrogen transporters, metabolism and mating type regulators. This is followed by the expression of genes involved in pheromone signalling and entry into meiosis, including *Ste11* and *Mei2*. STE11 is a transcription factor that is essential to the initiation of meiosis. It controls the transcription of several key genes, including *Mei2* and the mating type genes *mat1-Pc* and *mat1-Mc*, required for the initiation and continuation of meiosis (Mata et al., 2002). STE11 is an HMG-box protein, responsible for the expression of nitrogen responsive genes during starvation conditions. Ectopic expression of *Ste11* in vegetative growth conditions triggers mating and meiosis, while *Ste11* disruptions result in sterility (Sugimoto et al., 1991). This indicates an essential role for *Ste11* in *S. pombe* meiosis. STE11 binds DNA at TR (T-rich) boxes, present in varying copy numbers upstream of target genes including *matP*, *matM* , *Mei2* and *Ste11* itself (Sugimoto et al., 1991). The mating type genes are required for pheromone signalling which stimulates *Ste11* activity (Harigaya & Yamamoto, 2007). STE11 also binds to the TR box upstream of its gene, stimulating its own expression. This positive feedback loop reinforces the cell's commitment to meiosis (Kunitomo et al., 2000). *Ste11* activity is inhibited by CDK phosphorylation and since STE11 is highly unstable, the protein rapidly disappears if it is not able to stimulate its own expression (Kjærulff et al., 2007). This provides a means to tightly control expression of *Ste11*. CDK activity is low in the beginning of G1, and then increases through S and into G2. When CDK activity increases, STE11 is phosphorylated and degraded, this restricts STE11 function to G1 (Kjærulff et al., 2007). This is similar to the regulation in *S. cerevisiae* of IME1 by G1 cyclins in response to nutrient signals, which trigger arrest at the G1 phase (Colomina et al., 1999). This may indicate that CDK phosphorylation of a transcription factor plays a role in restricting meiosis initiation to the G1 phase in many organisms.

The stimulation of *Mei2*, *mat1-Pc* and *mat1-Mc* expression by STE11 leads to another level of meiotic control. Mating–type loci gene expression leads to pheromone production which stimulates *Mei3* expression. MEI3 inactivates *Pat1*, which functions to prevent the expression of both *Ste11* and *Mei2* during vegetative growth and in haploid cells (Yamamoto, 1996a). In this way, STE11 is responsible for the expression of both itself and *Mei2* in two different ways; directly through stimulation at the TR box, and indirectly through the pheromone response pathway which leads to the expression of *Mei3*. MEI2, expression leads to the induction of meiosis (Yamamoto, 1996a). Early genes are expressed after the initiation of meiosis, they are involved in S phase, chromosome pairing and recombination. Many of the genes expressed at this time contain an upstream element, the MluI box, which suggests they are controlled by the CDC10, RES2, REP1 transcription complex (Mata et al., 2002).

4.2.2 Commitment and continuation

Mei2 is critical for the mitotic-meiotic switch in fission yeast and for the commitment of the cell to meiosis (Y. Watanabe et al., 1988). Like *S. cerevisiae*, *S. pombe* makes the critical decision to enter meiosis before the premeiotic S phase (Marston & Amon, 2004). However, unlike budding yeast, once *Mei2* is active, *S. pombe* is fully committed to meiosis, and the cell cannot be induced to revert to mitosis (Y. Watanabe et al., 1988). It should be noted that *Mei2* is also required for meiosis I (reviewed in Yamamoto, 1996b). MEI2 contains three RNA recognition motifs and this RNA binding capability is essential for its function in stimulating meiotic initiation and continuation through meiosis I. The RNA that interacts with MEI2 to promote premeiotic DNA synthesis is currently unknown; however, it has been found that MEI2 must interact with meiRNA, the non-functional RNA product of *Sme2*, to successfully promote entry into meiosis I (Watanabe & Yamamoto, 1994, cited in Yamamoto, 1996b). meiRNA is required for the import of *Mei2* into the nucleus before meiosis I (Yamashita et al., 1998). *Mei2* is located in the cytoplasm of the cell during vegetative growth, but it condenses into a single spot within the nucleus during meiotic prophase. These dots can be identified in the nucleus even before premeiotic DNA synthesis and then they fade away after the first meiotic division (Yamashita et al., 1998). The formation of this dot requires MEI2 and meiRNA association, un-associated MEI2 and meiRNA remain in the cytoplasm. Once in the nucleus, MEI2 forms the dot and promotes meiosis I. meiRNA is then no longer required for the function of MEI2 in promoting meiosis I; however, MEI2 binding to other RNAs is crucial (Yamashita et al., 1998). Once in the nucleus, MEI2 promotes meiosis I by modifying the availability of meiosis specific mRNA transcripts. In vegetatively growing cells, meiosis specific mRNAs are selectively eliminated by the MMI1 RNA binding protein, which interacts with an RNA element termed the DSR (determinant of selective removal) (Harigaya et al., 2006). During meiosis, MMI1 changes its localization within the nucleus, from several spots to a single dot, which overlaps the MEI2 dot. It is believed that MEI2 sequesters MMI1, preventing it from eliminating meiosis specific genes, resulting in their stable expression (Harigaya et al., 2006). One of these stabilized genes is *Mei4*, a transcription factor involved in controlling middle meiosis genes and necessary for meiosis I (Harigaya et al., 2006; Yamamoto, 2010).

Mei4 is a meiosis specific transcription factor that binds to an element upstream of its target genes termed FLEX-D, which activates their transcription. It is part of the cascade that controls meiosis in *S. pombe* and *Mei4* mutants arrest in prophase I (Horie et al., 1998). *Mei4* is key to the expression of genes during middle meiosis in *S. pombe*, the genes involved in meiotic divisions, as well as its own expression. *Mei4* is autoregulated, and it possesses two FLEX-like sequences in its 5′ upstream region, so low levels of *Mei4* expression result in greater transcription (Abe & Shimoda, 2000). Middle genes include cell cycle regulators like *Cdc25*, kinases, components of the SPB and other genes required for progression through the cell cycle. Also represented were genes involved in cell morphogenesis, membrane trafficking, and possibly spore formation (Mata et al., 2002). Interestingly, two of the genes controlled by *Mei4*, *Mde3* and *Pit1*, are homologs to *S. cerevisiae Ime2*. These genes are involved in sporulation and asci formation, but they do not seem to delay meiotic progression (Abe & Shimoda, 2000). This makes sense, based on their different timing of expression, as *Ime2*'s role in early and middle meiosis requires it to be expressed in early meiosis, not middle meiosis like *Mde3* and *Pit1*. The upstream regions of more than half of these genes have elements similar to the *Mei4* binding motif and 90% of known *Mei4* target genes are found to be up-regulated in the middle phase of meiosis, (Mata et al., 2002). In this way, *Mei4* demonstrates some functional similarities to the *S. cerevisiae Ndt80*, which is also expressed during middle meiosis, where it regulates its own expression, promotes the expression of other middle meiosis genes and is essential for progression to the first meiotic divisions. Therefore, although they are not related proteins, NDT80 and MEI4 seem to fulfil functionally equivalent roles, indicating that this post-initiation/premeiotic division stage is a conserved component of meiotic completion.

Late meiosis genes in *S. pombe* are involved in spore formation and they are expressed after the meiotic divisions. These include stress response, cell cycle regulation and cell wall formation genes (Mata et al., 2002). Many late genes have a binding site for Atf transcription factors and over half of late genes are regulated by ATF21 and ATF31. These transcription factors are expressed during the middle phase of meiosis, and induce late gene expression (Mata et al., 2002). The conservation of transcriptional waves in *S. pombe* and *S. cerevisiae* as well as *C. cinerea* (Burns et al., 2010b) suggests this may be a wide spread mechanism to ensure that the orderly progression of meiosis.

4.3 *Ustilago maydis*

The *Ustilago maydis* meiotic transcriptional program has not been elucidated; however, initial data in this area is available in the form of transcript profiling during periods of sexual development and bioinformatic analysis comparing *U. maydis* genes to known meiotic genes in other organisms. Here we will review and update the results of these past analyses, reflect on what they suggest regarding transcriptional control of meiosis in *U. maydis* and propose future experiments.

Zahiri et al. (2005) used cDNA microarray hybridization experiments to investigate changes in gene expression during teliospore germination. They selected two time points for investigation: 4 hrs and 11 hrs post induction of germination. Transcript levels at these time points were compared to those in the dormant teliospore. To provide context, recall that *U. maydis* teliospores germinate at late prophase I (O'Donnell & McLaughlin, 1984). Therefore, by the time teliospores germinate, the early stages of meiosis are completed and the stage is set for completion of meiosis I. Early biochemical experiments showed that, during teliospore germination, total RNA increased steadily; however, protein synthesis did not proceed at a measurable level until approximately 6 hrs after inducing germination. From 6hrs onward, protein synthesis increased linearly with time (Tripathi & Gottlieb, 1974). Zahiri et al. (2005) identified genes whose transcript levels decreased upon germination and proposed that these transcripts were stored in the dormant spore and degraded as germination proceeded, possibly following translation. These transcripts were proposed to code proteins required early in the germination process, or for reinitiating meiosis before visible signs of germination were evident. Genes involved in early meiosis would be expected to be captured in this study, and Zahiri et al. (2005) identified those involved in recombination, DNA repair, transcription, translation, protein turnover and assembly, stress response, and metabolism. The interpretation of Zahiri et al. (2005) was that the array of genes found was consistent with change from a state of dormancy to one of physiological activity, and with the events of early meiosis, notably DNA recombination and repair. Upon re-examination, two of the genes in this category, *Rad51* and *Brh2*, in addition to their role in meiosis, are required for teliospore germination. Null mutants of *Rad51* and *Brh2* are unable to produce basidia (Kojic et al 2002). Transcript presence for these genes in the dormant teliospore, therefore, may or may not support the occurrence of the early stages of meiosis in the early stages of teliospore germination. Interestingly, *Rad51* transcript level decreases as germination proceeds, while the *Brh2* transcript level increases. *Brh2* expression is consistent with it having an ongoing role in the events of germination and/or meiosis. This data also showed evidence of waves of transcription. Zahiri et al. (2005) noted that genes upregulated at the later stage of germination and meiosis included: DNA repair, protein turnover, cell wall synthesis and metabolism. Cell wall synthesis genes are involved in basidium formation, which would fit with genes typically expected to be expressed during late meiosis. The increased expression of genes involved in protein turnover is believed to indicate the physiological transition and we now see that *Brh2*, a gene thought to be only involved in DNA repair, may have other roles during teliospore germination. Zahiri et al. (2005) proposed that the dormant teliospore is in a premeiotic state and that it begins meiosis immediately upon induction of germination. However, Donaldson and Saville (2008) proposed that DNA replication occurred before karyogamy, while the cell is still *in planta*. This means that the initial steps of meiosis up to prophase I would occur before the teliospore enters dormancy, with the cell arresting at the pachytene checkpoint (Donaldson & Saville, 2008). This would be consistent with the observation that as the basidium forms, the cell is already in late prophase I, clearing the pachytene checkpoint and beginning meiotic divisions only a short time after induction of germination, and before any new protein synthesis has been detected in the teliospore (Donaldson & Saville, 2008). The reinterpretation of the Zahiri et al. (2005) data, with the knowledge of alternate roles for some genes otherwise considered to be meiosis genes (Banuett, 2010), indicates that the Donaldson and Saville (2008) interpretation is more likely correct.

Donaldson and Saville (2008) performed comparative genomic analysis between *U. maydis*, and *S. cerevisiae*, *S. pombe*, and *N. crassa*. They identified 164 potential *U. maydis* orthologs to meiosis genes found in other fungi, of which 66 genes overlapped with the core meiotic genes conserved between *S. cerevisiae* and *S. pombe* (Mata et al., 2002). Potential orthologs were identified to several key meiotic genes, including: *Ime2*, *Ndt80*, *Ume6*, and *Ste11*. Notably absent, however, were *U. maydis* orthologs to *Mei4*, *Ime1*, *Atf21* and *Atf31* (Donaldson & Saville, 2008). Of the orthologs that were identified, the *Ime2* and *Ste11* orthologs, *Crk1* and *Prf1*, respectively, have been well characterized in *U. maydis*. Although both *Crk1* and *Prf1* are involved in mating and pathogenic development of *U. maydis* (discussed earlier in this chapter), a direct link to meiotic initiation has not been shown for either gene. This could indicate that these two proteins perform a different role in *U. maydis*; influencing mating, but not directly influencing meiosis; or that the role in meiosis could simply be obscured by the fact that *Crk1* and *Prf1* mutants prevent pathogenesis, arresting development before meiosis occurs. Distinguishing between these possibilities will require further investigation.

Ndt80 in *U. maydis* (hereinafter referred to as *UmNdt80*) is highly divergent from its *S. cerevisiae* ortholog. It was identified based on similarity with a *N. crassa* gene that had a low level of similarity to *S. cerevisiae Ndt80* (Donaldson & Saville, 2008; Borkovich et al., 2004). *U. maydis* strains in which *UmNdt80* is deleted are capable of mating, pathogenesis, tumour formation and teliospore production. However, *UmNdt80* mutant teliospores are tan coloured, in contrast to the dark brown teliospores formed by a wild-type *U. maydis* infection. These *UmNdt80* mutant teliospores are meiotically deficient, with up to 95% germinating as diploids, suggesting that they failed to complete meiosis (Doyle & Saville, unpublished). These initial analyses of *UmNdt80*, a potential meiotic control gene, suggest that it plays an essential role in teliospore formation and meiotic completion. *UmNdt80* appears to play a role after mating and initiation of pathogenesis is complete, affecting the later events of meiosis and spore formation. This is similar to the role of NDT80 in *S. cerevisiae*, which is interesting when one considers that NDT80 and UmNDT80 are highly divergent proteins, with only the NDT80-PhoG active site showing similarity. From this, we infer that *UmNdt80* functions as a transcription factor involved in meiotic progression and teliospore formation, but its expression is not regulated like *S. cerevisiae Ndt80*. *U. maydis* is a fungal pathogen that achieves meiotic competence and forms teliospores only *in planta*, it is clear that the environmental signals leading to meiosis and teliospore formation are different than those involved with meiosis in *S. cerevisiae*. Consistent with this, *U. maydis* lacks the main transcription factor (*Ime1*) that stimulates the expression of *Ndt80* in *S. cerevisiae* and the *U. maydis Ime2* ortholog, *Crk1* (Donaldson & Saville, 2008) does not have the direct link to meiosis exhibited by the *S. cerevisiae, Ime2*.

While the bioinformatic and functional analyses of transcription factors in *U. maydis* have revealed some interesting possibilities concerning meiotic control, further wet lab experiments are required. The focus of these experiments will be to understand the progression of gene expression during *U. maydis* meiosis and the role of UmNDT80 in this process. We will investigate the functions of *U. maydis* orthologs for *S. cerevisiae* genes involved in meiosis, including *Spo11*, and *Rim11*. We are also investigating the upstream control of *UmNdt80*, with a goal of working back to find transcription factors involved in the initiation of meiosis and teliospore formation. These, and other studies underway, will provide the tools to identify the environmental (plant) signals required for meiotic initiation.

5. Meiosis gene presence and function in fungi

In this section we provide an overview on how meiosis genes can be identified in organisms with a sequenced genome. We then provide a comparative analysis of the presence and absence of "core meiosis genes" in select pathogenic and non-pathogenic fungi and close with a transcriptional analysis of predicted *U. maydis* meiosis genes. The transcriptional analysis section contains a brief discussion of the possible post transcriptional events that control meiosis gene expression in *U. maydis*.

5.1 Identification of meiosis genes in fungi

The use of DNA microarrays facilitated the identification of a large number of genes involved in meiosis and sporulation in *S. cerevisiae* and *S. pombe*. Chu et al. (1998) identified over 1000 *S. cerevisiae* genes with differential expression (induced or repressed) during meiosis, representing nearly 16% of the budding yeast transcriptome. Genes were grouped based on their temporal expression into 7 distinct clusters; fine-tuning the previously identified early, middle, mid-late, and late meiotic- and sporulation-specific gene clusters (Mitchell, 1994). Primig et al. (2000) identified strain-dependent meiosis-specific genes by comparing the meiotic transcriptome of two yeast strains (SK1 and W303), which differ in their rates of sporulation. A core set of 900 genes with strain-independent meiotic expression was observed and the inventory of meiosis- and sporulation-specific genes in budding yeast was expanded with the identification of 650 previously unreported meiosis-specific genes. In both studies, the functions of uncharacterized genes were inferred as meiotic- or sporulation-specific, based on their temporal expression (Chu et al., 1998; Primig et al., 2000). Schlecht and Primig (2003) further defined a set of 75 core meiotic- and sporulation-specific genes for *S. cerevisiae*. Mata et al. (2002) identified approximately 1000 *S. pombe* genes with a fourfold increase in expression during meiotic growth conditions compared to vegetative growth conditions. Mata et al. (2002) compared the meiosis-specific fission yeast genes to the core strain-independent meiosis-specific genes (Primig et al., 2000) and identified 75 genes upregulated during meiosis in both budding and fission yeast. This core group of shared genes between the two diverged yeast species contained genes related to the process of meiosis and sporulation (*Rec8*, *Dmc1*, and *Hop2* for example) but lacked genes related to the timing of their expression; indicating the control of meiosis evolved after budding and fission yeast diverged from a common ancestor (Mata et al., 2002). Mata and Bahler (2003) expanded their analysis of *S. pombe* and observed that organism-specific genes (orphans) were over-represented during meiosis and sporulation, particularly in meiotic prophase. Mata and Bahler (2003) noted that meiotic structural proteins are poorly conserved, contrasting the identification of core recombination proteins among eukaryotes (Villeneuve & Hillers, 2001).

Information gained from the DNA microarray studies in budding yeast and fission yeast was used to identify meiotic orthologs in the filamentous ascomycete *Neurospora crassa* and the basidiomycete pathogen *U. maydis* (Borkovich et al., 2004; Donaldson & Saville, 2008). Putative orthologs for the majority (~77%) of core meiotic genes shared between *S. cerevisiae* and *S. pombe* were identified in *U. maydis*. The inability to detect all of the core meiotic genes indicates that either meiosis genes have diverged beyond sequence recognition using blastp-analysis in *U. maydis* relative to the yeasts, or there are basidiomycete- or *U. maydis*-specific meiosis genes that replace the functions of identified yeast genes (Donaldson & Saville, 2008).

5.2 The core eukaryotic meiosis-specific machinery

Much attention has been given to the identification of a conserved set of core meiotic proteins across eukaryotes. Villeneuve and Hillers (2001) identified meiotic recombination proteins shared between animals, plants and fungi; presumably stemming from a common ancestor. Key components of the conserved meiotic recombination machinery include: SPO11, RAD50/MRE11, DMC1, RAD51, MSH4/MSH5, and MLH1. While analyzing the *Giardia lamblia* genome, Ramesh et al. (2005) augmented the list of core meiotic proteins identified by Villeneuve and Hillers (2001) to include HOP1, HOP2, MND1, RAD52, MSH2, MSH6, MLH2, MLH3, and PMS1. While identifying meiotic homologs in *Trichomonas vaginalis*, Malik et al. (2007) added RAD1, MER3, SMC1-5, RAD18, RAD21, REC8, PDS5, and SCC3 to the list of core meiotic proteins conserved across a wide range of plants, animals and fungi. Of the 29 proteins conserved across eukaryotes, only SPO11, HOP1, HOP2, MND1, DMC1, MSH4, MSH5, MER3, and REC8 are reported to be functional solely during meiosis (Malik et al., 2007). Interestingly, when looking at the phylogenetic distribution of the 29 core meiotic proteins across the fungi studied by Malik et al. (2007), one can observe that the loss of key meiosis proteins is restricted to the 9 proteins only functional during meiosis (with the exception of MLH2 and MLH3 in some fungi).

Table 1 shows the presence and absence of these 9 proteins across the model yeasts *S. cerevisiae* and *S. pombe*, the model basidiomycete mushroom, *C. cinereus*, and fungal plant pathogens, including: *Gibberella zeae* (wheat head blight fungus), *Magnaporthe grisea* (rice blast fungus), *Sporisorium reilianum* (maize head smut), and *U. maydis* (common smut of corn). We will focus our discussion on the loss of HOP2, MND1, DMC1 and HOP1.

The absence of an ortholog to DMC1 is coincident with a loss of MND1, and HOP2 (reviewed in Neale & Keeney, 2006). *G. zeae, M. grisea, S. reilianum,* and *U. maydis* lack clear orthologs to DMC1, MND1, and HOP2 (Table 1). Loss of DMC1, MND1, and HOP2 is not unique to the fungal plant pathogens as it has been reported in *Drosophila melanogaster, Caenorhabditis elegans,* and *N. crassa* (Malik et al., 2007). Recently in a comparison of eight *Candida* genomes, these genes were absent in *C. guilliermondii* and *C. lusitaniae* (Butler et al., 2009). In an excellent review of the homologous recombination system in *U. maydis,* Holloman et al. (2008) concluded BRH2, RAD51, and REC2, a paralog to RAD51, efficiently mediate homologous pairing during meiosis in a RAD51-dependent manner (in the absence of DMC1, MND1, and HOP2). Additionally, given its capacity to promote DMC1-like DNA strand exchange, it was speculated that REC2 evolved as a substitute for DMC1 in *U. maydis* (Holloman et al., 2008). While the *U. maydis* REC2 is a weak ortholog to *S. cerevisiae* REC57 (e-value = 4-e4; reported in Holloman et al., 2008), there is a clear ortholog in *S. reilianum* (e-value = 0.0). Therefore, it would not be surprising for REC2 to compensate for the loss of DMC1 in *S. reilianum* as well.

A recognizable ortholog to HOP1 is absent in *G. zeae, M. grisea, S. reilianum,* and *U. maydis* (Table 1). This loss is not unique to the fungal plant pathogens, as *D. melanogaster, N. crassa, C. guilliermondii* and *C. lusitaniae* lack clear HOP1 orthologs (Butler et al., 2009; Malik et al., 2007). HOP1 is a structural component of the synaptonemal complex (SC), a tripartite structure which holds homologous chromosomes together during meiosis (reviewed in Loidl, 2006). The absence of clear orthologs to other synaptonemal complex proteins: ZIP1, ZIP2, ZIP3, and RED1, coupled with the inability to observe SCs microscopically (Fletcher, 1981), has brought into question whether SC formation occurs in *U. maydis* (Donaldson & Saville, 2008; Holloman et al., 2008). It should be noted that, with the exception of HOP1, SC proteins show great sequence divergence between organisms (Loidl, 2006). Additionally, SC formation may occur in *U. maydis* during teliospore development (*in planta*), or prior to teliospore germination when the thick cell wall interferes with SC visualization (Fletcher, 1981). Therefore, SC formation in *U. maydis* may occur. Conversely, SC formation may not occur in *U. maydis*, as is the case in *Aspergillus nidulans*, and *S. pombe* (Loidl, 2006). In *S. pombe*, linear elements (LinEs) are formed in the absence of SCs. It has been suggested that the low chromosome number in *S. pombe* enables an abridged version of the SC machinery to efficiently pair and recombine chromosomes during meiosis (Loidl, 2006). In this mechanism, an ortholog to *S. cerevisiae* RED1 (*S. pombe* REC10) is essential for the formation of LinEs, while HOP1, and MEK1 are active in their formation, but not required (Loidl, 2006). *U. maydis* contains an ortholog only to MEK1 (Donaldson & Saville, 2008), suggesting that it is unable to form LinEs.

Protein	S. cerevisiae[1]	S. pombe[1]	C. cinereus[2]	G. zeae[1]	M. grisea[1]	S. reilianum[3]	U. maydis[4]	S. cerevisiae function[5]
SPO11	+	+	+	+	+	+	+	Meiosis-specific protein that initiates meiotic recombination by catalyzing the formation of double-strand breaks in DNA via a transesterification reaction; required for homologous chromosome pairing and synaptonemal complex formation
HOP1	+	+	+	-	-	-	-	Meiosis-specific DNA binding protein that displays Red1p dependent localization to the unsynapsed axial-lateral elements of the synaptonemal complex; required for homologous chromosome synapsis and chiasma formation
HOP2	+	+	+	-	-	-	-	Meiosis-specific protein that localizes to chromosomes, preventing synapsis between nonhomologous chromosomes and ensuring synapsis between homologs; complexes with Mnd1p to promote homolog pairing and meiotic double-strand break repair
MND1	+	+	+	-	-	-	-	Protein required for recombination and meiotic nuclear division; forms a complex with Hop2p, which is involved in chromosome pairing and repair of meiotic double-strand breaks
DMC1	+	+	+	-	-	-	-	Meiosis-specific protein required for repair of double-strand breaks and pairing between homologous chromosomes; homolog of Rad51p and the bacterial RecA protein
MSH4	+	-	+	+	+	+	+	Protein involved in meiotic recombination, required for normal levels of crossing over; colocalizes with Zip2p to discrete foci on meiotic chromosomes, has homology to bacterial MutS protein
MSH5	+	-	+	+	+	+	+	Protein of the MutS family, forms a dimer with Msh4p that facilitates crossovers between homologs during meiosis; msh5-Y823H mutation confers tolerance to DNA alkylating agents; homologs present in C. elegans and humans
MER3	+	-	+	-	+	+	+	Meiosis specific DNA helicase involved in the conversion of double-stranded breaks to later recombination intermediates and in crossover control; catalyzes the unwinding of Holliday junctions; has ssDNA and dsDNA stimulated ATPase activity
REC8	+	+	+	+	+	+	+	Meiosis-specific component of sister chromatid cohesion complex; maintains cohesion between sister chromatids during meiosis I; maintains cohesion between centromeres of sister chromatids until meiosis II; homolog of S. pombe Rec8p

[1]Data retrieved from Malik et al. (2007). [2]Data retrieved from Burns et al. (2010a). [3]Data retrieved from the Munich Information center for Protein Sequences *Sporisorium reilianum* Database (Schirawski et al., 2010). [4]Data retrieved from Donaldson and Saville (2008). [5]Protein function retrieved from the *Saccharomyces* Genome Database (2011).

Table 1. Core meiosis-specific genes conserved among model fungi and select fungal plant pathogens. Putative loss of protein is highlighted in grey.

5.3 Core meiosis-specific gene expression in *U. maydis*

In order to gain insight into the timing of meiosis *in planta*, we investigated the expression of four core meiosis-specific genes in *U. maydis* (Table 1). We included a fifth gene, *Mre11*, a key component in the MRE11-RAD50-XRS2 (MRX) complex, which processes double strand break ends prior to homologous recombination (Holloman et al., 2008). Schlecht and Primig (2003) identified *Mre11* as being meiosis and sporulation specific in *S. cerevisiae*. Using a combination of statistical analysis and PCR, Ho et al. (2007) identified a conserved hypothetical gene (*Um01426*) to be highly expressed in the dormant teliospore, compared to haploid cells grown in rich media. This gene was included in our analysis to estimate the timing of teliospore maturation in the *in planta* time course expression analysis. Glyceraldehyde 3-phosphate dehydrogenase (*UmGapd*) is constitutively expressed in *U. maydis* and its expression was used to detect the presence of *U. maydis* cells *in planta*.

Golden Bantam seedlings were infected with compatible *U. maydis* haploids (FB1, FB2) and leaf samples were taken 2, 4, 6, 8, 10, and 14 days post infection (dpi). RNA was isolated from these six leaf time-points, as well as dikaryotic and forced diploid mycelia grown filamentously, teliospores isolated from mature tumours on infected ears of corn, and individual compatible haploids. Equal amounts of RNA were used as template for reverse transcriptase reactions primed with oligo-d(T)$_{16}$. The resulting cDNA was diluted and equal amounts were used as template for PCR. When possible, primers were designed to flank introns (*UmSpo11*, *Um01426*, and *UmGapd*), to clarify the difference in sizes of the amplified products between PCRs with cDNA as template and genomic DNA as template (Figure 2a). The results presented as "leaf" represent a combination of plant and fungal RNA isolated from infected leaves while the *U. maydis* cell type results are RT-PCR results from pure fungal RNA. Therefore, the results are displayed as two separate panels (leaf samples, or *U. maydis* cell-types; Figure 2a). Given that the amount of RNA going into first strand synthesis was equivalent in each panel, the resulting RT-PCR product viewed on the ethidium-bromide stained agarose gel may be interpreted as representing the relative expression for each transcript. Comparison between panels is not valid. We estimated the relative expression of the meiosis-specific transcripts for each gene individually, for the leaf and *U. maydis* cell-types (Figure 2b).

Banuett and Herskowitz (1996) provide an excellent framework for a comparison between the initiation of meiosis and homologous recombination in relation to hyphal development and teliospore formation *in planta*. It should be cautioned that differences in the timing of *U. maydis* development may arise due to the type of maize variety infected, and plant growth conditions (Banuett & Herskowitz, 1996). The total RNA isolated from leaf samples is expected to contain RNA from *U. maydis* cells at different stages of development since *U. maydis* development is asynchronous *in planta*. Additionally, changes in transcript levels does not necessarily imply changes in the respective protein levels, especially since meiosis genes are known to be post-transcriptionally regulated (Burns et al., 2010b). We will attempt to determine the timing of meiosis initiation and homologous recombination relative to the *in planta* development of *U. maydis*. Given that we used a different maize variety and infected corn seedlings 2 days later than Banuett and Herskowitz (1996), we might expect inconsistencies between the molecular and morphological studies. Nonetheless, it is interesting that the expression of *Um01426*, a gene highly expressed in mature teliospores (Ho et al., 2007) is upregulated 6-8 dpi, in line with the time that teliospore formation is initiated, and its expression peaks 14 dpi when mature teliospores are visible (Figure 2a,b; Banuett & Herskowitz, 1996).

a)

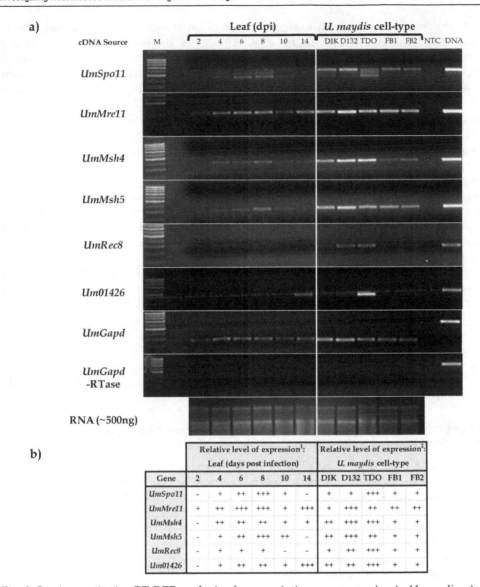

Fig. 2. Semi-quantitative RT-PCR analysis of core meiotic gene expression in *U. maydis*. a) The cellular origins of the cDNA templates were: Leaf samples taken at 2, 4, 6, 8, 10, and 14 days post infection (dpi); DIK, dikaryotic filamentous mycelia; D132, diploid filamentous mycelia; TDO, dormant teliospore; FB1, haploid cell (*a1 b1*); FB2 haploid cell (*a2 b2*); NTC, no template control; DNA, genomic DNA; M, FullRanger 100 bp DNA Ladder (Norgen Biotek). RNA quality was accessed by glyoxal agarose gel electrophoresis. b) Relative levels of gene expression were estimated on a gene-by-gene basis independently for the [1]leaf and [2]cell-type specific RT-PCR groupings; -, no expression; +, low expression; ++, mid expression; +++, high expression.

Spo11 expression has been used to signal the transition between non-meiotic and meiotic cells in *S. cerevisiae* and *S. pombe* (Burns et al., 2010b). We detected unspliced *UmSpo11* at 2 dpi where filaments have been observed on the leaf epidermal surface (Banuett & Herskowitz, 1996). A discussion on the implications of alternatively spliced *UmSpo11* will follow; however, if the spliced variant detected at 4 dpi (Figure 2a) yields a protein that is responsible for double strand break formation at a time where dikaryotic filaments penetrate through the stomata and branch within the plant cells, then the dikaryotic filament may be prepped to initiate the transition from mitotic to meiotic cells directly after karyogamy occurs (9-10 dpi; Banuett & Herskowitz, 1996). Premeiotic DNA replication in C. cinerea occurs prior to karyogamy (reviewed in Burns et al., 2010b). In U. maydis, the timing of premeiotic DNA replication, relative to karyogamy is unknown. The expression of many genes involved in homologous recombination (*UmMre11*, *UmMsh4*, *UmMsh5*, and *UmRec8*) prior to the time-points suggested for hyphal fragmentation and karyogamy is intriguing. We may have detected so-called "leaky" transcription, but due to the deleterious effects of leaky transcription, it is more likely that the transcript is present and the protein is not functional at the time we detected. In a comparison of the meiotic expression profiles of shared orthologs between *S. pombe*, *S. cerevisiae*, and *C. cinerea*, Burns et al. (2010b) note that the expression of *Spo11* and *Rec8* peak late in meiosis, prior to the first meiotic division, past the time-point when the protein is functional. We observed that, while the expression of core genes involved in homologous recombination in *U. maydis* are detectable 4 dpi, they peak in-between 6-10 dpi, coincident with teliospore formation, hyphal fragmentation and karyogamy (Figure 2a,b; Banuett & Herskowitz, 1996). Given this expression of the core meiotic genes responsible for homologous recombination, and using expression of *Um01426* (discussed above) as a reference, we propose cells are prepped for meiosis prior to teliospore maturation.

5.4 Post-transcriptional control of Spo11 and controlled meiotic splicing in *S. cerevisiae* and *S. pombe*

Under specific conditions in budding yeast, regulated splicing can affect translation by introducing frame shift mutations or nonsense codons, producing non-functional proteins (Juneau et al., 2007). Using high-density tiling arrays, Juneau et al. (2007) discovered 13 intronic meiosis-specific genes that undergo regulated splicing in *S. cerevisiae*. Using RT-PCR, it was observed that the transcripts *Ama1*, *Hfm1*, *Hop2*, *Mnd1*, *Rec107*, *Rec114*, *Rec102*, *Pch2*, *Spo22*, *Dmc1*, *Mei4*, *Sae3*, and *Spo1* spliced more efficiently (>84%) during sporulation compared to vegetative growth (Juneau et al., 2007). Therefore, regulated splicing of select meiotic-specific transcripts could help the cell overcome the deleterious effects of leaky meiotic gene expression during mitosis.

Similarly, meiosis-specific splicing has been noted in *S. pombe*. In a random sampling of 96 intronic meiosis-specific genes, Averbeck et al. (2005) used RT-PCR to identify 12 transcripts (including *Crs1*, *Meu13*, *Rec8*, *Spo4*, *Spo6*, and *Mfr1*) that underwent meiosis-specific splicing and correlated with the temporal waves of meiotic-gene expression (early, middle, late) observed during meiotic progression (Mata et al., 2002). Overexpression of *Crs1* (cyclin-like protein regulated via splicing), was toxic to vegetative cells, highlighting the requirement of mitotically growing cells to have tight regulation of meiosis-specific genes. Moldon et al. (2008) showed *Rem1* encodes two proteins with different functions, depending on the regulated splice form. During vegetative growth, full-length *Rem1* affects recombination in the premeiotic S phase, while spliced *Rem1* acts as a cyclin during meiosis I (whose expression is toxic in mitotic cells). Control of *Rem1* splicing is guided by two forkhead transcription factors, *Fkh2* and *Mei4*. During vegetative growth, FKH2 binds the *Rem1* promoter and *Rem1* is not spliced. During meiosis, MEI4 binds the *Rem1* promoter, and recruits the spliceosome, leading to a spliced variant of *Rem1* (Moldon et al., 2008). The prevalence of meiosis-specific splicing in *S. pombe* adds additional levels of meiotic posttranscriptional gene regulation.

Interestingly, in our analysis of the expression of *UmSpo11*, we detected three different PCR products of varying sizes ranging from ~400bp to ~600bp (Figure 2a). The ~600bp product co-migrated with the amplification using genomic DNA as template. We ruled out genomic DNA as a source of contamination due to the absence of amplification in PCR using RNA as template (minus reverse transcriptase) and the presence, in RT-PCR reactions of other intron containing genes (*Um01426*, and *UmGapd*), of only PCR products of a size consistent with the mature (spliced) transcripts. The primers designed for the *UmSpo11* PCR flanked an intron, and the expected amplicon size for processed transcripts, based on the current genome annotation, was ~350bp. No expressed sequence tag data is available for the *UmSpo11* region in question so it is possible that the current online annotation in the *Ustilago maydis* database for *UmSpo11* does not correctly predict the intron size (Mewes et al., 2008). Experiments are underway to determine whether or not the transcripts represented by the ~400bp or ~500bp amplicons contain an uninterrupted ORF. It is tempting to speculate that *UmSpo11* has transcripts of different sizes that code for proteins with distinct functions, and/or that the *UmSpo11* transcript undergoes meiosis-specific splicing. In mice and humans, two SPO11 isoforms are produced by alternative splicing (Romanienko & Camerini-Otero, 1999). The two isoforms in mice, SPO11β and SPO11α, are translated from a transcript containing all 13 exons and a transcript which skips the second exon, respectively (Bellani et al., 2010). Bellani et al. (2010) studied SPO11 isoform expression levels in mouse meiocytes and determined SPO11β initiates double strand breaks in early spermatocytes and SPO11α is present in pachytene/diplotene spermatocytes where it might act as a topoisomerase.

One final possibility is that one, or more of the *UmSpo11* amplicons represents a natural antisense transcript (NAT) and not an alternatively spliced transcript. NATs are polyadenylated in *U. maydis* (Ho et al., 2007) and therefore could be represented as a cDNA product in the presented reverse transcriptase reactions. If present, the NAT cDNA from the *UmSpo11* locus could then serve as template in the PCR. This is not without precedent as an antisense transcript overlaps the 3' end of the *Arabidopsis thaliana AtSpo11-2* locus (Hartung & Puchta, 2000). Three different isoforms of *AtSpo11-2* were observed, varying in their 3'UTR length and in the presence of an intron in the 3'UTR. While it is possible that the antisense transcript to *AtSpo11-2* elicits the RNA silencing pathway, it may be possible that the antisense transcript regulates the splicing of *AtSpo11-2* by masking the splice sites in the 3'UTR. Given the absence of a functional RNAi pathway in *U. maydis*, if one of the alternative splice forms seen for *UmSpo11* is an antisense transcript, it may act to regulate the splicing of *UmSpo11*. Experiments are underway to characterize all UmSpo11 transcripts.

It is noteworthy that Heimel et al. (2010) recently showed that *UmCib1*, a transcription factor required for pathogenic development, is predominantly unspliced during saprophytic growth, but undergoes splicing during biotrophic growth. Therefore, it is highly likely that other biological roles for alternate splicing in *U. maydis* remain to be discovered.

6. A regulatory role for fungal noncoding RNAs in meiosis

This section will begin with a broad overview of noncoding RNA (ncRNA) function in eukaryotes. There is very limited information on ncRNAs in other smuts and rusts; therefore, we will highlight the prevalence of meiosis-specific ncRNAs in *S. cerevisiae* and *S. pombe*, including putative functional roles for their expression. We will review specific examples of how ncRNAs function in controlling gene expression and conclude with knowledge that our laboratory is accumulating on ncRNAs and long natural antisense transcripts (NATs) in *U. maydis*.

6.1 Introduction to noncoding RNAs in eukaryotes

Recent estimates are that >90% of eukaryotic genomes are transcribed, but protein-coding transcripts only account for ~2-3% of eukaryotic genome transcription (reviewed in Costa, 2010). The difference is comprised of noncoding RNAs, some of which have a role in controlling gene expression. Thus, at least a portion of these so-called ncRNAs actually have a function that adds to the complexity of gene expression. Noncoding RNAs are divided into classes depending on their size and origin. For example, short ncRNAs (~20-31nt) include microRNAs (miRNAs), piwi-interacting RNAs (piRNAs), and small interfering RNAs (siRNAs). Long ncRNAs (>200nt) include, but are not limited to, long intergenic transcripts, and natural antisense transcripts (NATs). NATs are RNA molecules transcribed from the DNA strand complementary to that which codes the mRNA (reviewed in Costa, 2010; Faghihi & Wahlestedt 2009; Tisseur et al., 2011). Noncoding RNAs exert their function at the transcriptional or post-transcriptional level using a wide range of regulatory mechanisms including: RNA interference (RNAi), transcriptional interference, RNA masking (affecting mRNA-splicing, -transport, -polyadenylation, -translation, or -stability), RNA editing, X chromosome inactivation, imprinting, and chromatin remodelling (Costa, 2005; reviewed in Lavorgna et al., 2004; Munroe & Zhu, 2005). Notably, while most plants, animal and fungi contain functional RNA silencing machinery, phylogenetic analyses have revealed select ascomycetes (*S. cerevisiae* and *Candida lusitaniae*) and *U. maydis* lack the canonical RNAi machinery. The expression of ncRNAs, especially NATs in fungi that do not contain the canonical RNAi machinery sets the stage for discovering new mechanisms of controlling meiotic gene expression.

6.2 Global expression of meiosis-specific ncRNAs in *S. cerevisiae* and *S. pombe*

Yassour et al. (2011) created a strand-specific cDNA library using RNA isolated from *S. cerevisiae* cells during mid-log growth in rich media. Next-generation (Illumina) sequencing revealed transcript units that had an antisense orientation to 1,103 annotated transcripts. It was concluded that these antisense transcripts may be functional, given there was little evidence they arose from unterminated transcription, bi-directional transcription initiated from divergent promoters, or potential nucleosome-free regions (Yassour et al., 2010). A subset of sense-antisense transcript pairs was examined using strand-specific quantitative real-time PCR (qRT-PCR) and their expression patterns were inversely related, supporting a model where antisense transcripts interfere with sense transcript expression (Yassour et al., 2010). Additionally, the expression levels of six sense-antisense transcript pairs were studied across five yeast species using qRT-PCR. In some instances, both antisense transcript presence and differential expression were conserved, further supporting a functional role for antisense transcripts in *S. cerevisiae* (Yassour et al., 2010). 85 genes expressed at 8 hours post induction of sporulation were associated with antisense transcripts during *S. cerevisiae* mid-log phase growth. For example, the meiosis genes encoding *Ime4*, *Ndt80*, *Rec102*, *Gas2*, *Sps19*, *Slz1*, *Rim9*, and *Smk1* were all associated with long antisense transcripts. Overall, antisense units were prevalent in processes repressed during mid-log phase growth, leading Yassour et al. (2010) to hypothesize that NAT expression in *S. cerevisiae* may be involved in a global repression of stress, stationary phase, and meiosis genes when cells are grown in rich conditions. As a whole, in *S. cerevisiae*, and other yeast species, differentially expressed sense-antisense transcript pairs provide a means of controlling gene expression in response to environmental conditions and in some cases, enable the transition from mid-log phase- to meiotic-growth conditions.

Meiosis-specific genes were identified by T. Watanabe et al. (2001) using a subtractive cDNA library enriched for meiotic gene expression in heterozygous *S. pombe* diploid cells grown in nitrogen starvation conditions. This approach identified 31 *Meu* (meiotic expression upregulated) transcripts, 5 of which were ncRNAs. Notably, *Meu16* encodes an antisense ncRNA overlapping *Mde6* (a *Mei4*-dependent protein). The function of this antisense is not known. Additionally, individual RACE experiments uncovered ncRNAs at the *Rec7* (required for early steps of meiotic recombination) and *Spo6* (required for progression of meiosis II and sporulation) loci (Molnar et al., 2001; Nakamura et al., 2000). Specifically, three antisense transcripts (*Tos1*, *Tos2* and *Tos3*), of varying lengths, were discovered at the *Rec7* locus. It has been suggested that the *Tos* RNAs may indicate the location of dsDNA break formation (DSB; see below) at the *Rec7* locus (Wahls et al., 2008). *Spo6-L*, encodes a constitutively expressed bidirectional transcript in reverse orientation to the meiosis-specific *Spo6* gene. The function of *Spo6-L* remains unknown (Nakamura et al., 2000).

Subsequently, Wilhelm et al. (2008) studied the transcriptome of *S. pombe* from cells grown in rich media and from five stages of meiotic development under nitrogen limiting conditions.

Their investigation yielded 426 previously unannotated ncRNAs, including 58 ncRNAs (34 of which overlapped known genes in antisense orientation) upregulated under meiotic growth conditions. Many of the ncRNAs appear to be expressed in waves similar to meiosis-specific gene expression in *S. pombe* (Mata et al., 2002, 2007). Ni et al. (2010) created a strand-specific cDNA library using RNA isolated from *S. pombe* cells grown in rich media with or without heat shock. Next generation (Illumina) sequencing revealed ncRNA transcript units that did not overlap with previously defined protein-coding genes. Some loci in *S. pombe*, absent of protein-coding genes, had a comparable number of ncRNA transcripts on both strands, indicating that unlike *S. cerevisiae*, ncRNAs in *S. pombe* may function via the formation of double-stranded RNA, possibly eliciting chromatin remodelling, or posttranscriptional gene silencing. Additional differences that Ni et al. (2010) found between budding and fission yeast include: that antisense transcription is commonly driven through bidirectional promoters in S. *pombe* but not in S. *cerevisiae* and, while the majority of antisense transcripts in S. *pombe* are independently regulated, sense-antisense transcript levels are coordinated in S. *cerevisiae*. This may imply that antisense-mediated gene regulation in S. *pombe* occurs at the posttranscriptional level, or in *trans*, without affecting the relative levels of the sense transcript (Ni et al., 2010). It was observed that differentially expressed genes had a higher abundance of antisense transcripts than constitutively expressed genes, indicating that antisense RNAs may be involved in a targeted control of gene expression (Ni et al., 2010). In total, 2,409 S. *pombe* genes had overlapping antisense transcripts under normal or heat shock conditions. Gene ontology analysis revealed, like S. *cerevisiae*, S. *pombe* is enriched for antisense transcription at 68 loci involved in meiosis, meiotic chromosome segregation, meiotic recombination and ascospore formation. It was hypothesized that antisense transcripts may repress "leaky" meiotic genes under vegetative conditions at the transcriptional or posttranscriptional level (Ni et al., 2010). Using RNA-Seq, the transcriptomes of the fission yeasts S. *pombe*, S. *octosporus*, S. *cryophilus*, and S. *japonicus* were compared under growth conditions including log-phase, glucose starvation, early stationary phase, and heat shock (Rhind et al., 2011). This comparison identified conserved antisense transcripts among some of the fission yeasts. Additionally, for meiotic genes with antisense transcripts, the level of the antisense transcript was higher than the sense transcript (Rhind et al., 2011). For example, meiosis-upregulated genes (*Mug5*, *Mug7*, *Mug27*, *Mug28*, *Mug97*), MEI4 dependent genes (*Mde2*, *Mde3*, *Mde4*, *Mde7*), genes involved in meiotic recombination (*Rec7*, *Rec15*, *Rec24*, *Rec27*) and sporulation-specific genes (*Spo4*, *Spo6*) had higher levels of antisense transcript than sense transcript (Rhind et al., 2011, Supporting online material S21). Notably, antisense transcripts detected towards the core eukaryotic meiosis genes *Hop1* and *Dmc1*, were higher than sense transcript levels, while an antisense transcript level lower than that of the sense transcript was observed for *Rec8*. Strand-specific northern blotting was used to detect relative levels of S. *pombe Spo4*, *Spo6*, *Mde2*, *Mde7*, and *Mug8* sense-antisense transcript pairs. This revealed that antisense transcription of meiotic genes did not share an inverse relationship to sense transcription during log-phase growth, reinforcing the findings of Ni et al. (2010). Taken together, these findings suggest that antisense transcription during log-phase growth does not directly inhibit the transcription of sense transcripts as in S. *cerevisiae*, but elicits the formation of double-stranded RNA to recruit RNAi machinery to destroy "leaky" meiotic transcripts during mitotic growth (Rhind et al., 2011).

6.3 Rrp6 controls global expression of meiosis-specific ncRNAs in *S. cerevisiae*

Additional mechanisms of regulating the timing of ncRNA expression during meiosis in S. *cerevisiae* have been discovered. Degradation of ncRNAs in S. *cerevisiae* was linked to RRP6, a key component of the nuclear exosome complex responsible for RNA processing and degradation (Wyers et al., 2005). Deep transcriptome analyses identified cryptic unstable transcripts (CUTs) that accumulate in ΔRrp6 vegetative cells (Neil et al., 2009; Xu et al., 2009). Separately, high-resolution oligonucleotide tiling arrays identified meiotic unannotated transcripts (MUTs), ncRNAs that accumulate in meiotic but not fermenting or respiring cells, and rsSUTs, ncRNAs exhibiting peak expression during respiring or sporulating *MAT* **a**/α diploids (Lardenois et al., 2011). MUTs and rsSUTs were observed to overlap sense mRNAs in an antisense orientation, known promoter regions, or autonomously replicating sequences (ARSs). Lardenois et al. (2011) found that a subset of MUTs and rsSUTs, previously characterized as CUTs (Neil et al., 2009; Xu et al., 2009), are targeted by the nuclear exosome for degradation during vegetative growth conditions. Moreover, Rrp6 expression was observed across all cell-types; but RRP6 levels dropped dramatically as cells switched from mitotic to meiotic growth. During this transition, the decrease in RRP6 levels paralleled the increase in MUTs. Additionally, Δ*Rrp6* cells were unable to proceed with premeiotic DNA replication or to undergo meiosis and spore formation. The progression through meiosis may be facilitated by the accumulation of MUTs transcribed at mitotically active loci (such as *Chs2*, *Cln3*, and *Hug1*); MUT expression would interfere with the transcription of some mitotic genes during meiosis (Lardenois et al., 2011). Yassour et al. (2011) observed an increased level of antisense transcription in Δ*Rrp6* mutants, associated with a small decrease in complementary sense mRNA levels. Additionally, the transcription of the MUT, itself, may interfere with promoter regions required for proper transcription of select mRNAs or ARS elements required for DNA replication during specific cell-stages. For example, MUT expression was detected towards the *Cln2* promoter region. CLN2 is a repressor of IME1; therefore, *Cln2* levels decrease during meiosis, possibly facilitated through a promoter-interference mechanism where MUT transcription hinders promoter activity (Lardenois et al., 2011).

6.4 *S. cerevisiae IME4* and *ZIP1* ncRNA functions

S. cerevisiae MAT **a**/α diploids require full activation of *IME1* which is mediated by IME4 expression. Therefore, *Ime4* expression during nutrient starvation is pivotal in determining whether or not the cell initiates meiosis. Hongay et al. (2006) observed cell-type specific sense and antisense transcription at the *Ime4* locus. Haploids expressed an antisense transcript to *Ime4*, called *Rme2* (regulator of meiosis 2; Gelfand et al., 2011), and *MAT* **a**/α diploids expressed *Ime4* sense RNA. In *MAT* **a**/α diploids, the a1-α2 protein heterodimer silences expression of *Rme2*, enabling the expression of *Ime4*, full expression of *Ime1*, and entry into meiosis. Their findings were consistent with a transcriptional interference mechanism, since transcription of *Rme2* in haploid cells only interfered with *Ime4* sense expression in *cis* (Hongay et al., 2006). Similarly, *Zip2*, a meiosis-specific protein involved in synaptonemal complex formation, shows cell-type specific sense and antisense expression (Gelfand et al., 2011). Haploids express an antisense transcript to *Zip2*, called *Rme3* (regulator of meiosis 3) and *MAT* **a**/α diploids express *Zip2* sense RNA. The a1-α2 protein heterodimer silences expression of *Rme3* in *MAT* **a**/α diploids, enabling expression of *Zip2*. Gelfand et al. (2011) expanded on previous research to show that *Rme2* extension through the *Ime4* promoter region was not required for *Ime4* repression and *Rme3* does not extend through the entire *Zip2* ORF, indicating that both *Rme2* and *Rme3* do not interfere with TATA-binding proteins or polymerase binding in the promoter regions. Additionally, a 450 bp region within *Ime4* was essential for *Rme2*-mediated repression. This suggested that the 450 bp region may only be transcribed in a single direction at one time, or that extension of the transcript is terminated by specific protein complexes which bind this region, or chromatin remodelling occurs at this site. Overall, transcriptional interference may be prevalent in controlling yeast gene expression (Gelfand et al., 2011). Such mechanisms may be useful in fine-tuning condition-specific gene expression, especially when "leaky" expression of certain genes may be harmful to the cell.

6.5 *S. pombe* meiRNA function: The mei2-meiRNA complex

S. pombe MEI2 is an RNA-binding protein required for premeiotic DNA replication and the initiation of meiosis I (Y. Watanabe & Yamamoto, 1994; reviewed in Yamamoto et al., 2010). This protein has been discussed in detail in "4.2.2 Commitment and Continuation". In the context of this section, the role of a 0.5kb ncRNA transcribed at the *Sme2* locus (dubbed meiRNA) will be discussed. meiRNA binds to MEI2 and this ncRNA-protein complex is transported to the nucleus (Y. Watanabe & Yamamoto, 1994). Two scenarios have been described whereby meiRNA determines the nuclear localization of MEI2; meiRNA either interferes with MEI2 export from the nucleus, or facilitates MEI2 import into the nucleus (Sato et al., 2001). This nuclear MEI2-meiRNA complex binds MMI1, a protein that targets meiotic transcripts containing DSR motifs for degradation, during mitosis. This interaction interferes with MMI1, stabilizing transcripts required for the progression of S. pombe cells through meiosis I (Yamamoto et al., 2010). From this overview one can see that ncRNAs control the mitosis-meiosis switch in the divergent S. pombe and S. cerevisiae; however, they do so by very different mechanisms.

6.6 *S. pombe* meiotic hotspots

In *S. pombe*, the dsDNA breaks (DSBs) that initiate recombination, cluster to 194 prominent and 159 weak DSB peaks that favour intergenic regions (IGRs, Cromie et al., 2007). T. Watanabe et al. (2002) identified 68 polyadenylated ncRNAs from a random sampling of cDNAs originating from *S. pombe* cells in mitotic or meiotic growth phase. By cross referencing the aforementioned studies, it was determined that 24 polyadenylated ncRNAs were located entirely within a DSB peak (Wahls et al., 2008). Overall, Wahls et al. (2008) concluded that meiotic DSB hotspots preferentially form at loci that express long polyadenylated ncRNAs, many of which are expressed solely during meiosis. There are two mechanisms by which ncRNAs may guide meiotic recombination proteins to DSB hotspots: 1) the ncRNAs make DNA accessible to DSB formation through meiotically induced chromatin remodelling, and 2) ncRNA-DNA hybrids (R-loops) guide the meiotic recombination machinery to the DSB sites (Wahls et al., 2008).

6.7 *U. maydis* ncRNA/antisense transcription

RNA-seq has not been performed on *U. maydis* cell-types, but limited strand-specific expression data is available from the creation of EST libraries from various cell-types and nutritional conditions. To help facilitate *U. maydis* genome annotation, cDNA libraries were constructed from cell-types, including: germinating and dormant teliospores (Sacadura & Saville, 2003; Ho et al., 2007), filamentous diploids (Nugent et al., 2004) and dikaryons (Morrison et al., in preparation). Recall that teliospore formation and germination are temporally linked to meiosis in *U. maydis* (reviewed in Donaldson & Saville, 2008). Of the 319 uniESTs that did not match an annotated gene model in the *U. maydis* genome, 108 uniquely represented RNAs expressed in the dormant or germinating teliospores. This corresponds to 34% of the identified ncRNAs, while these two cDNA libraries account for only 17% of the total ESTs from all cell-types (Saville, unpublished). In total, ~250 NATs have been identified in *U. maydis*, including NATs expressed in the dormant and germinating teliospores (55 and 12, respectively). The function of the NATs and ncRNAs in *U. maydis* is under investigation. Ten teliospore-specific NATs, annotated during analysis of the dormant teliospore cDNA library, have been verified as teliospore-specific, using strand-specific RT-PCR (Ho et al., 2010). Unlike *S. cerevisiae* and *S. pombe*, the *U. maydis* NATs expressed in the dormant and germinating teliospore are not enriched for mitosis-specific genes and the NATs expressed during vegetative growth are not enriched for meiosis-specific genes. Additionally, inverse expression patterns have not been observed for sense-antisense transcript pairs; precluding transcriptional interference as the principal mechanism of action for NATs in *U. maydis*. Therefore, their function in the dormant and germinating teliospores appears to be unique to *U. maydis*.

7. Conclusion

This chapter provides an overview of meiotic events in the model fungal species *Saccharomyces cerevisiae*, *Schizosaccharomyces pombe* and *Coprinopsis cinerea* as a means of providing context for an exploration of meiosis in the model plant pathogen *Ustilago maydis*. Like the yeast fungi, *U. maydis* has set genetic requirements for entry into meiosis; it must be diploid and contain complementary alleles at the b mating type locus. With this genetic background, the fungus is able to accept an environmental signal that triggers entry into meiosis. This signal comes from the plant host and an exploration of the stages of pathogenic development led us to hypothesize that the stage before hyphal fragmentation, and after the cells become embedded in a mucilaginous matrix, is the time it must receive a signal from the plant to trigger entry into premeiotic S phase and undergo karyogamy. We uncover similarities between the role of the MAPK and cAMP/PKA pathways in mating and meiosis initiation in yeasts and the mating and pathogenesis signal transduction pathways in *U. maydis*. This is very relevant because of the requirement for growth within the host for *U. maydis* to become meiotically competent. These comparisons emphasized that the *U. maydis* genes *Crk1* and *Prf1*, which are orthologs of the major meiosis control genes *Ime2* in *S. cerevisiae* and *Ste11* in *S. pombe* respectively, provide a means whereby mating type and environmental signals could be transduced to influence meiosis. This led to a model of how meiosis is triggered in *U. maydis* and its linkage with teliospore development. We present an overview of waves in transcription in the yeasts and present evidence for potential waves of transcription in *U. maydis*. The identification of *U. maydis* meiosis genes by bioinformatic analyses is updated with an identification of the conserved absence of core meiosis genes in plant pathogenic fungi. We also present data that identifies *UmNdt80* as the first gene known to be required for meiosis completion in *U. maydis*. The timing of expression of six core meiosis genes in *U. maydis* is followed during *in planta* development. This uncovered support for the model that *U. maydis* enters meiosis very soon after karyogamy and then arrests during pachytene, when the teliospore matures and enters a dormant state. This information also identified transcriptional and posttranscriptional control of *Spo11* as potential key transitions in *U. maydis* meiotic progression. The final portion of the chapter highlights new data on the bioinformatic discovery of ncRNAs and NATs in *U. maydis* and their overrepresentation among ESTs in the teliospore and germinating teliospore libraries. In the context of the emerging role for these RNAs in controlling aspects of meiosis in *S. cerevisiae* and *S. pombe,* the discovery and confirmation of these RNAs in *U. maydis* is compelling. This chapter identifies several areas where further research will provide tremendous insight regarding meiosis initiation and progression in *U. maydis.*

During proofing, gametogenesis initiation (van Werven & Amon, 2011) and RNAi-independent roles for antisense transcripts in controlling meiotic genes (Chen & Neiman, 2011) in budding and fission yeasts were reviewed.

8. Acknowledgement

We would like to acknowledge Natalie Pearson for her excellent artwork in Figure 1, thank Laura Peers for fact finding and organizational support, and thank Christine Russell for assistance in proofing this manuscript. We also acknowledge funding from NSERC of Canada, and the Ontario Ministry of Research and Innovation's Ontario Research Fund-Research Excellence program.

9. References

Abe, H. & Shimoda, C. (2000). Autoregulated expression of Schizosaccharomyces pombe meiosis-specific transcription factor Mei4 and a genome-wide search for its target genes. *Genetics,* Vol.154, No.4, (April 2000), pp. 1497-1508, ISSN 0016-6731

Ahmed, N.T., Bungard, D., Shin, M.E., Moore, M. & Winter, E. (2009). The Ime2 protein kinase enhances the disassociation of the Sum1 repressor from middle meiotic promoters. *Molecular and cellular biology,* Vol.29, No.16, (August 2009), pp. 4352-4362, ISSN 1098-5549

Allen, A., Islamovic, E., Kaur, J., Gold, S., Shah, D. & Smith, T.J. (2011). Transgenic maize plants expressing the Totivirus antifungal protein, KP4, are highly resistant to corn smut. *Plant biotechnology journal,* (February 2011), ISSN 1467-7652

Alvarez, B. & Moreno, S. (2006). Fission yeast Tor2 promotes cell growth and represses cell differentiation. *Journal of cell science,* Vol.119, No.Pt 21, (November 2006), pp. 4475-4485, ISSN 0021-9533

Averbeck, N., Sunder, S., Sample, N., Wise, J.A. & Leatherwood, J. (2005). Negative control contributes to an extensive program of meiotic splicing in fission yeast. *Molecular cell,* Vol.18, No.4, (May 2005), pp. 491-498, ISSN 1097-2765

Banuett, F. (2010). *Ustilago maydis* and Maize: a Delightful Interaction, In: Cellular and Molecular Biology of Filamentous Fungi, K.A. Borkovich & D.J. Ebbole, (Ed.), 622-644, ASM Press, ISSN 978-1-55581-473-1, Washington, USA

Banuett, F. (2002). Pathogenic development in *Ustilago maydis*: A progression of morphological transitions that results in tumor formation and teliospore production, In: Molecular biology of fungal development, H.D. Osiewacz, (Ed.), 349-398, Marcel Dekker, ISSN 0824707443, New York, USA

Banuett, F. & Herskowitz, I. (1996). Discrete developmental stages during teliospore formation in the corn smut fungus, Ustilago maydis. *Development,* Vol.122, No.10, (October 1996), pp. 2965-2976, ISSN 0950-1991

Banuett, F. (1995).Genetics of *Ustilago maydis*, a fungal pathogen that induces tumors in maize.*Annual Review of Genetics,* Vol.29, (n.d.), pp. 179-208, ISSN 0066-4197

Banuett, F. & Herskowitz, I. (1989). Different a alleles of Ustilago maydis are necessary for maintenance of filamentous growth but not for meiosis. *Proceedings of the National Academy of Sciences of the United States of America*, Vol.86, No.15, (August 1989), pp. 5878-5882, ISSN 0027-8424

Barnes, C.W., Szabo, L.J., May, G. & Groth, J.V. (2004). Inbreeding levels of two Ustilago maydis populations. *Mycologia*, Vol.96, No.6, (November 2004), pp. 1236-1244, ISSN 0027-5514

Bauman, P., Cheng, Q.C. & Albright, C.F. (1998). The Byr2 kinase translocates to the plasma membrane in a Ras1-dependent manner. *Biochemical and biophysical research communications*, Vol.244, No.2, (March 1998), pp. 468-474, ISSN 0006-291X

Bellani, M.A., Boateng, K.A., McLeod, D. & Camerini-Otero, R.D. (2010). The expression profile of the major mouse SPO11 isoforms indicates that SPO11beta introduces double strand breaks and suggests that SPO11alpha has an additional role in prophase in both spermatocytes and oocytes. *Molecular and cellular biology*, Vol.30, No.18, (September 2010), pp. 4391-4403, ISSN 1098-5549

Boehm, E.W.A., Wenstrom, J.C., McLaughlin, D.J., Szabo, L.J., Roelfs, A.P. & Bushnell, W.R. (1992). An ultrastructural pachytene karyotype for Puccinia graminis f. sp. tritici. *Canadian Journal of Botany*, Vol.70, (n.d.), pp. 401-4113, ISSN 0008-4026

Borkovich, K.A., Alex, L.A., Yarden, O., Freitag, M., Turner, G.E., Read, N.D., Seiler, S., Bell-Pedersen, D., Paietta, J., Plesofsky, N., Plamann, M., Goodrich-Tanrikulu, M., Schulte, U., Mannhaupt, G., Nargang, F.E., Radford, A., Selitrennikoff, C., Galagan, J.E., Dunlap, J.C., Loros, J.J., Catcheside, D., Inoue, H., Aramayo, R., Polymenis, M., Selker, E.U., Sachs, M.S., Marzluf, G.A., Paulsen, I., Davis, R., Ebbole, D.J., Zelter, A., Kalkman, E.R., O'Rourke, R., Bowring, F., Yeadon, J., Ishii, C., Suzuki, K., Sakai, W. & Pratt, R. (2004). Lessons from the genome sequence of Neurospora crassa: tracing the path from genomic blueprint to multicellular organism. *Microbiology and molecular biology reviews*, Vol.68, No.1, (March 2004), pp. 1-108, ISSN 1092-2172

Brefort, T., Doehlemann, G., Mendoza-Mendoza, A., Reissmann, S., Djamei, A. & Kahmann, R. (2009). Ustilago maydis as a Pathogen. *Annual Review of Phytopathology*, Vol.47, (n.d.), pp. 423-445, ISSN 0066-4286

Burdon, J.J. & Roelfs, A.P. (1985). The effect of sexual and asexual reproduction on the isozyme structure of populations of Puccinia graminis. *Phytopathology*, Vol.75, (n.d.), pp. 1068-1073, ISSN 0031-949X

Burns, C., Pukkila, P.J. & Miriam, Z.E. (2010a). Meiosis, In: Cellular and Molecular Biology of Filamentous Fungi, K.A. Borkovich & D.J. Ebbole, (Ed.), 81-95, ASM Press, ISSN 978-1-55581-473-1, Washington, USA

Burns, C., Stajich, J.E., Rechtsteiner, A., Casselton, L., Hanlon, S.E., Wilke, S.K., Savytskyy, O.P., Gathman, A.C., Lilly, W.W., Lieb, J.D., Zolan, M.E. & Pukkila, P.J. (2010b). Analysis of the Basidiomycete Coprinopsis cinerea reveals conservation of the core meiotic expression program over half a billion years of evolution. *PLoS genetics*, Vol.6, No.9, (September 2010), pp. e1001135, ISSN 1553-7404

Butler, G., Rasmussen, M.D., Lin, M.F., Santos, M.A., Sakthikumar, S., Munro, C.A., Rheinbay, E., Grabherr, M., Forche, A., Reedy, J.L., Agrafioti, I., Arnaud, M.B., Bates, S., Brown, A.J., Brunke, S., Costanzo, M.C., Fitzpatrick, D.A., de Groot, P.W., Harris, D., Hoyer, L.L., Hube, B., Klis, F.M., Kodira, C., Lennard, N., Logue, M.E., Martin, R., Neiman, A.M., Nikolaou, E., Quail, M.A., Quinn, J., Santos, M.C., Schmitzberger, F.F., Sherlock, G., Shah, P., Silverstein, K.A., Skrzypek, M.S., Soll, D., Staggs, R., Stansfield, I., Stumpf, M.P., Sudbery, P.E., Srikantha, T., Zeng, Q., Berman, J., Berriman, M., Heitman, J., Gow, N.A., Lorenz, M.C., Birren, B.W., Kellis,

M. & Cuomo, C.A. (2009). Evolution of pathogenicity and sexual reproduction in eight Candida genomes. *Nature*, Vol.459, No.7247, (June 2009), pp. 657-662, ISSN 1476-4687

Capitol Commodity Services. (February, 2011). U.S. crop values jumped 22% in 2010, 04.09.2011, Available from <http://www.ccstrade.com/futures-news/grains/story/u-s-crop-values-jumped-22-in-2010-800410457/>

Celerin, M., Merino, S.T., Stone, J.E., Menzie, A.M. & Zolan, M.E. (2000). Multiple roles of Spo11 in meiotic chromosome behavior. *The EMBO journal*, Vol.19, No.11, (June 2000), pp. 2739-2750, ISSN 0261-4189

Chen, H.M. & Neiman, A.M. (2011). A conserved regulatory role for antisense RNA in meiotic gene expression in yeast. *Current opinion in microbiology*, Vol.14, No.6, (December 2011), pp. 655-659, ISSN 1879-0364

Chu, S., DeRisi, J., Eisen, M., Mulholland, J., Botstein, D., Brown, P.O. & Herskowitz, I. (1998). The transcriptional program of sporulation in budding yeast. *Science*, Vol.282, No.5389, (October 1998), pp. 699-705, ISSN 0036-8075

Chu, S. & Herskowitz, I. (1998). Gametogenesis in yeast is regulated by a transcriptional cascade dependent on Ndt80. *Molecular cell*, Vol.1, No.5, (April 1998), pp. 685-696, ISSN 1097-2765

Chung, K.S., Won, M., Lee, S.B., Jang, Y.J., Hoe, K.L., Kim, D.U., Lee, J.W., Kim, K.W. & Yoo, H.S. (2001). Isolation of a novel gene from Schizosaccharomyces pombe: stm1+ encoding a seven-transmembrane loop protein that may couple with the heterotrimeric Galpha 2 protein, Gpa2. *The Journal of biological chemistry*, Vol.276, No.43, (October 2001), pp. 40190-40201, ISSN 0021-9258

Colomina, N., Gari, E., Gallego, C., Herrero, E. & Aldea, M. (1999). G1 cyclins block the Ime1 pathway to make mitosis and meiosis incompatible in budding yeast. *The EMBO journal*, Vol.18, No.2, (January 1999), pp. 320-329, ISSN 0261-4189

Cooper, K.F. & Strich, R. (2002). Saccharomyces cerevisiae C-type cyclin Ume3p/Srb11p is required for efficient induction and execution of meiotic development. *Eukaryotic cell*, Vol.1, No.1, (February 2002), pp. 66-74, ISSN 1535-9778

Costa, F.F. (2010). Non-coding RNAs: Meet thy masters. *BioEssays: news and reviews in molecular, cellular and developmental biology*, Vol.32, No.7, (July 2010), pp. 599-608, ISSN 1521-1878

Costa, F.F. (2005). Non-coding RNAs: new players in eukaryotic biology. *Gene*, Vol.357, No.2, (September 2005), pp. 83-94, ISSN 0378-1119

Covitz, P.A. & Mitchell, A.P. (1993). Repression by the yeast meiotic inhibitor RME1. *Genes & development*, Vol.7, No.8, (August 1993), pp. 1598-1608, ISSN 0890-9369

Covitz, P.A., Herskowitz, I. & Mitchell, A.P. (1991). The yeast RME1 gene encodes a putative zinc finger protein that is directly repressed by a1-alpha 2. *Genes & development*, Vol.5, No.11, (November 1991), pp. 1982-1989, ISSN 0890-9369

Cromie, G.A., Hyppa, R.W., Cam, H.P., Farah, J.A., Grewal, S.I. & Smith, G.R. (2007). A discrete class of intergenic DNA dictates meiotic DNA break hotspots in fission yeast. *PLoS genetics*, Vol.3, No.8, (August 2007), pp. e141, ISSN 1553-7404

Donaldson, M.E. & Saville, B.J. (2008). Bioinformatic identification of Ustilago maydis meiosis genes. *Fungal genetics and biology*, Vol.45, Suppl 1, (August 2008), pp. S47-53, ISSN 1096-0937

Donzeau, M. & Bandlow, W. (1999). The yeast trimeric guanine nucleotide-binding protein alpha subunit, Gpa2p, controls the meiosis-specific kinase Ime2p activity in response to nutrients. *Molecular and cellular biology*, Vol.19, No.9, (September 1999), pp. 6110-6119, ISSN 0270-7306

Doyle, C.E., Donaldson, M.E., Morrison, E.N. & Saville, B.J. (2011). Ustilago maydis transcript features identified through full-length cDNA analysis. *Molecular genetics and genomics,* Vol.286, No.2, (August 2011), pp. 143-159, ISSN 1617-4623

Faghihi, M.A. & Wahlestedt, C. (2009). Regulatory roles of natural antisense transcripts. *Nature reviews molecular cell biology,* Vol.10, No.9, (September 2009), pp. 637-643, ISSN 1471-0080

Feldbrugge, M., Kamper, J., Steinberg, G. & Kahmann, R. (2004). Regulation of mating and pathogenic development in Ustilago maydis.*Current opinion in microbiology,* Vol.7, No.6, (December 2004), pp. 666-672, ISSN 1369-5274

Fletcher, H.L. (1981). A search for synaptonemal complexes in Ustilago maydis. *Journal of cell science,* Vol.50, (August 1981), pp. 171-180, ISSN 0021-9533

Gallego, C., Gari, E., Colomina, N., Herrero, E. & Aldea, M. (1997). The Cln3 cyclin is down-regulated by translational repression and degradation during the G1 arrest caused by nitrogen deprivation in budding yeast. *The EMBO journal,* Vol.16, No.23, (December 1997), pp. 7196-7206, ISSN 0261-4189

Garrido, E., Voss, U., Muller, P., Castillo-Lluva, S., Kahmann, R. & Perez-Martin, J. (2004). The induction of sexual development and virulence in the smut fungus Ustilago maydis depends on Crk1, a novel MAPK protein. *Genes & development,* Vol.18, No.24, (December 2004), pp. 3117-3130, ISSN 0890-9369

Garrido, E. & Perez-Martin, J. (2003). The crk1 gene encodes an Ime2-related protein that is required for morphogenesis in the plant pathogen Ustilago maydis. *Molecular microbiology,* Vol.47, No.3, (February 2003), pp. 729-743, ISSN 0950-382X

Gelfand, B., Mead, J., Bruning, A., Apostolopoulos, N., Tadigotla, V., Nagaraj, V., Sengupta, A.M. & Vershon, A.K. (2011). Regulated antisense transcription controls expression of cell-type-specific genes in yeast. *Molecular and cellular biology,* Vol.31, No.8, (April 2011), pp. 1701-1709, ISSN 1098-5549

Gold, S.E., Brogdon, S.M., Mayorga, M.E. & Kronstad, J.W. (1997). The Ustilago maydis regulatory subunit of a cAMP-dependent protein kinase is required for gall formation in maize. *The Plant cell,* Vol.9, No.9, (September 1997), pp. 1585-1594, ISSN 1040-4651

Govin, J. & Berger, S.L. (2009). Genome reprogramming during sporulation. *The International journal of developmental biology,* Vol.53, No.2-3, (n.d.), pp. 425-432, ISSN 1696-3547

Guttmann-Raviv, N., Martin, S. & Kassir, Y. (2002). Ime2, a meiosis-specific kinase in yeast, is required for destabilization of its transcriptional activator, Ime1. *Molecular and cellular biology,* Vol.22, No.7, (April 2002), pp. 2047-2056, ISSN 0270-7306

Harigaya, Y. & Yamamoto, M. (2007). Molecular mechanisms underlying the mitosis-meiosis decision. *Chromosome research: an international journal on the molecular, supramolecular and evolutionary aspects of chromosome biology,* Vol.15, No.5, (n.d.), pp. 523-537, ISSN 0967-3849

Harigaya, Y., Tanaka, H., Yamanaka, S., Tanaka, K., Watanabe, Y., Tsutsumi, C., Chikashige, Y., Hiraoka, Y., Yamashita, A. & Yamamoto, M. (2006). Selective elimination of messenger RNA prevents an incidence of untimely meiosis. *Nature,* Vol.442, No.7098, (July 2006), pp. 45-50, ISSN 1476-4687

Hartmann, H.A., Kruger, J., Lottspeich, F. & Kahmann, R. (1999). Environmental signals controlling sexual development of the corn Smut fungus Ustilago maydis through the transcriptional regulator Prf1. *The Plant cell,* Vol.11, No.7, (July 1999), pp. 1293-1306, ISSN 1040-4651

Hartmann, H.A., Kahmann, R. & Bolker, M. (1996). The pheromone response factor coordinates filamentous growth and pathogenicity in Ustilago maydis. *The EMBO journal*, Vol.15, No.7, (April 1996), pp. 1632-1641, ISSN 0261-4189

Hartung, F. & Puchta, H. (2000). Molecular characterisation of two paralogous SPO11 homologues in Arabidopsis thaliana. *Nucleic acids research*, Vol.28, No.7, (April 2000), pp. 1548-1554, ISSN 1362-4962

Heimel, K., Scherer, M., Schuler, D. & Kamper, J. (2010). The Ustilago maydis Clp1 protein orchestrates pheromone and b-dependent signaling pathways to coordinate the cell cycle and pathogenic development. *The Plant cell*, Vol.22, No.8, (August 2010), pp. 2908-2922, ISSN 1532-298X

Hepworth, S.R., Friesen, H. & Segall, J. (1998). NDT80 and the meiotic recombination checkpoint regulate expression of middle sporulation-specific genes in Saccharomyces cerevisiae. *Molecular and cellular biology*, Vol.18, No.10, (October 1998), pp. 5750-5761, ISSN 0270-7306

Higuchi, T., Watanabe, Y. & Yamamoto, M. (2002). Protein kinase A regulates sexual development and gluconeogenesis through phosphorylation of the Zn finger transcriptional activator Rst2p in fission yeast. *Molecular and cellular biology*, Vol.22, No.1, (January 2002), pp. 1-11, ISSN 0270-7306

Ho, E.C., Donaldson, M.E. & Saville, B.J. (2010). Detection of antisense RNA transcripts by strand-specific RT-PCR, In: RT-PCR Protocols,2nd Edition, N. King, (Ed.), 125-138, Humana Press, ISSN 1940-6029, New York, USA

Ho, E.C., Cahill, M.J. & Saville, B.J. (2007). Gene discovery and transcript analyses in the corn smut pathogen Ustilago maydis: expressed sequence tag and genome sequence comparison. *BMC genomics*, Vol.8, No.1, (September 2007), pp. 334, ISSN 1471-2164

Holloman, W.K., Schirawski, J. & Holliday, R. (2008). The homologous recombination system of Ustilago maydis. *Fungal genetics and biology*, Vol.45, Suppl 1, (August 2008), pp. S31-9, ISSN 1096-0937

Hongay, C.F., Grisafi, P.L., Galitski, T. & Fink, G.R. (2006). Antisense transcription controls cell fate in Saccharomyces cerevisiae. *Cell*, Vol.127, No.4, (November 2006), pp. 735-745, ISSN 0092-8674

Honigberg, S.M. & Purnapatre, K. (2003). Signal pathway integration in the switch from the mitotic cell cycle to meiosis in yeast. *Journal of cell science*, Vol.116, No.Pt 11, (June 2003), pp. 2137-2147, ISSN 0021-9533

Honigberg, S.M. & Lee, R.H. (1998). Snf1 kinase connects nutritional pathways controlling meiosis in Saccharomyces cerevisiae. *Molecular and cellular biology*, Vol.18, No.8, (August 1998), pp. 4548-4555, ISSN 0270-7306

Horie, S., Watanabe, Y., Tanaka, K., Nishiwaki, S., Fujioka, H., Abe, H., Yamamoto, M. & Shimoda, C. (1998). The Schizosaccharomyces pombe mei4+ gene encodes a meiosis-specific transcription factor containing a forkhead DNA-binding domain. *Molecular and cellular biology*, Vol.18, No.4, (April 1998), pp. 2118-2129, ISSN 0270-7306

Horst, R.J., Doehlemann, G., Wahl, R., Hofmann, J., Schmiedl, A., Kahmann, R., Kamper, J., Sonnewald, U. & Voll, L.M. (2010a). Ustilago maydis infection strongly alters organic nitrogen allocation in maize and stimulates productivity of systemic source leaves. *Plant physiology*, Vol.152, No.1, (January 2010), pp. 293-308, ISSN 1532-2548

Horst, R.J., Doehlemann, G., Wahl, R., Hofmann, J., Schmiedl, A., Kahmann, R., Kamper, J. & Voll, L.M. (2010b). A model of Ustilago maydis leaf tumor metabolism. *Plant signaling & behavior*, Vol.5, No.11, (November 2010), pp. 1446-1449, ISSN 1559-2324

Huerta-Espino, J., Singh, R.P., German, S., McCallum, B.D., Park, R.F., Chen, W.Q., Bhardway, S.C. & Goyeau, H. (2011). Global status of wheat leaf rust caused by Puccinia triticina. *Euphytica*, Vol.179, (n.d.), pp. 143-160, ISSN 0014-2336

Isshiki, T., Mochizuki, N., Maeda, T. & Yamamoto, M. (1992). Characterization of a fission yeast gene, gpa2, that encodes a G alpha subunit involved in the monitoring of nutrition. *Genes & development*, Vol.6, No.12B, (December 1992), pp. 2455-2462, ISSN 0890-9369

Jin, Y. (2011). Role of Berberis spp. as alternate hosts in generating new races of Puccinia graminis and P. striiformis. *Euphytica*, Vol.179, (n.d.), pp. 105-108, ISSN 0014-2336

Juneau, K., Palm, C., Miranda, M. & Davis, R.W. (2007). High-density yeast-tiling array reveals previously undiscovered introns and extensive regulation of meiotic splicing. *Proceedings of the National Academy of Sciences of the United States of America*, Vol.104, No.5, (January 2007), pp. 1522-1527, ISSN 0027-8424

Kaffarnik, F., Muller, P., Leibundgut, M., Kahmann, R. & Feldbrugge, M. (2003). PKA and MAPK phosphorylation of Prf1 allows promoter discrimination in Ustilago maydis. *The EMBO journal*, Vol.22, No.21, (November 2003), pp. 5817-5826, ISSN 0261-4189

Kahmann, R. & Kamper, J. (2004). *Ustilago maydis*: how its biology relates to pathogenic development. *New Phytologist*, Vol.164, (n.d.), pp. 31-42, ISSN 1469-8137

Kamper, J., Kahmann, R., Bolker, M., Ma, L.J., Brefort, T., Saville, B.J., Banuett, F., Kronstad, J.W., Gold, S.E., Muller, O., Perlin, M.H., Wosten, H.A., de Vries, R., Ruiz-Herrera, J., Reynaga-Pena, C.G., Snetselaar, K., McCann, M., Perez-Martin, J., Feldbrugge, M., Basse, C.W., Steinberg, G., Ibeas, J.I., Holloman, W., Guzman, P., Farman, M., Stajich, J.E., Sentandreu, R., Gonzalez-Prieto, J.M., Kennell, J.C., Molina, L., Schirawski, J., Mendoza-Mendoza, A., Greilinger, D., Munch, K., Rossel, N., Scherer, M., Vranes, M., Ladendorf, O., Vincon, V., Fuchs, U., Sandrock, B., Meng, S., Ho, E.C., Cahill, M.J., Boyce, K.J., Klose, J., Klosterman, S.J., Deelstra, H.J., Ortiz-Castellanos, L., Li, W., Sanchez-Alonso, P., Schreier, P.H., Hauser-Hahn, I., Vaupel, M., Koopmann, E., Friedrich, G., Voss, H., Schluter, T., Margolis, J., Platt, D., Swimmer, C., Gnirke, A., Chen, F., Vysotskaia, V., Mannhaupt, G., Guldener, U., Munsterkotter, M., Haase, D., Oesterheld, M., Mewes, H.W., Mauceli, E.W., DeCaprio, D., Wade, C.M., Butler, J., Young, S., Jaffe, D.B., Calvo, S., Nusbaum, C., Galagan, J. & Birren, B.W. (2006). Insights from the genome of the biotrophic fungal plant pathogen Ustilago maydis. *Nature*, Vol.444, No.7115, (November 2006), pp. 97-101, ISSN 1476-4687

Kassir, Y., Adir, N., Boger-Nadjar, E., Raviv, N.G., Rubin-Bejerano, I., Sagee, S. & Shenhar, G. (2003).Transcriptional regulation of meiosis in budding yeast. *International review of cytology*, Vol.224, (n.d.), pp. 111-171, ISSN 0074-7696

Kato, T.,Jr, Okazaki, K., Murakami, H., Stettler, S., Fantes, P.A. & Okayama, H. (1996). Stress signal, mediated by a Hog1-like MAP kinase, controls sexual development in fission yeast. *FEBS letters*, Vol.378, No.3, (January 1996), pp. 207-212, ISSN 0014-5793

Kjaerulff, S., Andersen, N.R., Borup, M.T. & Nielsen, O. (2007). Cdk phosphorylation of the Ste11 transcription factor constrains differentiation-specific transcription to G1. *Genes & development*, Vol.21, No.3, (February 2007), pp. 347-359, ISSN 0890-9369

Klosterman, S.J., Perlin, M.H., Garcia-Pedrajas, M., Covert, S.F. & Gold, S.E. (2007). Genetics of morphogenesis and pathogenic development of Ustilago maydis. *Advances in Genetics*, Vol.57, (n.d.), pp. 1-47, ISSN 0065-2660

Kojic, M., Kostrub, C.F., Buchman, A.R. & Holloman, W.K. (2002). BRCA2 homolog required for proficiency in DNA repair, recombination, and genome stability in Ustilago maydis. *Molecular cell*, Vol.10, No.3, (September 2002), pp. 683-691, ISSN 1097-2765

Kronstad, J.W. (2008). Mining the genome of the biotrophic fungal pathogen Ustilago maydis. *Fungal genetics and biology*, Vol.45 Suppl 1, (August 2008), pp. S1-2, ISSN 1096-0937

Kruger, J., Loubradou, G., Wanner, G., Regenfelder, E., Feldbrugge, M. & Kahmann, R. (2000). Activation of the cAMP pathway in Ustilago maydis reduces fungal proliferation and teliospore formation in plant tumors. *Molecular plant-microbe interactions*, Vol.13, No.10, (October 2000), pp. 1034-1040, ISSN 0894-0282

Kues, U. (2000). Life history and developmental processes in the basidiomycete Coprinus cinereus. *Microbiology and molecular biology reviews*, Vol.64, No.2, (June 2000), pp. 316-353, ISSN 1092-2172

Kunitomo, H., Higuchi, T., Iino, Y. & Yamamoto, M. (2000). A zinc-finger protein, Rst2p, regulates transcription of the fission yeast ste11(+) gene, which encodes a pivotal transcription factor for sexual development. *Molecular biology of the cell*, Vol.11, No.9, (September 2000), pp. 3205-3217, ISSN 1059-1524

Lardenois, A., Liu, Y., Walther, T., Chalmel, F., Evrard, B., Granovskaia, M., Chu, A., Davis, R.W., Steinmetz, L.M. & Primig, M. (2011). Execution of the meiotic noncoding RNA expression program and the onset of gametogenesis in yeast require the conserved exosome subunit Rrp6. *Proceedings of the National Academy of Sciences of the United States of America*, Vol.108, No.3, (January 2011), pp. 1058-1063, ISSN 1091-6490

Lavorgna, G., Dahary, D., Lehner, B., Sorek, R., Sanderson, C.M. & Casari, G. (2004). In search of antisense. *Trends in biochemical sciences*, Vol.29, No.2, (February 2004), pp. 88-94, ISSN 0968-0004

Lee, N., D'Souza, C.A. & Kronstad, J.W. (2003). Of smuts, blasts, mildews, and blights: cAMP signaling in phytopathogenic fungi. *Annual Review of Phytopathology*, Vol.41, (n.d.), pp. 399-427, ISSN 0066-4286

Loidl, J. (2006). S. pombe linear elements: the modest cousins of synaptonemal complexes. *Chromosoma*, Vol.115, No.3, (June 2006), pp. 260-271, ISSN 0009-5915

Lu, B.C. (2000). The control of meiosis progression in the fungus Coprinus cinereus by light/dark cycles. *Fungal genetics and biology*, Vol.31, No.1, (October 2000), pp. 33-41, ISSN 1087-1845

Malik, S.B., Pightling, A.W., Stefaniak, L.M., Schurko, A.M. & Logsdon, J.M.,Jr. (2007). An expanded inventory of conserved meiotic genes provides evidence for sex in Trichomonas vaginalis. *PloS one*, Vol.3, No.8, (August 2007), pp. e2879, ISSN 1932-6203

Marston, A.L. & Amon, A. (2004). Meiosis: cell-cycle controls shuffle and deal. *Nature reviews. Molecular cell biology*, Vol.5, No.12, (December 2004), pp. 983-997, ISSN 1471-0072

Martinez-Espinoza, A.D., Garcia-Pedrajas, M.D. & Gold, S.E. (2002). The Ustilaginales as plant pests and model systems. *Fungal genetics and biology*, Vol.35, No.1, (February 2002), pp. 1-20, ISSN 1087-1845

Mata, J., Wilbrey, A. & Bahler, J. (2007).Transcriptional regulatory network for sexual differentiation in fission yeast. *Genome biology*, Vol.8, No.10, (October 2007), pp. R217, ISSN 1465-6914

Mata, J. & Bahler, J. (2003). Correlations between gene expression and gene conservation in fission yeast. *Genome research,* Vol.13, No.12, (December 2003), pp. 2686-2690, ISSN 1088-9051

Mata, J., Lyne, R., Burns, G. & Bahler, J. (2002). The transcriptional program of meiosis and sporulation in fission yeast. *Nature genetics,* Vol.32, No.1, (September 2002), pp. 143-147, ISSN 1061-4036

Matsuo, T., Otsubo, Y., Urano, J., Tamanoi, F. & Yamamoto, M. (2007). Loss of the TOR kinase Tor2 mimics nitrogen starvation and activates the sexual development pathway in fission yeast. *Molecular and cellular biology,* Vol.27, No.8, (April 2007), pp. 3154-3164, ISSN 0270-7306

Mewes, H.W., Dietmann, S., Frishman, D., Gregory, R., Mannhaupt, G., Mayer, K.F., Munsterkotter, M., Ruepp, A., Spannagl, M., Stumpflen, V. & Rattei, T. (2008). MIPS: analysis and annotation of genome information in 2007. *Nucleic acids research,* Vol.36, No.Database issue, (January 2008), pp. D196-201, ISSN 1362-4962

Mims, C.W. & Richardson, E.A. (2005). Light and electron microscopy of teliospores and teliospore germination in the rust fungus Coleosporium ipomoeae. *Canadian Journal of Botany,* Vol.83, No.5, (May 2005), pp. 451-458, ISSN 0008-4026

Mitchell, A.P. (1994). Control of meiotic gene expression in Saccharomyces cerevisiae. *Microbiological reviews,* Vol.58, No.1, (March 1994), pp. 56-70, ISSN 0146-0749

Mitchell, A.P., Driscoll, S.E. & Smith, H.E. (1990). Positive control of sporulation-specific genes by the IME1 and IME2 products in Saccharomyces cerevisiae. *Molecular and cellular biology,* Vol.10, No.5, (May 1990), pp. 2104-2110, ISSN 0270-7306

Mochizuki, N. & Yamamoto, M. (1992). Reduction in the intracellular cAMP level triggers initiation of sexual development in fission yeast. *Molecular & general genetics,* Vol.233, No.1-2, (May 1992), pp. 17-24, ISSN 0026-8925

Moldon, A., Malapeira, J., Gabrielli, N., Gogol, M., Gomez-Escoda, B., Ivanova, T., Seidel, C. & Ayte, J. (2008). Promoter-driven splicing regulation in fission yeast. *Nature,* Vol.455, No.7215, (October 2008), pp. 997-1000, ISSN 1476-4687

Molnar, M., Parisi, S., Kakihara, Y., Nojima, H., Yamamoto, A., Hiraoka, Y., Bozsik, A., Sipiczki, M. & Kohli, J. (2001). Characterization of rec7, an early meiotic recombination gene in Schizosaccharomyces pombe. *Genetics,* Vol.157, No.2, (February 2001), pp. 519-532, ISSN 0016-6731

Muller, P., Leibbrandt, A., Teunissen, H., Cubasch, S., Aichinger, C. & Kahmann, R. (2004). The Gbeta-subunit-encoding gene bpp1 controls cyclic-AMP signaling in Ustilago maydis. *Eukaryotic cell,* Vol.3, No.3, (June 2004), pp. 806-814, ISSN 1535-9778

Muller, P., Weinzierl, G., Brachmann, A., Feldbrugge, M. & Kahmann, R. (2003). Mating and pathogenic development of the Smut fungus Ustilago maydis are regulated by one mitogen-activated protein kinase cascade. *Eukaryotic cell,* Vol.2, No.6, (December 2003), pp. 1187-1199, ISSN 1535-9778

Munroe, S.H. & Zhu, J. (2006). Overlapping transcripts, double-stranded RNA and antisense regulation: a genomic perspective. *Cellular and molecular life sciences,* Vol.63, No.18, (September 2006), pp. 2102-2118, ISSN 1420-682X

Nakamura, T., Kishida, M. & Shimoda, C. (2000). The Schizosaccharomyces pombe spo6+ gene encoding a nuclear protein with sequence similarity to budding yeast Dbf4 is required for meiotic second division and sporulation. *Genes to cells: devoted to molecular & cellular mechanisms,* Vol.5, No.6, (June 2000), pp. 463-479, ISSN 1356-9597

Neale, M.J. & Keeney, S. (2006). Clarifying the mechanics of DNA strand exchange in meiotic recombination. *Nature*, Vol.442, No.7099, (July 2006), pp. 153-158, ISSN 1476-4687

Neil, H., Malabat, C., d'Aubenton-Carafa, Y., Xu, Z., Steinmetz, L.M. & Jacquier, A. (2009). Widespread bidirectional promoters are the major source of cryptic transcripts in yeast. *Nature*, Vol.457, No.7232, (February 2009), pp. 1038-1042, ISSN 1476-4687

Ni, T., Tu, K., Wang, Z., Song, S., Wu, H., Xie, B., Scott, K.C., Grewal, S.I., Gao, Y. & Zhu, J. (2010). The prevalence and regulation of antisense transcripts in Schizosaccharomyces pombe. *PloS one*, Vol.5, No.12, (December 2010), pp. e15271, ISSN 1932-6203

Nielsen, O. (1993). Signal transduction during mating and meiosis in S. pombe. *Trends in cell biology*, Vol.3, No.2, (February 1993), pp. 60-65, ISSN 0962-8924

Nugent, K.G., Choffe, K. & Saville, B.J. (2004). Gene expression during Ustilago maydis diploid filamentous growth: EST library creation and analyses. *Fungal genetics and biology*, Vol.41, No.3, (March 2004), pp. 349-360, ISSN 1087-1845

Obara, T., Nakafuku, M., Yamamoto, M. & Kaziro, Y. (1991). Isolation and characterization of a gene encoding a G-protein alpha subunit from Schizosaccharomyces pombe: involvement in mating and sporulation pathways. *Proceedings of the National Academy of Sciences of the United States of America*, Vol.88, No.13, (July 1991), pp. 5877-5881, ISSN 0027-8424

O'Donnell, K.L. & McLaughlin, D.J. (1984). Ultrastructure of Meiosis in *Ustilago maydis*. *Mycologia*, Vol.76, No.3, (May 1984), pp. 468-485, ISSN 0027-5514

Ordonez, M.E. & Kolmer, J.A. (2009). Differentiation of molecular genotypes and virulence phenotypes of Puccinia triticina from common wheat in North America. *Phytopathology*, Vol.99, No.6, (June 2009), pp. 750-758, ISSN 0031-949X

Pak, J. & Segall, J. (2002a). Regulation of the premiddle and middle phases of expression of the NDT80 gene during sporulation of Saccharomyces cerevisiae. *Molecular and cellular biology*, Vol.22, No.18, (September 2002), pp. 6417-6429, ISSN 0270-7306

Pak, J. & Segall, J. (2002b). Role of Ndt80, Sum1, and Swe1 as targets of the meiotic recombination checkpoint that control exit from pachytene and spore formation in Saccharomyces cerevisiae. *Molecular and cellular biology*, Vol.22, No.18, (September 2002), pp. 6430-6440, ISSN 0270-7306

Pawlowski, W.P., Sheehan, M.J. & Ronceret, A. (2007). In the beginning: the initiation of meiosis. *BioEssays: news and reviews in molecular, cellular and developmental biology*, Vol.29, No.6, (June 2007), pp. 511-514, ISSN 0265-9247

Peng, Z., Wang, W., Schettino, A., Leung, B. & McLeod, M. (2003). Inactivation of Ran1/Pat1 kinase bypasses the requirement for high-level expression of mei2 during fission yeast meiosis. *Current genetics*, Vol.43, No.3, (June 2003), pp. 178-185, ISSN 0172-8083

Piekarska, I., Rytka, J. & Rempola, B. (2010). Regulation of sporulation in the yeast Saccharomyces cerevisiae. *Acta Biochimica Polonica*, Vol.57, No.3, (n.d.), pp. 241-250, ISSN 1734-154X

Pierce, M., Benjamin, K.R., Montano, S.P., Georgiadis, M.M., Winter, E. & Vershon, A.K. (2003). Sum1 and Ndt80 proteins compete for binding to middle sporulation element sequences that control meiotic gene expression. *Molecular and cellular biology*, Vol.23, No.14, (July 2003), pp. 4814-4825, ISSN 0270-7306

Primig, M., Williams, R.M., Winzeler, E.A., Tevzadze, G.G., Conway, A.R., Hwang, S.Y., Davis, R.W. & Esposito, R.E. (2000). The core meiotic transcriptome in budding yeasts. *Nature genetics*, Vol.26, No.4, (December 2000), pp. 415-423, ISSN 1061-4036

Pukkila, P.J., Yashar, B.M. & Binninger, D.M. (1984). Analysis of meiotic development in Coprinus cinereus. *Symposia of the Society for Experimental Biology*, Vol.38, (n.d.), pp. 177-194, ISSN 0081-1386

Purnapatre, K., Gray, M., Piccirillo, S. & Honigberg, S.M. (2005). Glucose inhibits meiotic DNA replication through SCFGrr1p-dependent destruction of Ime2p kinase. *Molecular and cellular biology*, Vol.25, No.1, (January 2005), pp. 440-450, ISSN 0270-7306

Ramberg, J.E. & McLaughlin, D.J. (1980). Ultrastructural study of promycelial development and basidiospore initiation in Ustilago maydis. *Canadian Journal of Botany*, Vol.58, No.14, (n.d.), pp. 1548-1561, ISSN 0008-4026

Ramesh, M.A., Malik, S.B. & Logsdon, J.M.,Jr. (2005). A phylogenomic inventory of meiotic genes; evidence for sex in Giardia and an early eukaryotic origin of meiosis. *Current biology*, Vol.15, No.2, (January 2005), pp. 185-191, ISSN 0960-9822

Regenfelder, E., Spellig, T., Hartmann, A., Lauenstein, S., Bolker, M. & Kahmann, R. (1997). G proteins in Ustilago maydis: transmission of multiple signals? *The EMBO journal*, Vol.16, No.8, (April 1997), pp. 1934-1942, ISSN 0261-4189

Rhind, N., Chen, Z., Yassour, M., Thompson, D.A., Haas, B.J., Habib, N., Wapinski, I., Roy, S., Lin, M.F., Heiman, D.I., Young, S.K., Furuya, K., Guo, Y., Pidoux, A., Chen, H.M., Robbertse, B., Goldberg, J.M., Aoki, K., Bayne, E.H., Berlin, A.M., Desjardins, C.A., Dobbs, E., Dukaj, L., Fan, L., FitzGerald, M.G., French, C., Gujja, S., Hansen, K., Keifenheim, D., Levin, J.Z., Mosher, R.A., Muller, C.A., Pfiffner, J., Priest, M., Russ, C., Smialowska, A., Swoboda, P., Sykes, S.M., Vaughn, M., Vengrova, S., Yoder, R., Zeng, Q., Allshire, R., Baulcombe, D., Birren, B.W., Brown, W., Ekwall, K., Kellis, M., Leatherwood, J., Levin, H., Margalit, H., Martienssen, R., Nieduszynski, C.A., Spatafora, J.W., Friedman, N., Dalgaard, J.Z., Baumann, P., Niki, H., Regev, A. & Nusbaum, C. (2011). Comparative functional genomics of the fission yeasts. *Science*, Vol.332, No.6032, (May 2011), pp. 930-936, ISSN 1095-9203

Roeder, G.S. & Bailis, J.M. (2000). The pachytene checkpoint. *Trends in genetics*, Vol.16, No.9, (September 2000), pp. 395-403, ISSN 0168-9525

Roelfs, A.P. (1982). Effects of barberry eradication on stem rust in the United States. *Plant Disease*, Vol.66, No.2, (n.d.), pp. 177-181, ISSN 0191-2917

Romanienko, P.J. & Camerini-Otero, R.D. (1999). Cloning, characterization, and localization of mouse and human SPO11. *Genomics*, Vol.61, No.2, (October 1999), pp. 156-169, ISSN 0888-7543

Rubin-Bejerano, I., Mandel, S., Robzyk, K. & Kassir, Y. (1996). Induction of meiosis in Saccharomyces cerevisiae depends on conversion of the transcriptional represssor Ume6 to a positive regulator by its regulated association with the transcriptional activator Ime1. *Molecular and cellular biology*, Vol.16, No.5, (May 1996), pp. 2518-2526, ISSN 0270-7306

Sacadura, N.T. & Saville, B.J. (2003). Gene expression and EST analyses of Ustilago maydis germinating teliospores. *Fungal genetics and biology*, Vol.40, No.1, (October 2003), pp. 47-64, ISSN 1087-1845

Saccharomyces Genome Database. (August, 2011). Saccharomyces Genome Database, 17.08.2011, Available from <http://www.yeastgenome.org/>

Sagee, S., Sherman, A., Shenhar, G., Robzyk, K., Ben-Doy, N., Simchen, G. & Kassir, Y. (1998). Multiple and distinct activation and repression sequences mediate the regulated transcription of IME1, a transcriptional activator of meiosis-specific genes in Saccharomyces cerevisiae. *Molecular and cellular biology*, Vol.18, No.4, (April 1998), pp. 1985-1995, ISSN 0270-7306

Sato, M., Shinozaki-Yabana, S., Yamashita, A., Watanabe, Y. & Yamamoto, M. (2001). The fission yeast meiotic regulator Mei2p undergoes nucleocytoplasmic shuttling. *FEBS letters*, Vol.499, No.3, (June 2001), pp. 251-255, ISSN 0014-5793

Schirawski, J., Mannhaupt, G., Munch, K., Brefort, T., Schipper, K., Doehlemann, G., Di Stasio, M., Rossel, N., Mendoza-Mendoza, A., Pester, D., Muller, O., Winterberg, B., Meyer, E., Ghareeb, H., Wollenberg, T., Munsterkotter, M., Wong, P., Walter, M., Stukenbrock, E., Guldener, U. & Kahmann, R. (2010). Pathogenicity determinants in smut fungi revealed by genome comparison. *Science*, Vol.330, No.6010, (December 2010), pp. 1546-1548, ISSN 1095-9203

Schlecht, U. & Primig, M. (2003). Mining meiosis and gametogenesis with DNA microarrays. *Reproduction*, Vol.125, No.4, (April 2003), pp. 447-456, ISSN 1470-1626

Schumann, G.L. & D'Arcy, J.D. (2009). What Are the Causes of Plant Diseases?, In: *Essential Plant Pathology, Second Edition*, pp. 21-39, American Phytopathological Society Press, ISSN 978-0-89054-381-8, Minnesota, USA

Shah, J.C. & Clancy, M.J. (1992). IME4, a gene that mediates MAT and nutritional control of meiosis in Saccharomyces cerevisiae. *Molecular and cellular biology*, Vol.12, No.3, (March 1992), pp. 1078-1086, ISSN 0270-7306

Shenhar, G. & Kassir, Y. (2001). A positive regulator of mitosis, Sok2, functions as a negative regulator of meiosis in Saccharomyces cerevisiae. *Molecular and cellular biology*, Vol.21, No.5, (March 2001), pp. 1603-1612, ISSN 0270-7306

Shin, M.E., Skokotas, A. & Winter, E. (2010). The Cdk1 and Ime2 protein kinases trigger exit from meiotic prophase in Saccharomyces cerevisiae by inhibiting the Sum1 transcriptional repressor. *Molecular and cellular biology*, Vol.30, No.12, (June 2010), pp. 2996-3003, ISSN 1098-5549

Shiozaki, K. & Russell, P. (1996). Conjugation, meiosis, and the osmotic stress response are regulated by Spc1 kinase through Atf1 transcription factor in fission yeast. *Genes & development*, Vol.10, No.18, (September 1996), pp. 2276-2288, ISSN 0890-9369

Shuster, E.O. & Byers, B. (1989). Pachytene arrest and other meiotic effects of the start mutations in Saccharomyces cerevisiae. *Genetics*, Vol.123, No.1, (September 1989), pp. 29-43, ISSN 0016-6731

Simchen, G. (2009). Commitment to meiosis: what determines the mode of division in budding yeast? *BioEssays: news and reviews in molecular, cellular and developmental biology*, Vol.31, No.2, (February 2009), pp. 169-177, ISSN 1521-1878

Singh, R.P., Hodson, D.P., Huerta-Espino, J., Jin, Y., Bhavani, S., Njau, P., Herrera-Foessel, S., Singh, P.K., Singh, S. & Govindan, V. (2011). The Emergence of Ug99 Races of the Stem Rust Fungus is a Threat to World Wheat Production. *Annual Review of Phytopathology*, Vol.49, (September 2011), pp. 465-481, ISSN 0066-4286

Snetselaar, K.M. & Mims, C.W. (1992). Sporidial fusion and infection of maize seedlings by the smut fungus *Ustilago maydis*. *Mycologia*, Vol.84, (n.d.), pp. 193-203, ISSN 0027-5514

Sugimoto, A., Iino, Y., Maeda, T., Watanabe, Y. & Yamamoto, M. (1991). Schizosaccharomyces pombe ste11+ encodes a transcription factor with an HMG motif that is a critical regulator of sexual development. *Genes & development*, Vol.5, No.11, (November 1991), pp. 1990-1999, ISSN 0890-9369

Takeda, T., Toda, T., Kominami, K., Kohnosu, A., Yanagida, M. & Jones, N. (1995). Schizosaccharomyces pombe atf1+ encodes a transcription factor required for sexual development and entry into stationary phase. *The EMBO journal*, Vol.14, No.24, (December 1995), pp. 6193-6208, ISSN 0261-4189

Thevelein, J.M. & de Winde, J.H. (1999). Novel sensing mechanisms and targets for the cAMP-protein kinase A pathway in the yeast Saccharomyces cerevisiae. *Molecular microbiology*, Vol.33, No.5, (September 1999), pp. 904-918, ISSN 0950-382X

Tisseur, M., Kwapisz, M. & Morillon, A. (2011). Pervasive transcription - Lessons from yeast. *Biochimie*, Vol.93, No.11, (November 2011), pp. 1889-1896, ISSN 1638-6183

Tripathi, R.K. & Gottlieb, D. (1974). Sequential biosynthesis of ribonucleic acids during germination of teliospores of Ustilago maydis. *Mycologia*, Vol.66, No.3, (May-Jun 1974), pp. 413-421, ISSN 0027-5514

Tung, K.S., Hong, E.J. & Roeder, G.S. (2000). The pachytene checkpoint prevents accumulation and phosphorylation of the meiosis-specific transcription factor Ndt80. *Proceedings of the National Academy of Sciences of the United States of America*, Vol.97, No.22, (October 2000), pp. 12187-12192, ISSN 0027-8424

U.S. Department of Agriculture. (August, 2011). Wheat Data: Yearbook Tables, World and U.S. wheat production, exports, and ending stocks, 06.09.2011, Available from <http://www.ers.usda.gov/data/wheat/YBtable04.asp>

U.S. Grains Council.(November, 2010). Corn, 06.09.2011, Available from <http://www.grains.org/corn>

Valbuena, N. & Moreno, S. (2010). TOR and PKA pathways synergize at the level of the Ste11 transcription factor to prevent mating and meiosis in fission yeast. *PloS one*, Vol.5, No.7, (July 2010), pp. e11514, ISSN 1932-6203

van Werven, F.J. & Amon, A. (2011). Regulation of entry into gametogenesis. *Philosophical Transactions of the Royal Society B: Biological Sciences*, Vol.366, No.1584, (December 2011), pp. 3521-3531, ISSN 0962-8436

Vershon, A.K. & Pierce, M. (2000). Transcriptional regulation of meiosis in yeast. *Current opinion in cell biology*, Vol.12, No.3, (June 2000), pp. 334-339, ISSN 0955-0674

Villeneuve, A.M. & Hillers, K.J. (2001). Whence meiosis? *Cell*, Vol.106, No.6, (September 2001), pp. 647-650, ISSN 0092-8674

Voegele, R.T. & Mendgen, K. (2003). Rust haustoria: nutrient uptake and beyond. *New Phytologist*, Vol.159, (n.d.), pp. 93-100, ISSN 1469-8137

Wahls, W.P., Siegel, E.R. & Davidson, M.K. (2008). Meiotic recombination hotspots of fission yeast are directed to loci that express non-coding RNA. *PloS one*, Vol.3, No.8, (August 2008), pp. e2887, ISSN 1932-6203

Watanabe, T., Miyashita, K., Saito, T.T., Nabeshima, K. & Nojima, H. (2002). Abundant poly(A)-bearing RNAs that lack open reading frames in Schizosaccharomyces pombe. *DNA research: an international journal for rapid publication of reports on genes and genomes*, Vol.9, No.6, (December 2002), pp. 209-215, ISSN 1340-2838

Watanabe, T., Miyashita, K., Saito, T.T., Yoneki, T., Kakihara, Y., Nabeshima, K., Kishi, Y.A., Shimoda, C. & Nojima, H. (2001). Comprehensive isolation of meiosis-specific genes identifies novel proteins and unusual non-coding transcripts in Schizosaccharomyces pombe. *Nucleic acids research*, Vol.29, No.11, (June 2001), pp. 2327-2337, ISSN 1362-4962

Wilhelm, B.T., Marguerat, S., Watt, S., Schubert, F., Wood, V., Goodhead, I., Penkett, C.J., Rogers, J. & Bahler, J. (2008). Dynamic repertoire of a eukaryotic transcriptome surveyed at single-nucleotide resolution. *Nature*, Vol.453, No.7199, (June 2008), pp. 1239-1243, ISSN 1476-4687

Watanabe, Y. & Yamamoto, M. (1994). S. pombe mei2+ encodes an RNA-binding protein essential for premeiotic DNA synthesis and meiosis I, which cooperates with a novel RNA species meiRNA. *Cell*, Vol.78, No.3, (August 1994), pp. 487-498, ISSN 0092-8674

Watanabe, Y., Lino, Y., Furuhata, K., Shimoda, C. & Yamamoto, M. (1988). The S. pombe mei2 gene encoding a crucial molecule for commitment to meiosis is under the regulation of cAMP. *The EMBO journal*, Vol.7, No.3, (March 1988), pp. 761-767, ISSN 0261-4189

Wilkinson, M.G., Samuels, M., Takeda, T., Toone, W.M., Shieh, J.C., Toda, T., Millar, J.B. & Jones, N. (1996). The Atf1 transcription factor is a target for the Sty1 stress-activated MAP kinase pathway in fission yeast. *Genes & development*, Vol.10, No.18, (September 1996), pp. 2289-2301, ISSN 0890-9369

Willer, M., Hoffmann, L., Styrkarsdottir, U., Egel, R., Davey, J. & Nielsen, O. (1995). Two-step activation of meiosis by the mat1 locus in Schizosaccharomyces pombe. *Molecular and cellular biology*, Vol.15, No.9, (September 1995), pp. 4964-4970, ISSN 0270-7306

Wyers, F., Rougemaille, M., Badis, G., Rousselle, J.C., Dufour, M.E., Boulay, J., Regnault, B., Devaux, F., Namane, A., Seraphin, B., Libri, D. & Jacquier, A. (2005). Cryptic pol II transcripts are degraded by a nuclear quality control pathway involving a new poly(A) polymerase. *Cell*, Vol.121, No.5, (June 2005), pp. 725-737, ISSN 0092-8674

Xiao, Y. & Mitchell, A.P. (2000). Shared roles of yeast glycogen synthase kinase 3 family members in nitrogen-responsive phosphorylation of meiotic regulator Ume6p. *Molecular and cellular biology*, Vol.20, No.15, (August 2000), pp. 5447-5453, ISSN 0270-7306

Xu, Z., Wei, W., Gagneur, J., Perocchi, F., Clauder-Munster, S., Camblong, J., Guffanti, E., Stutz, F., Huber, W. & Steinmetz, L.M. (2009). Bidirectional promoters generate pervasive transcription in yeast. *Nature*, Vol.457, No.7232, (February 2009), pp. 1033-1037, ISSN 1476-4687

Yamamoto, M. (2010). The selective elimination of messenger RNA underlies the mitosis-meiosis switch in fission yeast. *Proceedings of the Japan Academy:Series B, Physical and biological sciences*, Vol.86, No.8, (n.d.), pp. 788-797, ISSN 1349-2896

Yamamoto, M. (1996a). The molecular control mechanisms of meiosis in fission yeast. *Trends in biochemical sciences*, Vol.21, No.1, (January 1996), pp. 18-22, ISSN 0968-0004

Yamamoto, M. (1996b). Regulation of meiosis in fission yeast. *Cell structure and function*, Vol.21, No.5, (October 1996), pp. 431-436, ISSN 0386-7196

Yamashita, A., Watanabe, Y., Nukina, N. & Yamamoto, M. (1998). RNA-assisted nuclear transport of the meiotic regulator Mei2p in fission yeast. *Cell*, Vol.95, No.1, (October 1998), pp. 115-123, ISSN 0092-8674

Yassour, M., Pfiffner, J., Levin, J.Z., Adiconis, X., Gnirke, A., Nusbaum, C., Thompson, D.A., Friedman, N. & Regev, A. (2010). Strand-specific RNA sequencing reveals extensive regulated long antisense transcripts that are conserved across yeast species. *Genome biology*, Vol.11, No.8, (n.d.), pp. R87, ISSN 1465-6914

Zahiri, A.R., Babu, M.R. & Saville, B.J. (2005). Differential gene expression during teliospore germination in Ustilago maydis. *Molecular genetics and genomics*, Vol.273, No.5, (June 2005), pp. 394-403, ISSN 1617-4615

Zaman, S., Lippman, S.I., Zhao, X. & Broach, J.R. (2008). How Saccharomyces responds to nutrients. *Annual Review of Genetics*, Vol.42, (n.d.), pp. 27-81, ISSN 0066-4197

Permissions

The contributors of this book come from diverse backgrounds, making this book a truly international effort. This book will bring forth new frontiers with its revolutionizing research information and detailed analysis of the nascent developments around the world.

We would like to thank Dr. Andrew Swan, for lending his expertise to make the book truly unique. He has played a crucial role in the development of this book. Without his invaluable contribution this book wouldn't have been possible. He has made vital efforts to compile up to date information on the varied aspects of this subject to make this book a valuable addition to the collection of many professionals and students.

This book was conceptualized with the vision of imparting up-to-date information and advanced data in this field. To ensure the same, a matchless editorial board was set up. Every individual on the board went through rigorous rounds of assessment to prove their worth. After which they invested a large part of their time researching and compiling the most relevant data for our readers. Conferences and sessions were held from time to time between the editorial board and the contributing authors to present the data in the most comprehensible form. The editorial team has worked tirelessly to provide valuable and valid information to help people across the globe.

Every chapter published in this book has been scrutinized by our experts. Their significance has been extensively debated. The topics covered herein carry significant findings which will fuel the growth of the discipline. They may even be implemented as practical applications or may be referred to as a beginning point for another development. Chapters in this book were first published by InTech; hereby published with permission under the Creative Commons Attribution License or equivalent.

The editorial board has been involved in producing this book since its inception. They have spent rigorous hours researching and exploring the diverse topics which have resulted in the successful publishing of this book. They have passed on their knowledge of decades through this book. To expedite this challenging task, the publisher supported the team at every step. A small team of assistant editors was also appointed to further simplify the editing procedure and attain best results for the readers.

Our editorial team has been hand-picked from every corner of the world. Their multi-ethnicity adds dynamic inputs to the discussions which result in innovative outcomes. These outcomes are then further discussed with the researchers and contributors who give their valuable feedback and opinion regarding the same. The feedback is then collaborated with the researches and they are edited in a comprehensive manner to aid the understanding of the subject.

Apart from the editorial board, the designing team has also invested a significant amount of their time in understanding the subject and creating the most relevant covers. They scrutinized every image to scout for the most suitable representation of the subject and create an appropriate cover for the book.

The publishing team has been involved in this book since its early stages. They were actively engaged in every process, be it collecting the data, connecting with the contributors or procuring relevant information. The team has been an ardent support to the editorial, designing and production team. Their endless efforts to recruit the best for this project, has resulted in the accomplishment of this book. They are a veteran in the field of academics and their pool of knowledge is as vast as their experience in printing. Their expertise and guidance has proved useful at every step. Their uncompromising quality standards have made this book an exceptional effort. Their encouragement from time to time has been an inspiration for everyone.

The publisher and the editorial board hope that this book will prove to be a valuable piece of knowledge for researchers, students, practitioners and scholars across the globe.

List of Contributors

N. Songsasen
Department of Reproductive Sciences, Center for Species Survival, Smithsonian Conservation Biology Institute, Front Royal, Virginia

Sylvie Bilodeau-Goeseels and Nora Magyara
Agriculture and Agri-Food Canada, Lethbridge Research Centre, Canada

James J. Faust, Madhavi Kalive, Anup Abraham and David G. Capco
Arizona State University, USA

Kyung-Ah Lee, Eun-Young Kim, Hyun-Seo Lee, Su-Yeon Lee and Eun-Ah Kim
Department of Biomedical Science, College of Life Science, CHA University, Korea

Se-Jin Yoon
Department of Genetics, Stanford University, USA

Jeehyeon Bae
Department of Pharmacy, College of Pharmacy, CHA University, Korea

Yuko Onohara and Sadaki Yokota
Nagasaki International University, Japan

Maria Elena Dell'Aquila, Nicola Antonio Martino, Manuel Filioli Uranio, and Lucia Rutigliano
Dept. Animal Production, University of Bari "Aldo Moro", Bari, Italy

Yoon Sung Cho
Assisted Procreation Unit - Ospedale Accreditato Clinica Santa Maria, Bari, Italy

Katrin Hinrichs
Dept. Physiology and Pharmacology, Texas A&M University, College Station, USA

L. Xanthopoulou and H. Ghevaria
University College London, UK

Hiba Waldman Ben-Asher and Jeremy Don
The Mina & Everard Goodman Faculty of Life Sciences, Bar-Ilan University, Israe

Maria Suely Pagliarini, Maria Lúcia Carneiro Vieira and Cacilda Borges do Valle
University of Maringá, Department of Cell Biology and Genetics, Brazil
University of Sao Paulo, College of Agriculture 'Luiz de Queiroz', Department of Genetics, Embrapa Beef Cattle, Brazil

Cynthia Ross Friedman
Department of Biological Sciences, Thompson Rivers University, Canada

Hua-Feng Wang
Beijing Urban Ecosystem Research Station, China
State Key Laboratory of Urban and Regional Ecology, China
Research Center for Eco-Environmental Sciences, Chinese Academy of Sciences, China

Filipe Ressurreição, Augusta Barão, Wanda Viegas and Margarida Delgado
CBAA, Instituto Superior de Agronomia, Technical University of LisbonTapada da Ajuda, Portugal

Jorge E. A. Mariath
Plant Anatomy Laboratory, Biosciences Institute, Electron Microscopy Centre, Federal University of Rio Grande do Sul-RS, Brazil

B. J. Saville, M. E. Donaldson and C. E. Doyle
Trent University, Canada

Printed in the USA
CPSIA information can be obtained
at www.ICGtesting.com
JSHW011453221024
72173JS00005B/1051